# ANCIENT EGYPT

Oxford

# ANCIENT EGYPT

*General Editor:*
*David P. Silverman*

New York
Oxford University Press
1997

Ancient Egypt

Published in the United States of America by Oxford University Press, Inc.
198 Madison Avenue, New York, N.Y. 10016
Oxford is a registered trademark of Oxford University Press



Conceived, created, and designed by
Duncan Baird Publishers, London, England

Library of Congress
Cataloging-in-Publication Data

Ancient Egypt / David Silverman, general editor.
p. cm.
Includes bibliographical references and index.
ISBN 0-19-521270-3
1. Egypt--Civilization--To 30BCE I. Silverman, David P.
DT61.A59 1995b
923--dc21 96-37171
CIP

*Editor:* Peter Bently
*Co-ordinating editor:* Daphne Bien Tebbe
*Assistant editors:* Ingrid Court-Jones, Judy Dean
*Designers:* Paul Reid, Lucie Penn
*Picture research:* Cecilia Weston-Baker
*Commissioned artwork:* Nigel Brookes and Abigail Horner of Ian Fleming & Associates Ltd; Stephen Conlin pp.182–3, 208–9
*Decorative borders:* Neil Packer

Typeset in Erhardt MT 11/15pt
Color reproduction by Colourscan, Singapore
Printed in China by Imago Publishing Limited

NOTE
The abbreviations CE and BCE are used throughout this book:
CE Common Era (the equivalent of AD)
BCE Before the Common Era (the equivalent of BC)
Captions to pages 1–5 appear on p.256

Printing (last digit) 9 8 7 6 5 4 3 2 1



# CONTENTS

# INTRODUCTION

To the modern intelligence "Ancient Egypt" presents a conundrum: this land of the past has thousands of familiar visual cues that make it spring to life with a coherent character unequalled by any other civilization of antiquity, yet its full depth and breadth still remain to be discovered. For almost two millennia before the publication of Champollion's decipherment of hieroglyphs in 1822, those fascinated by the creations of this long-dead culture could interpret them only by guesswork and hypothesis. They could recognize the images that comprised the hieroglyphs, but could not read them. They could marvel at the engineering skills involved in building a giant pyramid, but were ignorant of the multiple levels of significance that such a structure possessed. They could explore temple and tomb architecture, but could only theorize about its relevance to the religion, life and death of these ancient peoples. They could gaze upon statues of hybrid creatures, but remained unaware of their exact role. Mummies, unearthed by the thousand, were a constant source of interest, but their true significance was overshadowed by the strange belief that powdered mummy, if consumed by the ill, had curative powers. Even after more reliable information was in circulation, there were those who turned a blind eye to the facts, instead succumbing to the intoxication of legend, rumour and half-truth. Such speculations have survived to the

*The main pylon gateway of the Temple of Luxor. One obelisk stands here still, but the matching obelisk of the pair was shipped to Paris in the 19th century and erected in the Place de la Concorde, where it stands today – a testament to the enduring popularity of the great civilization of the Nile Valley.*

present day in a whole corpus of popular mythology with virtually no basis in fact. The curse of the pharaohs, the preservative powers of pyramids, the extraterrestrial origins of Egyptian culture, are just some of the themes that still thread their way through literature for the mass market.

"Egyptomania" is not a new phenomenon. Shortly after the discovery of the Rosetta Stone, ancient Egyptian motifs became a source of embellishment in the architecture and decorative arts of Europe – a trend that eventually reached the United States. In the nineteenth century, many new banks, hotels, libraries and other public and commercial buildings were erected in the "Egyptian" style, with papyrus and lotus column ornamentation – though perhaps the most apt use for this mode of ornament was in cemeteries, with gateways in the form of temple pylons, and mausoleums fashioned as miniature temples. Ancient monuments were even shipped to New York and European capitals as adornments for streets, squares and parks. After the discovery of the tomb of Tutankhamun in 1922, a new wave of Egyptomania broke on both sides of the Atlantic. Pendants in the form of scarab beetles and the Eye of Horus became *haute couture* accessories. After another fifty years or so, the phenomenon recurred, the impetus coming this time from the great travelling exhibition of Tutankhamun's treasures. Decorative arts, makeup and hairstyles showed the renewed popularity of Egyptian imagery.

The high visibility of ancient Egypt to the popular imagination should not be allowed to eclipse the perhaps less spectacular but more factual and more important work undertaken by scholars. The last two centuries have seen the gradual development of reputable scientific investigation in the field. When Champollion's work became available, philologists began to provide information about what the inscriptions really said. Eventually anthropologists, art historians, cultural historians, religious historians and other academicians took part in a concerted investigative effort, and soon these scholars were able to piece together much of the puzzle. In the last few decades botanists, physical anthropologists, geneticists, radar technologists, computer experts and a variety of other specialists have joined the quest. Egyptologists are asking about all levels of Egyptian society, not just the royal and élite classes. They are looking into socio-political and socio-economic issues, among others, and in their studies are applying contemporary hypotheses (such as anthropological and economic models and literary and art historical theories). Work by many of the eminent scholars investigating these areas is presented in the following chapters. Perhaps we may never solve all of the Sphinx's riddles, but each day of scholarship and research brings us closer to an understanding of one of the ancient world's greatest civilizations.

*David P. Silverman, Curator-in-Charge,*
*Egyptian Section, University of Pennsylvania Museum*

*Part I*

# THE EGYPTIAN WORLD

*A crop inspection by the "Chief Grain Measurer", a government tax assessor whose title alone remains in the fragment at top left. His clerk stands by the barley and two chariot drivers await their masters, perhaps the assessor and Nebamun, owner of the tomb from which this wall painting comes. New Kingdom, ca. 1400BCE.*



OPPOSITE: *The agricultural Nile floodplain gives way abruptly to deserts and rugged hills, as illustrated in this view of Deir el-Bahari, an area of Western Thebes. The mortuary temples of the pharaohs Hatshepsut and Nebhepetre Mentuhotep II are in the foreground.*

● CHAPTER 1

# THE GIFT OF THE NILE

"Egypt," wrote the Greek historian Herodotus in the fifth century BCE, "is so to speak the gift of the Nile." According to the priests, he said, Egypt was nothing but marsh before the land was created by layer upon layer of silt deposited by the great river. Modern geographers might differ in their account of Egypt's physical origins, but the central role of the river in the life of the country is as evident today as it was in ancient times.

▲

## THE RIVER IN THE SAND

Egypt lies at the northern end of the longest river in the world: the Nile, which rises in the East African highlands and flows into the Mediterranean more than four thousand miles (6,500km) away. The rhythm of the river was the most important feature of life in ancient Egypt. Until this century, when huge dams have been built to control the Nile's flow, monsoon rains in Ethiopia caused it to swell along its lower reaches and inundate the surrounding countryside every year from June to October. Most of Egypt's population were farmers, who stood idle during the inundation, unless they were called up to work on public monuments such as the king's tomb. When the Nile receded, it deposited rich silt, ensuring that the farmers always planted in fertile soil. Except for those years when the flood was disastrously high or low, Egyptians were secure in their knowledge that the river would guarantee them enough to eat (see pp.12–13).

The Nile in Egypt has two main parts: the Valley and the Delta, corresponding to the ancient divisions of the country into Upper and Lower Egypt. The Valley, some 660 miles (1,060km) long, is a remarkable canyon that is an offshoot of the African Great Rift Valley. The floodplain occupies 4,250 square miles (11,000km$^2$) and ranges in width from just one and a quarter miles (2km) at Aswan to eleven miles (17km) at el-Amarna.

At present, the Nile splits near Cairo into two branches that flow into the sea at Rosetta (Rashid) and Damietta (Dumyat). These are all that remain of several branches that existed until medieval times (see map, opposite). The silt left by the branches formed a broad triangle of fertile land that covers some 8,500 square miles (22,000km$^2$). The Greeks called this land the "Delta" because its shape reminded them of the inverted fourth letter of their alphabet (Δ). The Delta is fifty-seven feet (17m) above sea level near Cairo and is fringed in the coastal regions by lagoons, wetlands, lakes and sand dunes. In parts of the eastern Delta there are conspicuous low hills known as "turtle backs". These sandy "islands" in the surrounding silty plain were rarely submerged by the annual inundation and in Predynastic times (to ca. 4000BCE) villages and burial grounds

ABOVE: *The* shaduf, *a device for lifting water out of the Nile and emptying it into irrigation trenches, dates from pharaonic times and is still in use today in parts of rural Egypt.*

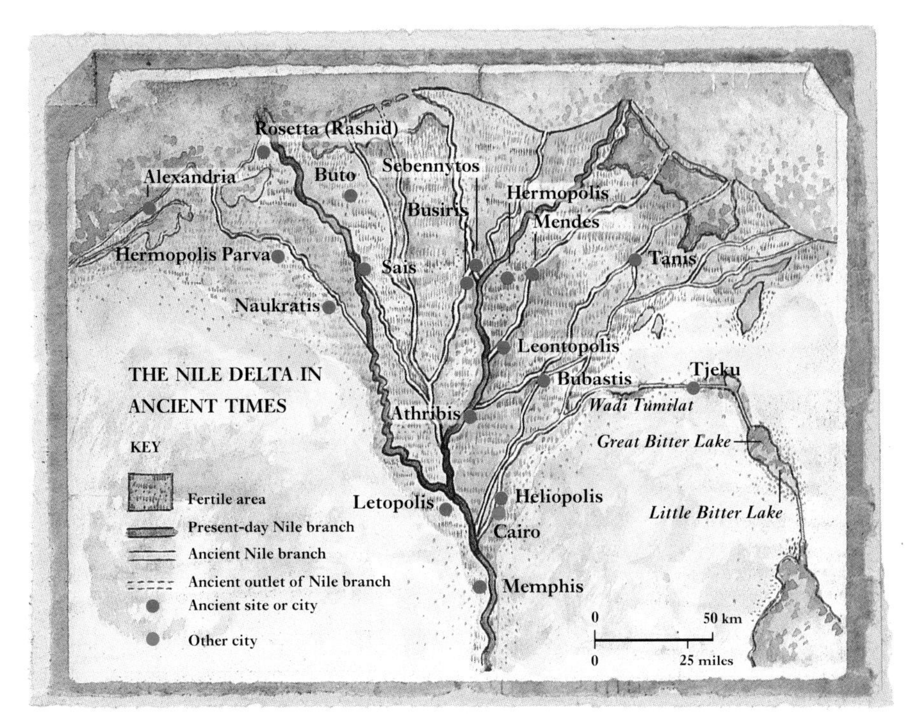


*A map of the Nile Delta in ancient times, showing the possible course of its numerous distributaries. According to Greek and Roman historians, there were once at least five and possibly as many as sixteen Nile branches. Changes in the hydrographic regimes of the Nile between the 10th and 12th centuries* CE *were responsible for the reduction of the branches to just two, flowing into the sea at Rosetta and Damietta (see illustration, below).*

became established on their slopes. From the Old Kingdom (ca. 2625–2130BCE) onward, the apex of the Delta was close to Memphis, the ancient capital. It is now fifteen miles (25km) north of Cairo.

The Nile divides the eastern margin of the Sahara into the Western Desert (also known as the Eastern Sahara and the Libyan Desert) and the Eastern Desert. The Western Desert covers about two-thirds of Egypt, and its most striking features are a series of rocky desert plateaux and sandy depressions, in which nestle lush oases (see map, p.50). The Eastern Desert, characterized by the prominent Red Sea Hills, was important in pharaonic times for its minerals (see map, p.65). The Sinai, essentially an extension of the Eastern Desert across the Gulf of Suez, was also a major source of minerals, especially copper. Wheat, barley, sheep and goats were domesticated in the Near East at least two thousand years before they appeared in the Nile Valley. Herders in the deserts of Palestine and the Sinai were probably driven to seek refuge in the Delta by great droughts seven thousand years ago.

The Western Desert, which was not always as dry as it is today, has yielded the oldest evidence of humankind in Egypt. Tools at least half a million years old have been found close to long-vanished rivers and springs, and the first domesticated cattle in Africa may have been tended ca. 9000BCE near ephemeral lakes (*playas*) in the southwest of the Western Desert, not far from the present-day border with Sudan. It may well be in these areas that the roots of Egypt's civilization lie: here, herders took the first steps toward complex social organization and developed fundamental elements of Egyptian society and religion, before the increasing aridity of the region forced them to drift toward the Nile Valley (see pp.106–7).

*A colour-enhanced Landsat satellite view of Egypt and the Nile: the red areas represent the fertile floodplain. The meanders of the Nile have drifted over time. During the Old Kingdom, the main channel was close both to Memphis and to the limestone quarries near Tura on the opposite side. Over the last millennium, the river has shifted eastward: for example, most of the west bank of Cairo accrued between the 10th and 14th centuries.*

# THE NURTURING WATERS

**MATERIAL RESOURCES**
**Abundant food supplies were supplemented by other important economic resources. Flax was used to make fine linen garments and rope. Now rare in Egypt, papyrus grew in great thickets in the marshes and swamps. The stalks were sliced into strips to make sheets (a sheet consisted of one layer of horizontal strips placed over a layer of vertical strips). The two layers were then beaten together to make the best writing material available until the arrival of paper in Arab times. Other reeds and grasses were turned into mats and baskets. Nile mud provided clay for pottery and sun-dried bricks.**

**Egyptian sycamore, fig and acacia were employed in shipbuilding, but better-quality timber had to be imported. Lebanese cedar was used for ships, fine chests and coffins.**

The civilization of Egypt and its spectacular achievements were based throughout its history on the prosperity of a mainly agrarian economy. The country's verdant green fields and bountiful food resources depended on the fertile soil of the Nile floodplain and the annual summer flood, which commenced in mid-June and lasted until mid-October (see p.10). As soon as the waters began to recede, the farmers returned to their sodden fields to sow their seeds. The crops were ready for harvest from February to early June, when the Nile was at its lowest level.

Egyptian agriculture involved the cultivation of a wide range of plants, the most common being emmer-wheat and barley, staples from which Egyptians made bread, cakes and a nutritious type of beer that was frequently flavoured with spices, honey or dates. The predominantly cereal diet was supplemented by fava beans, lentils and peas (good sources of protein); and other vegetables grown included lettuces, cucumbers, leeks, onions and radishes. Among the most popular fruits, grown in orchards, were melons, dates, sycamore figs and pomegranates. Grapes were also cultivated and were used to make both red and white wines. Oils were extracted from flax and the castor-oil plant (*Ricinus communis*), as well as from sesame in Ptolemaic times. The Egyptians also grew a wide variety of herbs for medicinal purposes.

Poultry and livestock had an important place in the economy. Geese were a common sight along the canals and villages that lined the Nile, and at the time of the annual inundation, migratory water birds flocked to

*The Nile at Beni Hasan in Upper Egypt, about 15 miles (25km) north of el-Minya. The agricultural floodplain contrasts with the desert that begins abruptly on either side.*

Egypt from afar. Pintail ducks, in particular, were caught in nets and snared in traps. Farmers also kept sheep, goats, cattle and pigs. Donkeys were Egypt's main beasts of burden and chief mode of transport on land. Horses were introduced only during the New Kingdom (after ca. 1539BCE); camels and buffaloes did not appear in the Egyptian landscape until a thousand years later, during the Persian occupation.

The Nile itself was a source of an abundance of fish such as tilapia and catfish, both of which were found close to the banks of the river in the muddy waters between the reeds. Nile perch (*Lates nilotica*) was a favoured catch in the irrigation ditches that were dug to channel water from the Nile to the fields.

OVERLEAF: *The view south across the Nile toward the heights of Beni Hasan (compare illustration, opposite page). The cliff face at Beni Hasan contains around 40 Middle Kingdom rock-cut tombs, including several belonging to local nomarchs (provincial governors) of the 11th and 12th dynasties.*

## THE UNPREDICTABLE NILE

The flood discharge from the Nile varied enormously from year to year. When floods were very low, there might be severe food shortages, but excessively high floods would wreak catastrophic destruction in villages and fields. Moreover, the floods sometimes arrived too late or too early, and the floodwaters might not retreat until after planting was supposed to have started. A brief inundation could mean that the waters receded quickly, making it difficult to get enough water to the fields before planting time.

Conditions become particularly difficult when "bad" floods recurred over several successive years. There were periods when low floods and high floods alternated annually, causing major disruption to planting and harvesting schedules. Repeated low floods also led to the silting of major transport canals and the disappearance of many Delta branches of the Nile (see p.11).

Ancient records are scanty, but those of the Nilometer (Nile flood gauge) at Rhoda near Cairo, over the last thirteen hundred years, reveal that from the early tenth century CE to the late fourteenth century CE the floods were still highly variable, with the periods 930 to 1070 and 1180 to 1350 marked by severe droughts. Times of drought were accompanied by outbreaks of pestilence and civil disorder, and it is known that some people resorted to cannibalism. It is not known whether such effects occurred in pharaonic times, but according to one theory, poor floods contributed to the demise of the Old Kingdom.

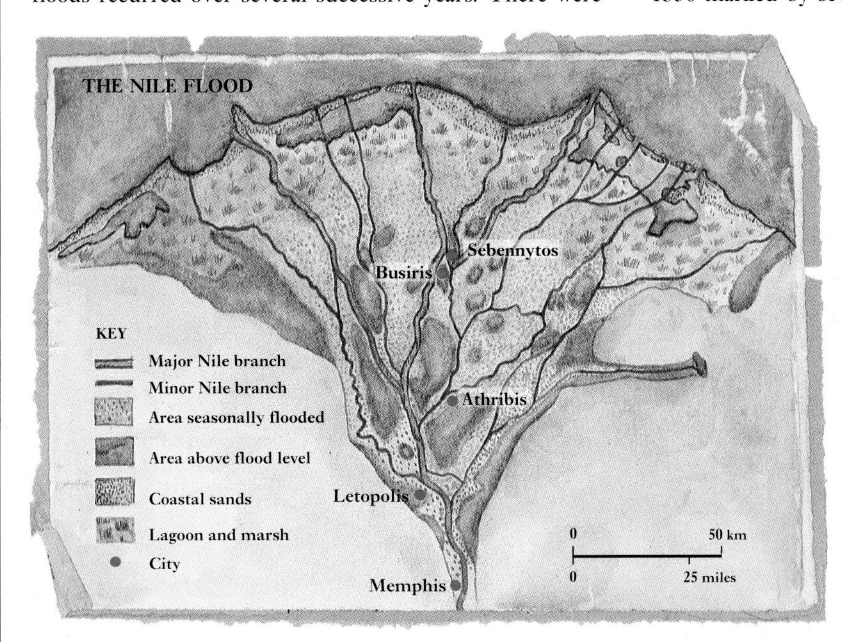


*The Nile Delta in ancient times, showing the area subjected to flooding. The often catastrophic variability of the Nile inundation was the chief motivation behind the construction this century of the Aswan Dam and, especially, the Aswan High Dam, completed in 1971.*

# THE GREAT HIGHWAY

The Nile was at once Egypt's richest source of sustenance and its main communications artery. It flowed from south to north at an average speed of four knots (7.4kph) during the season of inundation, which meant that the voyage from Thebes to Memphis, a distance of around 550 miles (885km), would have taken approximately two weeks. Navigation was faster during the inundation because the water was on average about twenty-five to thirty-three feet (7.5–10m) deep. In contrast, during the season of drought, when the water level was low, the speed of the current was much slower, about one knot (1.8kph), and the same trip would have taken at least two months. At the Nile's lowest point, in June, the water was no more than seven feet (2m) at Aswan compared with just under eighteen feet (5.5m) near Memphis.

The trip from north to south would have been extremely slow before the invention of sails (probably ca. 3350BCE or a little later) to take advantage of the northerly and northwesterly winds blowing off the Mediterranean. At all times of the year the great bend near Qena, where the Nile flows from west to east and then back from east to west, slows down river travel considerably. Night sailing was generally avoided because of the danger of running aground on one of the many sandbanks and low sandy islands (see illustration, p.12).

In the late Predynastic or Naqada II period (ca. 3500–3100BCE), Egyptian boats developed from craft made of reed bundles into big ships constructed from wood planks. Early rock art suggests that some boats were over fifty feet (15m) long and could carry a crew of thirty-two. Multi-oared boats existed before this time, in the early fourth millennium BCE. Clay models of boats found at Merimde Beni Salama in the Delta date back to the fifth millennium BCE.

By the Early Dynastic Period, Egyptian boatbuilding had attained high standards. At Abydos, boat pits (see p.172) associated with a First-Dynasty funerary complex of ca. 3000BCE have revealed a fleet of twelve boats between fifty and sixty feet (15–18m) long. But perhaps the greatest discovery from this period is that of a barque of the pharaoh Khufu, builder of the Great Pyramid (see p.158). Buried in pieces next to the pyramid, it was recently reassembled and measured an impressive 144 feet (43.8m) in length.

From the earliest times, boats were used to transport people

**CANALS**
**Extinct branches of the Nile in the eastern Delta once played a key role in the import of goods into Egypt from the Near East. When the branches began to silt up, they were re-excavated as artificial canals. A major canal east of the Delta is depicted in reliefs showing King Sety I (ca. 1290–1279BCE) crossing the border into Asia. Later, Necho II (ca. 610–595BCE) dug a canal to connect the Nile with the Red Sea (see map, opposite). This waterway was maintained and deepened by the Persians and the Ptolemies. The Greek historian Herodotus remarked that two large ships could navigate the canal side by side. As early as the Sixth Dynasty (ca. 2350–2170BCE), the Egyptians also dug a canal at the First Cataract to ease the movement of boats through the rapids. However, during low water, boats had to be hauled out of the river and dragged on land past the cataract. Another canal at the cataract was excavated in the reign of Senwosret III (ca. 1836–1818BCE).**

*Mourners with a mummy aboard a wooden model boat of unknown provenance; it was placed in a tomb ca. 1900BCE to symbolize a voyage to the sanctuary of Osiris at Abydos.*

*A scene from the Book of the Dead of the priest Chensumose showing him sailing on the waters of the underworld in a vessel resembling a small Nile skiff rigged to catch the Nile's prevailing northerly winds. Twenty-First Dynasty (ca. 1075–945BCE).*

between villages during the inundations, to ferry them across the river, and to transport cattle, grain and other commodities. They were also deployed in military campaigns. From the Fifth Dynasty onward, Egyptian shipwrights were making sailing boats capable of ocean navigation.

Together with the donkey – the principal overland transport – boats made possible the economic and political integration of the country. The capitals of the nomes, or provinces (see p.27), were linked with the national capital by boats and barges that carried local revenues to the royal storehouses. The emergence of a royal state in Egypt may have been linked with the coordination of grain collection and other relief activities developed as part of a strategy to deal with unexpected crop failures in a particular district. In pharaonic times, grain from several districts stored in a central granary would be sent by river to an area hit by famine.

Artificial harbours and ports to accommodate large cargo boats were an essential feature of the riverine landscape. Towns took advantage of the deeper side of the Nile channel close to the shore to establish ports. They also built rock jetties that extended a short way into the river, perhaps in response to changes in the course of the Nile. The site of a huge harbour at Medinet Habu in Western Thebes, built during the reign of Amenhotep III (ca. 1390–1353BCE), is marked by huge elongated mounds created by the earth from the harbour's excavation.

Other large harbours are known from Memphis and the Delta city of Tanis. The port at Tanis was used by Thutmose III (ca. 1479–1425BCE) to connect Memphis with the eastern Delta.

# A LANDSCAPE OF THE MIND

The unpredictability of the Nile floods (see p.13) exercised a powerful hold on the Egyptian imagination. The period just before the inundation, when the river was so low that in places a person could cross on foot, was a time of apprehension: when the flood came it was frequently wild and dangerous. The Egyptians could not tame the river, but they sought to prevent its worst effects by managing the landscape to take advantage of natural conditions – for example, by strengthening natural levees to form embankments. At times of low floods, artificial canals carried water to the thirsty uplands of the floodplain. Flood basins were managed so that water could flow from one basin to another, enabling areas up and down the valley to have sufficient water in time for planting. The desire for order which permeated the Egyptians' world-view was surely derived in no small measure from the chaotic presence of the river in their midst.

To the Egyptians, every being, including Pharaoh and the gods, had to abide by the fundamental cosmic principle of *ma'at,* personified as Ma'at, the goddess of order, justice and goodness. The cosmic order was also embodied in the movement of the god Re, the sun, the other prominent natural element whose rhythms regulated Egyptian lives. The sun god was believed to be ferried daily across the sky in a boat, and to return through the underworld on a barque to a point below the eastern horizon (see pp.118–19). Such mythological vessels recalled the ferryboats that plied between the banks of the Nile.

On earth, order was maintained by the pharaoh, the manifestation of the god Horus, son of Osiris and Isis (see pp.134–5). According to allusions in early religious texts, and later literary and artistic references, Osiris taught the people how to take advantage of the Nile by giving them

*The weighing of the heart (see p.137) before the god Osiris, whose face was depicted green (now discoloured by time) because of his association with the growth of crops and the annual renewal of the land by the Nile. In this scene from the Book of the Dead of Nefer-Is (ca. 350BCE), he is presented with offerings representing Egypt's bounty.*

the arts of cultivation and civilization. He was slain by his brother, Seth, who was identified with the forces of evil and chaos. After death, Osiris returned to life as king of the underworld, where he ordained the life-giving waters of the annual inundation.

Egyptian concepts of time were based on the daily rising and setting of the sun and the three-part cycle of the Nile: drought; the season in between; and the season of inundation. Cosmic space was delimited by the four corners: the south (the source of the Nile), the north (where the pole star shone), the east (where the sun rose) and the west (where it set). Time and space were thus linked to the two most important elements in Egyptian cosmology; and these elements, in turn, were linked in the cosmic order with life, death and rebirth.

## DEITIES OF THE NILE

Although Osiris ordained the annual inundation, the god most associated with the river itself was Hapi, depicted as a human figure with a large belly and pendulous breasts. This corpulence represented the bounties of the Nile, whose waters flowed to nurture Egypt. Hymns addressed to the Nile spoke of its bounty, expressing joy at its coming, and sorrow at the plight of Egypt when the Nile floods failed. The inundation was ritually greeted with thanks and jubilation in honour of Hapi, its patron divinity. The god is depicted with a papyrus plant, another symbol of the benefits of the Nile, sprouting from the top of his head.

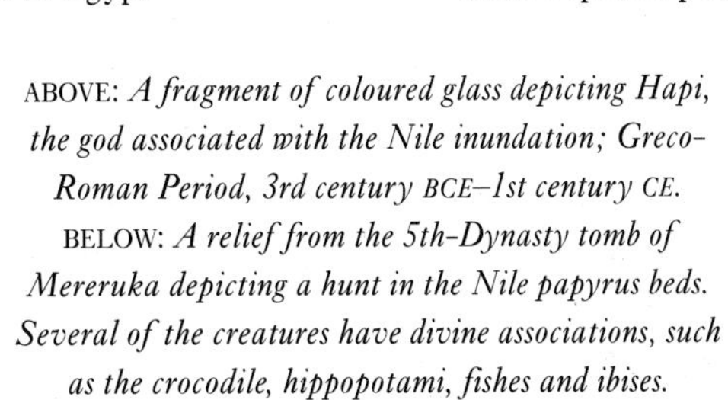

ABOVE: *A fragment of coloured glass depicting Hapi, the god associated with the Nile inundation; Greco-Roman Period, 3rd century BCE–1st century CE.*
BELOW: *A relief from the 5th-Dynasty tomb of Mereruka depicting a hunt in the Nile papyrus beds. Several of the creatures have divine associations, such as the crocodile, hippopotami, fishes and ibises.*

The Nile was a river of creative forces. Its source was believed to be in the underworld, where it was connected to a subterranean stream. From the underworld it issued to the surface between granite rocks close to the First Cataract near Elephantine in the far south. As the fount of Egypt's fertility, the (supposed) source of the Nile was linked to the ram-headed creator god Khnum, who was believed to have fashioned humankind from Nile mud on a potter's wheel. Satis, the consort of Khnum in the south, together with her companion Anuket, were revered as the dispensers of cool water. Satis was often depicted pouring water onto the earth to endow it with life. Unlike Khnum, she was shown in human form wearing the crown of Upper Egypt with two gazelle horns.

Nile creatures, such as the hippopotamus, the crocodile and fish, were venerated as gods of fertility. Heket, a frog, was revered as a goddess of childbirth, as was the hippopotamus goddess, Taweret. In the story of Isis and Osiris, Heket was said to have assisted Isis in bringing the murdered Osiris briefly back to life, in order that he could father the god Horus (see main text).

● CHAPTER 2

# THREE KINGDOMS AND THIRTY-FOUR DYNASTIES

The Egyptians were by no means the first people to acquire the prerequisites of civilization – the arts of farming and urban living – but once these had become established, the land of the Nile developed a culture of extraordinary durability. For most of its ancient history, Egypt was under the sway of the pharaohs, all-powerful monarchs of divine status. For three thousand years, the name of Pharaoh commanded awe not only among Egyptians but also throughout the civilized lands of Africa and the Near East.

▲



ABOVE: *Part of a roll-call of the names of the royal predecessors of Ramesses II, from his cult temple at Abydos. Such "king-lists" are an important source for our knowledge of ancient Egyptian chronology, although some pharaohs, such as the "heretic" Akhenaten (see pp.128–9) and his immediate successors, are regularly omitted.*

## "MOST ANCIENT EGYPT"

Egypt began its march to civilization rather late compared with some regions of the Near East. Yet once it had taken root, the great civilization of the Nile proved to be the most durable of all, spanning more than three thousand years from the appearance of the first unified kingdom to the final eclipse of ancient Egyptian culture in the early Christian era.

For most of its ancient history, Egypt was ruled by kings, or pharaohs, who in ancient times were grouped into thirty-one dynasties (see sidebar, opposite). Egyptologists now tend to count the Macedonian and Ptolemaic dynasties as numbers thirty-two and thirty-three, and they have also added a thirty-fourth, the so-called Dynasty "0", to account for a handful of very early kings. The dynasties in turn are subdivided into several periods, three of which are regarded as the peaks of Egyptian civilization: the "Old Kingdom" (the earliest pyramid age); the "Middle Kingdom" (virtually synonymous with a single great dynasty, the Twelfth) and the "New Kingdom" (the age of the great warrior-pharaohs, such as Thutmose III and Ramesses II).

Features of civilized life, such as agriculture and towns, only appear in Egypt in the sixth millennium BCE, some two thousand years later than in Anatolia, Mesopotamia and Syria-Palestine. This may be owing to Egypt's rich natural resources rather than any cultural retardedness: the savannas adjoining the Nile Valley remained home to an abundance of plants and animals until these areas became desert, by ca. 2000BCE. The seeds of

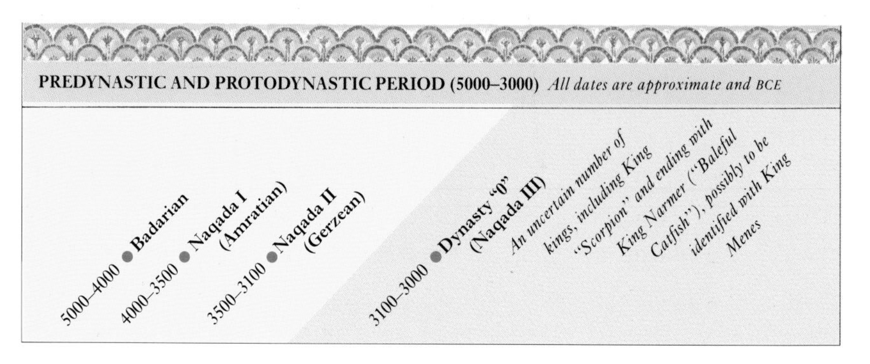

*One of two pairs of colossal statues of the pharaoh Ramesses II (see p.35) that flank the entrance to the temple of Abu Simbel in Lower Nubia. Ramesses' reign (ca. 1279–1213BCE) – the longest of Egypt's imperial age – is one of the best-documented periods of Egyptian history. Like many of his surviving monuments, the temple of Abu Simbel incorporates accounts of the events of his reign.*

Egyptian civilization lie in a number of late Neolithic cultures that emerged ca. 5000BCE and, over the next thousand years or so, developed into distinctive regional cultures in Upper and Lower Egypt. In the late fourth millennium BCE, the autonomy of the northern culture was eroded by the rise of an aggressive rival in Upper Egypt. The development of this southern culture is traced through a number of stages named after archaeological sites: Badarian, Naqada I (or Amratian), Naqada II (or Gerzean) and Naqada III (or Dynasty "0"). Collectively these make up the "Predynastic" and "Protodynastic" periods (see timechart).

The Naqada II period saw the growth of a prosperous and unified culture in Upper Egypt, with political power consolidated in towns such as Hierakonpolis (see p.69), Naqada and This. Classic Egyptian concepts of divine authority began to evolve, including the ruler's identification with the sky god Horus. By the later Predynastic Period, the southern kingdom's cultural penetration of Lower Egypt would be followed, gradually but inevitably, by a political takeover of the north. (See also pp.106–7.)

**WRITING EGYPTIAN HISTORY**

**The framework used today to describe Egypt's ancient history is not entirely a construct of modern scholarship, but is based on sources from antiquity. The original division into dynasties derives from a history of Egypt by Manetho, an Egyptian priest who wrote in Greek in the third century BCE, perhaps for King Ptolemy I. The history, which identified thirty-one dynasties before the Ptolemies, is mostly lost, but other ancient authors occasionally cite it and a précis of its contents is preserved by early Christian writers.**

**Manetho's history clearly depends upon authentic materials, such as king-lists (see illustration, opposite), historical monuments and literary texts based on actual events. However, the work poses other problems apart from the abbreviated form in which it survives. Manetho uses his sources uncritically and there are mistakes in the text as it has been transmitted. For the most part, scholars now use Manetho only to supplement the more substantial and reliable archaeological records.**

**The broader division of Egyptian history into three "kingdoms" (see main text) is also ancient in origin. It is implied in a list of royal ancestors in the mortuary temple of Ramesses II (ca. 1279–1213BCE) in Western Thebes, at the head of which are three great kings who unified Egypt – Menes (Dynasty "0" or First Dynasty; see p.23), Nebhepetre Mentuhotep II (Eleventh Dynasty) and Ahmose (the founder of the Eighteenth Dynasty).**

# THE FIRST NATION-STATE

**THE BATTLES OF HORUS AND SETH**

**King Peribsen's apparent abandonment of Horus in favour of Seth (see main text) was once interpreted as a historical prototype for the antagonism between Horus and Seth in Egyptian religious thinking. The theologians of Heliopolis developed a pantheon in which the ancient sky god, Horus ("He who is on high"), with whom every pharaoh was identified, became the son of the god Osiris and his sister and consort, Isis. Osiris was king of Egypt during "the period of the god", a primeval era when divinities were said to have ruled on earth. But Seth murdered Osiris and took the throne; Horus grew to manhood and eventually defeated the usurper: their struggle, known as *The Contendings of Horus and Seth* and first described in the Pyramid Texts (see p.188), became one of the epic themes in Egyptian literature (see also pp.134–5).**

The consolidation of the Egyptian state occurred gradually over a period spanning the later Predynastic Period and the early First Dynasty. A late tradition claims that the first two dynasties were based at This, or Thinis, in Upper Egypt, and it is true that, from Dynasty "0" to the end of the Second Dynasty, the Upper Egyptian kings who conquered the north were buried at Abydos, not far from This. These early rulers are known as "Horus-kings" because their names are written in a frame, or *serekh*, incorporating an image of a palace façade (𓉔) surmounted by a falcon (𓅃) identified with the god Horus. The name in a *serekh* is called the "Horus-name". The most conspicuous steps toward Egyptian unity were taken around the beginning of the First Dynasty under the Horus-kings Narmer ("Baleful Catfish") and Aha ("Fighter"). Aha transferred the centre of government to the vicinity of Memphis. (See also pp.106–7.)

Arguably the greatest achievement of Egypt's early rulers was to forge, not only a powerful state, but also a national consciousness across widely separated regions with strong local customs. The main instrument of this nation-building was the royal government, which reserved the highest offices for members of the king's family, while it also became increasingly dependent on a bureaucracy of able commoners. The tombs of Early Dynastic officials reveal both the rewards of royal service and a civilization becoming ever more culturally unified.

Many of the classic institutions of Egyptian government and society were evolving during this Early Dynastic Period, but it is difficult to get a complete picture of developments. Surviving annals of the late Old Kingdom are fragmentary and say nothing at all of the ideological adjustments which, other evidence suggests, the regime was forced to make from time to time. For example, late in the Second Dynasty, King Peribsen (ruled ca. 2700BCE) took on the identity of the god Seth in addition to, or in place of, that of Horus. Peribsen's move may simply have been an attempt to appeal to worshippers of Seth in Upper Egypt and does not seem to have provoked any enduring hostility, because Peribsen's cult survived into the

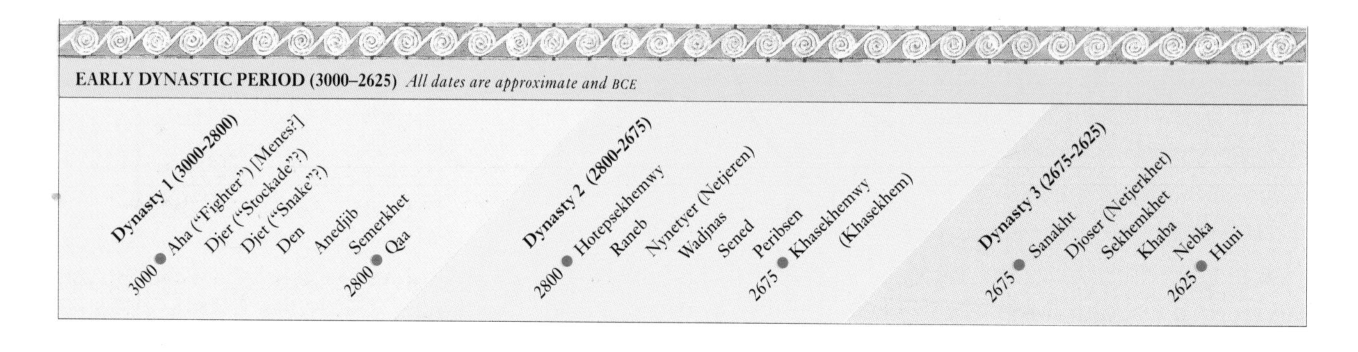

Fourth Dynasty. In any case, a compromise was clearly reached under Khasekhemwy, or Khasekhem, the last king of the dynasty, who proclaimed that "the two gods who are in [the king] are at peace". His *serekh* is unique in having the image of Seth alongside that of Horus; his successors reverted to the Horus falcon alone.

At the start of the Third Dynasty, the royal cemetery was transferred from Abydos to Memphis, perhaps partly to enhance the latter city's status as the national capital. Certainly, the Step Pyramid complex of King Djoser (ca. 2650BCE) at Saqqara (see pp.178–9) is not only the grandest monument to divine monarchy to date but, with its stone models of Upper and Lower Egyptian shrines, it is also the most explicit reflection of the king's status as "Lord of the Two Lands".

## MENES, THE LEGENDARY UNIFIER OF EGYPT

*The late Protodynastic "Narmer Palette" depicts King Narmer, who is often identified with Menes, wearing the crown of Upper Egypt, crushing a people of the Delta. On the other side of the slate he is shown wearing the Lower Egyptian crown.*

In Egyptian tradition, the unification of the "Two Lands" is credited to a legendary ruler called Meni ("Min" and "Menes" in Greek sources), who is said to have founded the capital city of Memphis on the border between Upper and Lower Egypt. In the fifth century BCE, Egyptian priests had assured the Greek historian Herodotus (*Histories*, Book Two) that Min was the first king of Egypt, and had reclaimed the land around the city that he founded, Memphis. When Manetho wrote his history two centuries later (see p.21), he confidently identified Menes as the first human king of Egypt, adding that "he made a foreign expedition and won renown, but was killed by a hippopotamus".

But the tradition of Menes as the first king of all Egypt goes back only to the fifteenth century BCE, when Eighteenth-Dynasty monuments call the legendary unifier "Meni". He is commonly identified with the early king Narmer, who appears on a ceremonial slate palette triumphing over a people of the Delta and wearing the traditional crowns of the two Egyptian kingdoms (see illustration). However, the name or title Meni is recorded neither for Narmer nor any other "Horus-king" of early Egypt, although the name Men is attested for other individuals of the period. It may be a king's personal name, as opposed to the official "Horus name", like Narmer or Aha (see main text). Another royal title was the "Two Ladies" name. An ivory label of Aha shows him before a structure inscribed with the name "Two Ladies *Men*". Even if Men does not refer to Aha, as the king who historically made Memphis the capital of Egypt he already has quite a strong claim to be identified with the Menes of legend.

It has been suggested that Menes might have been purely a historical construct: the Egyptian *Men-i* can mean "So-and-so-who-once-came". According to yet another theory, the name might represent a deliberate reversal of the syllables of the god Amun, or Amen, who in Egypt's imperial age came to be seen as the divine "father" of every pharaoh.

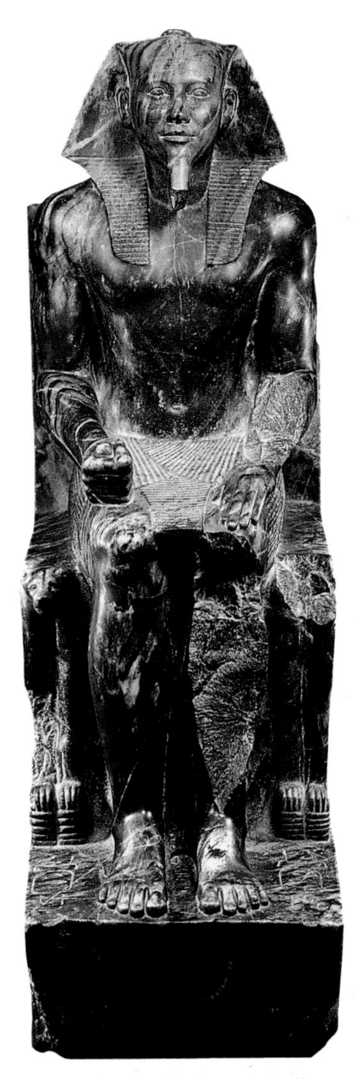

*By the 4th Dynasty, Egyptian sculptors had achieved a level of mastery of stone that they were never to surpass. Probably the finest Old Kingdom sculpture yet discovered is this superbly modelled diorite statue of Khafre, or Chephren, who built the second of the Giza pyramids (see pp.184–5).*

# THE ZENITH OF THE UNITARY STATE

Early Egyptian civilization reached a peak of efficiency and splendour during the Old Kingdom (the Fourth to Eighth dynasties). Royal power, reflected in the great pyramid complexes (see pp.168–91), would never be greater than it was in this period, and Egypt's international prestige, proudly vaunted in official records, is also reflected in archaeological finds in Asia, Nubia and the deserts flanking the Nile Valley.

Under King Sneferu (ca. 2625–2585BCE), the founder of the Fourth Dynasty, the royal tomb became a true pyramid, perhaps symbolizing a ramp of sunbeams which would lead the pharaoh to his ultimate divine destiny in the heavens. The sovereign now embodied not only Horus, but also the sun god, Re, and the formal royal title *Sa-Re*, "Son of Re", was in use by the middle of the Fourth Dynasty. A vogue for temples to the sun can be seen in the royal cemeteries of the Fifth Dynasty (see p.188), while the "Pyramid Texts", an anthology of spells carved on the inner walls of pyramids from the late Fifth Dynasty onward, repeatedly state that "the king belongs to the sky".

Royal power, centred on the royal residence near each king's pyramid complex, was implemented by a government that increasingly came to be in the hands of trusted commoners. The highest offices became closed to royal relatives during the Fourth Dynasty and were filled by lower-born individuals who sometimes bore the honorific title "King's Son". At the head of the administration was a chief minister (or sometimes two ministers) who presided over the departments of government: the granary and treasury, public works, the judiciary and the civil service. Officials were rotated in different jobs around the country, as dictated by necessity, ability and the royal will. Instead of a standing army, local militias were put into the field when required. Temples were endowed by royal charter, but in times of need they were liable to have property or manpower requisitioned by other official departments unless specifically exempted.

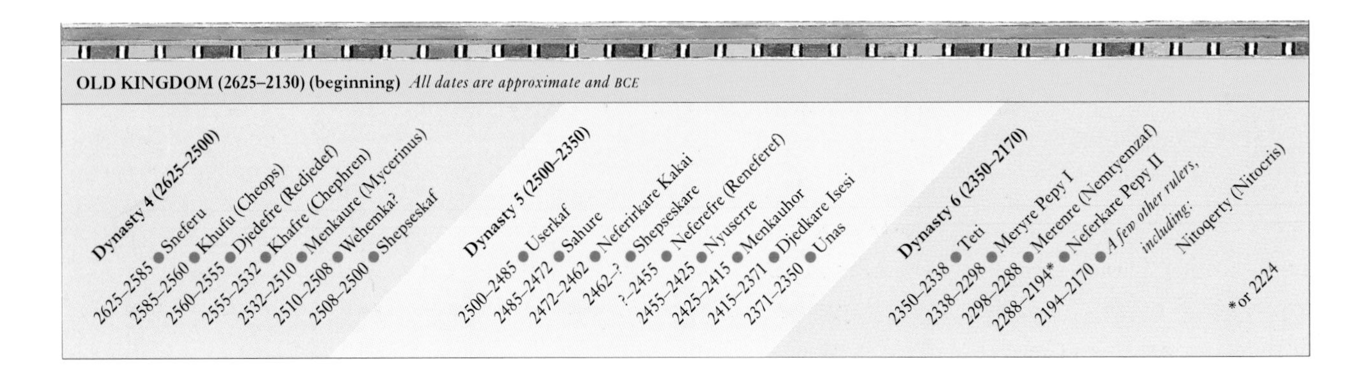

## KHUFU: THE MAKING OF A LEGEND

The person of King Khufu (known to the Greeks as Cheops) is virtually eclipsed by his monumental resting place, the Great Pyramid at Giza. He is hardly a shadowy figure, even in the shattered records of his own time, but most of what we know about his reign (nearly all of it through records of his family and officials) comes from tombs in the vicinity of that colossal monument. However, little in the way of hard historical fact emerges from these sources. It was once believed that some intrigue surrounded the reburial of Khufu's mother at Giza after her tomb was robbed – it was thought that courtiers conspired to keep the full truth from the king. But this story is no longer beyond dispute. An alleged feud within the royal family, involving Khufu's sons and successors, Djedefre and Khafre (or Chephren), is also doubted today. Even the king's features would escape us, were it not for a single tiny statuette discovered in two pieces in the temple of Osiris at Abydos in Upper Egypt (see illustration).

*This ivory statuette – shown actual size – is the only image so far discovered of Khufu, the builder of the Great Pyramid.*

Khufu's commanding monument has thrown its shadow over the king's historical reputation. Later Egyptian literature presents him as a grim, authoritarian figure, most notably in a popular tale recounted in the Westcar Papyrus. In this story, a magician, Djedi, who can reputedly bring the dead back to life, is presented to the king: "His Person [Khufu] said: 'Bring me a prisoner from prison, that his punishment may be inflicted [that is, the prisoner would be killed and brought back to life].' " Djedi protests: " 'O my sovereign lord – live, prosper, be healthy! ... It is not ordained that such things be done to [humankind].' " Khufu relents and Djedi demonstrates his magic on a goose and an ox.

By the fifth century BCE, native tradition claimed that Khufu became a tyrant owing to his obsession with finishing the Great Pyramid. The priests told the Greek historian Herodotus that to raise money for the project, Khufu seized temple property throughout Egypt and even forced his own daughter to work in a brothel. The idea of a tyrannical Khufu continues to resonate today, when tourists are still entertained with a composite of such ancient stories.

As with the Early Dynastic Period, few concrete events of the Old Kingdom can be gleaned from contemporary records. It is now doubted that there was a quarrel among the sons of Khufu (see box), as was once believed. Similarly, a popular tradition of a rivalry between the last kings of the Fourth Dynasty and their successors in the Fifth is also now discounted. The autobiographies in the tombs of Sixth-Dynasty officials provide few hard facts but a more coherent sense of developments. In Upper Egypt, certain families became entrenched in office as hereditary "nomarchs", rulers of the "nomes" (provinces; see p.27). Two members of one such family even became queens of Pepy I (ca. 2338–2298BCE) and the mothers of his two successors, Merenre and Pepy II. Further challenges to the pharaohs arose with the formation of small states in Nubia, which led to growing friction between Egypt and its southern neighbour during the long reign of Pepy II.

# THE FIRST CHALLENGE TO UNITY

*A statue from Thebes of the pharaoh Nebhepetre Mentuhotep II, the greatest ruler of the 11th Dynasty, who reunited Egypt and ruled for half a century (ca. 2008–1957BCE). He wears the Red Crown of Lower Egypt.*

By the later years of the Old Kingdom, Egypt had become engulfed in a crisis that had been developing for a long time. After the extravagance of pyramid-building at the height of the Fourth Dynasty, there was a decrease both in the size of royal tombs and in the care with which they were built. The financing of mortuary monuments for royal and private patrons must have strained the resources of a centralized economy, and documents from pyramid cities in the Fifth Dynasty show endowments being reduced and redistributed. Such economic problems were probably linked to a progressive drying-out of the climate which had been taking place all over the Near East in the later third millennium BCE. The need to cope with the effects of lower inundations of the Nile may be one reason for the decentralization of power that occurred in the Sixth Dynasty, when dynasties of "nomarchs" become entrenched in the provinces (see box, opposite). Their descendants, a few generations later, would speak of water shortages and having to protect their people from famine.

The crisis probably came to a head ca. 2200BCE, during the long reign of Neferkare Pepy II (see pp.24–5), the last significant ruler of the Old Kingdom. For some time after his death, kings continued to reign in Memphis (the last rulers of the Sixth Dynasty, followed by the Seventh and Eighth dynasties). Most of them are very obscure, and on the rare occasions when they emerge into the light, they are clearly overshadowed by the powerful southern nomarchs, who effectively governed Upper Egypt. With the passing of the last kings in Memphis (ca. 2130BCE), the nomarch of Herakleopolis, in the twentieth Upper Egyptian nome (see map, opposite), claimed royal power as Akhtoy I. His descendants in the Ninth and Tenth dynasties were supported by Lower Egypt and the other nomarchs of Middle Egypt, but within one or two generations they were challenged by the rise (ca. 2081BCE) of the Eleventh Dynasty, based at Thebes. A prolonged contest between these two regimes ensued, with the

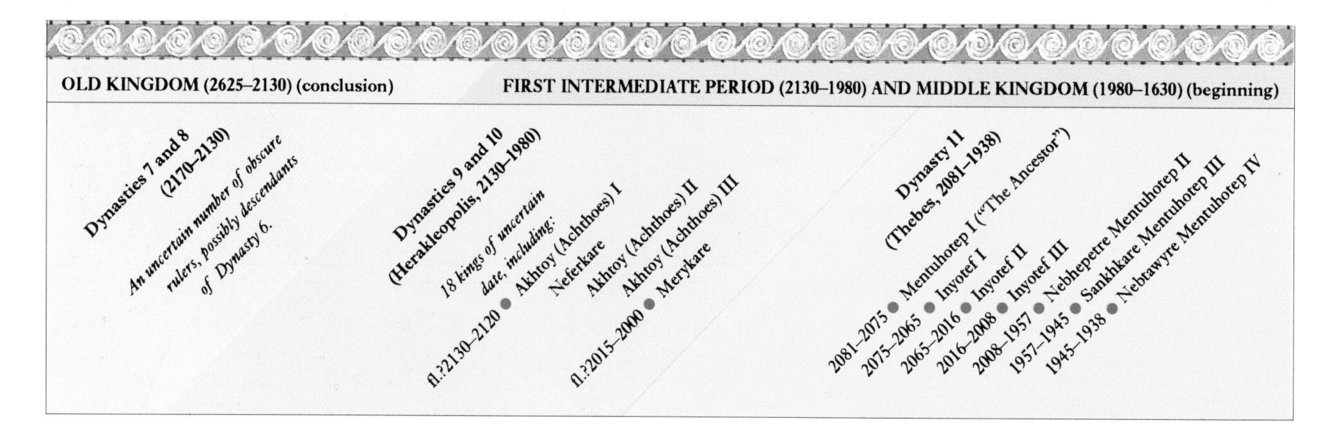

Herakleopolitans unable to eliminate their rival and the Thebans incapable of advancing far beyond the southern boundary of the Thinite nome (probably the eighth Upper Egyptian nome; see map). But under Nebhepetre Mentuhotep II, Thebes harnessed the resources of Nubia, which had been outside Egyptian control since the late Old Kingdom, to subvert most Herakleopolitan partisans in Middle Egypt. Nebhepetre finally reunited Egypt when he overthrew the Herakleopolitans ca. 1980BCE.

Under the Eleventh Dynasty, Egypt recovered quickly from the civil wars which had long divided the country. However, in spite of building projects and other public gestures, the dynasty's focus remained too southern to satisfy everyone in the reunited "Two Lands". The royal residence remained in the confines (then relatively provincial) of Thebes, and complaints of neglect in the border regions of the northeastern Delta ring true (although admittedly they are contained in a later, avowedly propagandist work). Amenemhet, chief minister to Nebtawyre Mentuhotep IV (ca. 1945–1938BCE), successfully bid for power and emerged ca. 1938BCE as the founder of a new dynasty. With Amenemhet I began the "Middle Kingdom", which later Egyptians would regard as a worthy successor to the glories of the Old Kingdom at its height.

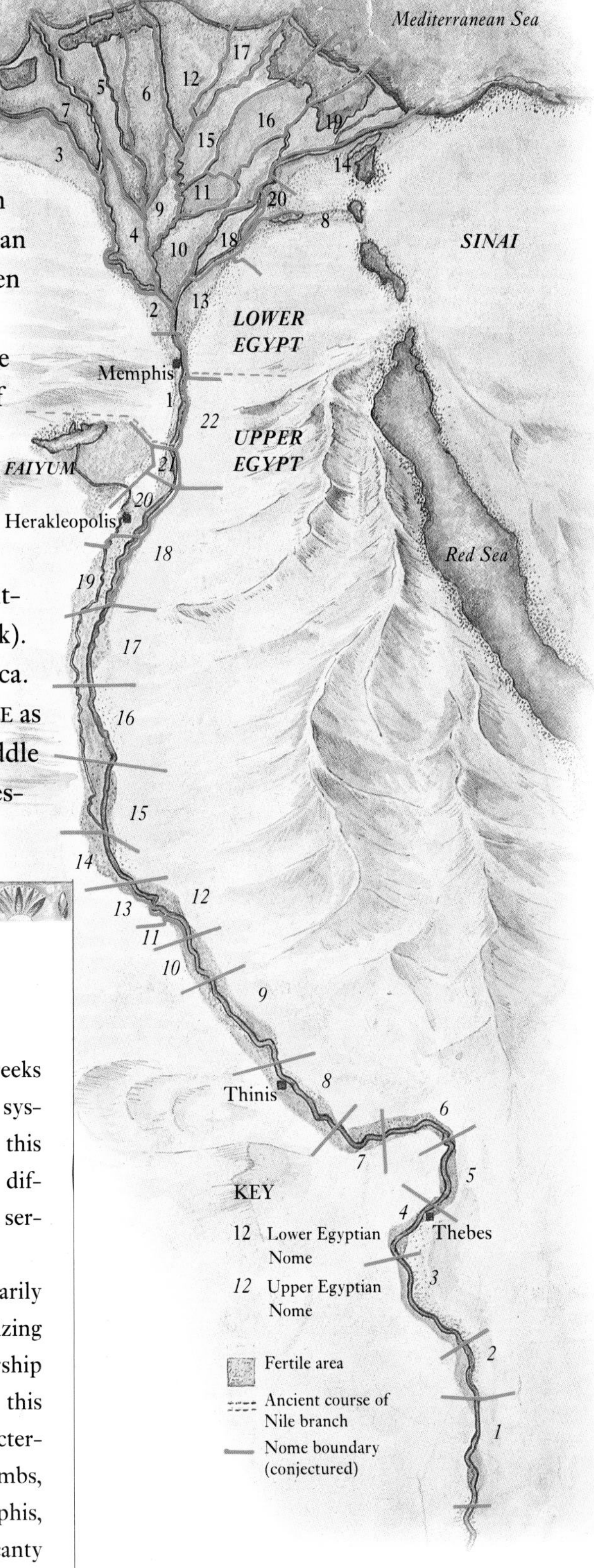


*Ancient Egypt was divided into 42 nomes or provinces. There were 22 nomes in Upper Egypt and 20 in Lower Egypt; the Faiyum and other oases were not administered as part of the nome system. The boundaries between the nomes are conjectural.*

## NOMES AND NOMARCHS

Ancient Egypt was divided into provinces (or "nomes", as the Greeks called them) under provincial governors, or "nomarchs". The nome system is first attested early in the Old Kingdom, in the Third Dynasty. At this period, a typical nomarch seems to have spent his early career serving in different provinces or in other branches of the civil administration. As a royal servant, he would be buried in the cemetery of the capital, Memphis.

However, from the Fifth Dynasty some nomarchs began to serve primarily in one nome, where they lived and were eventually buried. This decentralizing trend, together with the hereditary right of ruling families to the governorship of their provinces, became widespread in the later Sixth Dynasty. From this time on, more and more governors adopted what became the most characteristic nomarch title: "Great Chief of Nome X". Their impressive tombs, increasingly located in Upper and Middle Egypt instead of around Memphis, reflect their power and practical independence, in contrast to the scanty records left by the contemporary kings of the later Sixth to Eighth dynasties. These grand monuments highlight something of an anomaly in early Egypt: royal officials who became, for a time, a hereditary nobility.

# THE GLORIOUS TWELFTH DYNASTY

*The kings of the 12th Dynasty were renowned for their great public works. Amenemhet III (ca. 1818–1772BCE), depicted here in the guise of a god of the Nile, undertook one of the most important projects, a massive reclamation scheme in the Faiyum oasis. The shrinking of Lake Moeris (modern Lake Qarun) greatly increased the area of cultivable land in this rich offshoot of the Nile Valley.*

The beginning of the Twelfth Dynasty, and of the Middle Kingdom, under Amenemhet I was marked by notable achievements on the one hand and apparent growth pains on the other. At home, Amenemhet I's promised commitment to the north took shape in a new fortress, called "The Walls of the Ruler", which protected the Delta's eastern border. The royal residence returned north, to a site south of old Memphis called Itj-Tawy, "[Amenemhet is] The Seizer of the Two Lands", probably modern el-Lisht. Amenemhet also confirmed the rights of the nomarchs, who were allowed to remain as a hereditary nobility in Middle Egypt after changing sides toward the end of the civil war (see pp.26–7). The new king dealt with them firmly but diplomatically.

However, trouble seems to have come from within the royal family itself. Literary works such as *The Tale of Sinuhe* (the supposed autobiography of a courtier who fled to Asia to escape political tumult in Egypt) and *The Instruction of King Amenemhet I* (which claims to be the king's last testament) portray both the pharaoh and his son, Senwosret I, as besieged by disloyalty within the palace. This evidence may suggest either that Amenemhet I was assassinated, or that he made his son co-regent for the balance of his reign (ca. 1919–1909BCE).

Whatever the precise nature of this crisis, the dynasty weathered the storm and went on to establish a remarkable record for length, stability and accomplishment. Its eight kings ruled on average for more than twenty-two years each and, once past the troubled transition to Senwosret I, nothing marred the orderly succession from father to son until the end of the dynasty, when Amenemhet IV's early death brought his sister Sobekneferu to the throne. She was one of the few women to rule as king.

In foreign affairs, the Twelfth Dynasty continued the Old Kingdom policy of keeping the outside world at a safe distance. In Nubia, Amenemhet I built the first of a chain of fortresses which, under his successors, dominated the region of the Nile's second cataract. "The Walls of the

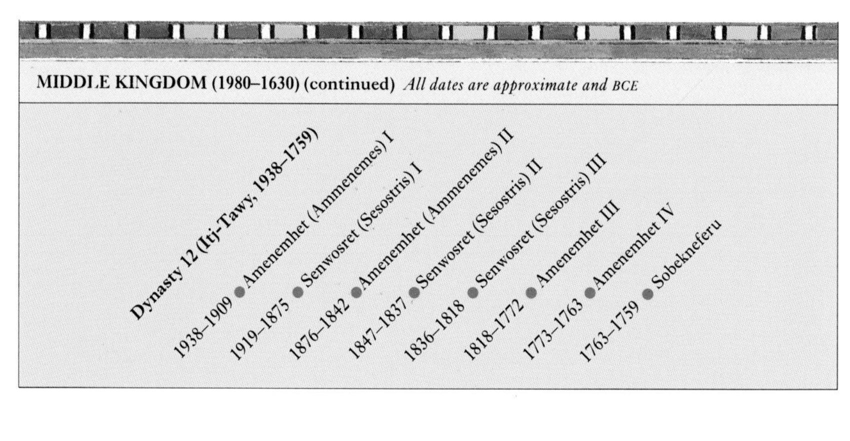

## THE LEGEND OF "SESOSTRIS"

In late antiquity, Egyptian priests regaled Greek and Roman visitors with tales of the fabulous exploits of a pharaoh called "Sesostris". His conquests, they said, had ranged from deep inside Africa to the Near East, and even into Scythia (southwestern Russia) which no later conqueror – not even Darius I of Persia or Alexander the Great – had been able to subdue.

This image of "Sesostris" is manifestly an amalgam of several warrior pharaohs in Egyptian history. However, he can ultimately be traced to the three Twelfth-Dynasty kings called Senwosret. Foreign affairs were a notable feature of each reign. Senwosret I extended Egypt's southern frontiers and sent forays against the Libyans, while Senwosret II expanded Egypt's trading links with Nubia and the states of Western Asia. Senwosret III (see illustration, p.220) campaigned in person in Western Asia. From the chain of forts on Egypt's southern border, begun by his predecessors and completed by him, he also launched numerous campaigns into Nubia. He apparently achieved sufficient success in these expeditions to win enduring fame in the southern lands as a god.

Already deified by the late Middle Kingdom, Senwosret III was still worshipped by his great successors, the warrior-pharaohs of the Eighteenth and Nineteenth Dynasties, including Thutmose III and Ramesses II (whose own achievements, ironically, contributed to the later Sesostris legend). The vivid and personal narrative in which Senwosret III relates his exploits survives on a stone tablet that was specially commissioned by the king to record them.

*This cobra of gold and semi-precious stones was part of the regalia of King Senwosret II, one of the three kings whose reputations formed the core of the legend of Sesostris. The image of the female cobra, known as the* uraeus, *rearing up as if to protect the sovereign and the land, formed part of the pharaonic regalia from Old Kingdom times (see p.109).*

Ruler" served a similar purpose on Egypt's northeastern border. Contemporary records show extensive interaction (mainly in the form of trade and diplomatic exchanges) between Egypt and Western Asia, although it is not known whether Egypt was in touch with the nearest superpower, the Babylonian empire of Hammurabi and his successors.

At home, the central government gradually gained ground against implicit rivals, such as the nomarchs of Middle Egypt. This process, which was complete by the reign of Amenemhet III, is best explained by the royal family's patient policy of encouraging members of these noble families to enter the king's service. A new centralized system emerged which divided the country into two major administrative units, essentially Upper and Lower Egypt, each under a chief minister. The system survived – at least in principle – until ca. 1000BCE and was among the lasting achievements of a peaceful and prosperous age which occupied a favoured place in the memory of later Egyptians.

# A SECOND UNRAVELLING

Egypt remained united under the pharaohs of the Thirteenth Dynasty, who continued to reside at Itj-Tawy. Power appears to have rotated among factions of the bureaucracy, giving rise to a string of short-lived royal families – the dynasty numbers at least fifty-five kings. In fact, the civil service that had been built on behalf of the royal power seems to have ended up by absorbing it.

This regime ultimately failed to match the Twelfth Dynasty's strength at home and abroad. After allowing discipline at the southern forts to deteriorate, the government eventually withdrew its garrisons, and not long afterward the forts were reoccupied by the rising Nubian state of Kush. In the north, parts of Lower Egypt were becoming heavily settled by an immigrant Asiatic population, and an independent line of kings

## THE WARS OF KING KAMOSE

Kamose's exploits against the Hyksos and the Kushites were inscribed in unusually vivid and personal detail on two stelae set up inside the temple of Amun at Karnak (see pp.208–9). The war narrative begins at a meeting of the king's privy council, at which Kamose contemptuously rejects his advisors' urgings to continue his predecessors' policy of cooperation with the Hyksos regime. Kamose sails north, plundering the towns of Hyksos vassals that "had betrayed Egypt their mistress".

Kamose then beards his Hyksos opponent in his capital of Avaris, "as if a kite were preying on the territory of Avaris. I caught sight of his women on the top of his palace ... as they peeped out of their loopholes on their walls, like the young of lizards". The ships in the harbour are plundered of "all their costly woods and all the fine products of Syria". But the greatest humiliation for King Apopi is the capture of his messenger, on his way south to rouse the Nubians against Kamose. Apopi's message to his ally is brusque: "Do you see what Egypt has done against me? ... Come, travel north. Do not grow pale! Behold, he is here in my grasp," he boasts, somewhat prematurely. Kamose in turn boasts that "I caused [the letter from Apopi] to be taken back to him ... so that my victory should invade his heart and his limbs should be paralyzed."

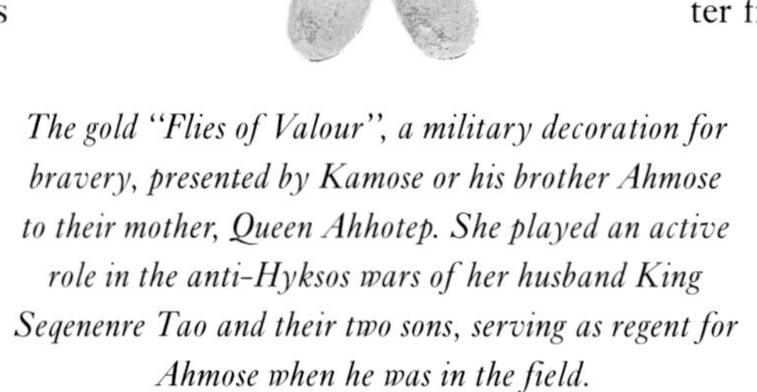

*The gold "Flies of Valour", a military decoration for bravery, presented by Kamose or his brother Ahmose to their mother, Queen Ahhotep. She played an active role in the anti-Hyksos wars of her husband King Seqenenre Tao and their two sons, serving as regent for Ahmose when he was in the field.*

Kamose is wise enough not to overstretch himself and withdraws to Thebes. He arrives with the rising waters of the Nile flood and receives a warm reception: "women and men came out to see me, every woman embracing her companion". The king then repairs to Karnak to offer thanks to Amun. Such a thanksgiving is the usual, even stereotypical, ending to a war narrative in ancient Egypt.

(referred to as the Fourteenth Dynasty) seems to have arisen in the western Delta during the later Thirteenth Dynasty. A text known as *The Admonitions of Ipuwer*, dated toward the end of this period, describes a demoralized and rebellious country with its institutions in disarray, while "foreigners have become [native] people everywhere".

According to Manetho (see p.21), into this unstable mix came invaders from the east. These Semitic newcomers are called the "Hyksos", a Greek rendering of the Egyptian *Heka Khaswt* ("Rulers of Foreign Lands"). Their regime (the Fifteenth Dynasty) replaced the Thirteenth and Fourteenth Dynasties in most of the country. Only in the south did a significant territory remain united under native rule: a kingdom based at Thebes coalesced around a new regime termed the Seventeenth Dynasty. Even this rump state eventually had to acknowledge the suzerainty of the Hyksos, who ruled from the city of Avaris in the northeastern Delta. Egypt's new Asiatic government was also the pre-eminent power in the Near East, with trade and diplomatic relations that extended to the Minoan kings on Crete.

*A grey granite sphinx, originally sculpted as a representation of Amenemhet III of the 12th Dynasty (ca. 1818–1872BCE; see also illustration, p.28). It was later usurped by a Hyksos ruler, who had his own name carved on the statue.*

The Hyksos dealt a further blow to their Theban rivals when they became firmly allied with the Kushite kingdom. As a consequence, the Theban regime was hemmed in by enemies for close to a century. In the 1540s, however, King Seqenenre Tao confronted the Hyksos and their allies. He fell in battle ca. 1543BCE, but his successor Kamose was more successful. He recovered the Nubian forts and led a bold raid as far as the outskirts of Avaris (see box). Kamose died suddenly without issue and the final stages of the war of liberation were delayed until his brother Ahmose had come of age. Ahmose eventually triumphed, capturing Avaris and chasing the Hyksos into Canaan. For his reunification of Egypt, later Egyptians conferred on this son of the Seventeenth Dynasty the honour of beginning a new dynasty – the Eighteenth – and a new era: the New Kingdom.

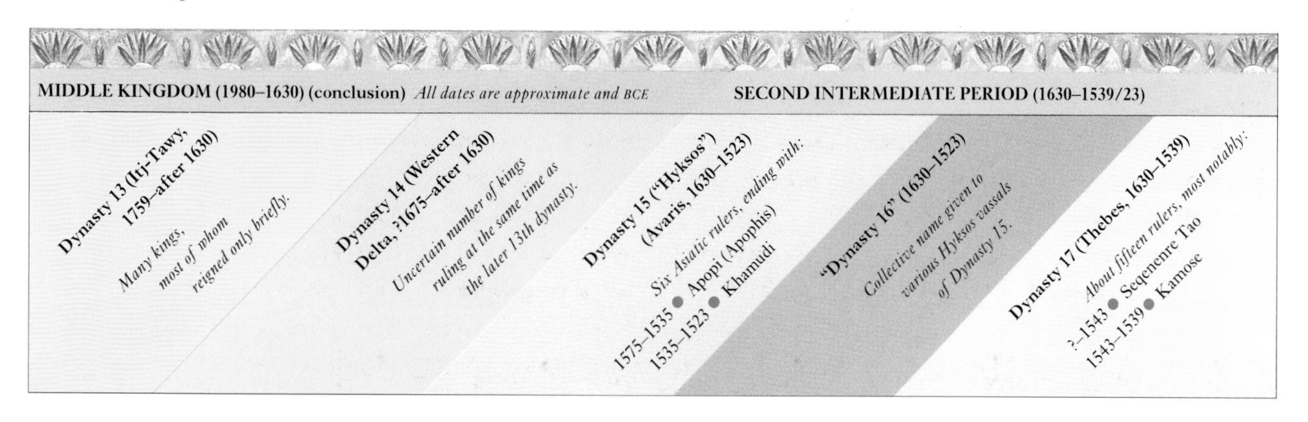

# FROM RECOVERY TO EMPIRE

*The deified King Amenhotep I came to be revered as the guardian of the burial grounds of Western Thebes. He is portrayed here on the inside of the coffin of a priest of Amun-Re, Djedhoriufankh, who lived during the 21st Dynasty (ca. 1000BCE). The central scene shows the pharaoh, who is depicted as a mummy, in the protective embrace of two images of the sun god Re-Harakhte, represented in the form of a hawk surmounted by the solar disc.*

The early rulers of the new Eighteenth Dynasty set Egypt – gradually and not entirely by design – on a course of imperial expansion. They pushed deeper into Nubia than ever before, extinguishing the Kushite kingdom and carving out an enormous province that stretched to beyond the fourth cataract and the great bend in the southern Nile. To the north, in Asia, a region that was more alien in language and culture than Nubia and politically more diverse, Egyptian imperial power was slower to develop. Early Eighteenth-Dynasty kings, such as Ahmose, the founder of the dynasty, and Thutmose I, preferred to keep Western Asia at bay, as their predecessors had done.

With Thutmose III, Egypt began to build an Asiatic empire of vassal states, but it took two more generations for imperial borders to stabilize, when Thutmose IV and Mitanni, a powerful Syrian empire, reached an accord. This entente was disrupted when the rising Hittite empire, based in Asia Minor (modern Turkey), overthrew Mitanni during the reign of Amenhotep IV, or Akhenaten. Three generations of warfare between Egypt and the Hittites ensued, until the two superpowers came to agreement under Ramesses II (ca. 1279–1213BCE; see p.35). The Egyptian empire shrank in the process, but remained a great power.

At home, the succession was not always smooth. Following the death of Amenhotep I, Ahmose's successor, the throne passed to a commoner, Thutmose I, for reasons that are not clear. His son and grandson, also called Thutmose, both married princesses who were probably related to Ahmose, but the kings themselves were born to non-royal women. When Thutmose II died suddenly (ca. 1479BCE), the regency for the infant Thutmose III was assumed by his stepmother Hatshepsut, his father's chief queen and a descendant of Ahmose. She eventually made herself king and reigned for nearly two decades alongside her stepson, presumably until her death (see p.89). Later in his reign, Thutmose III desecrated her monuments and defamed her memory.

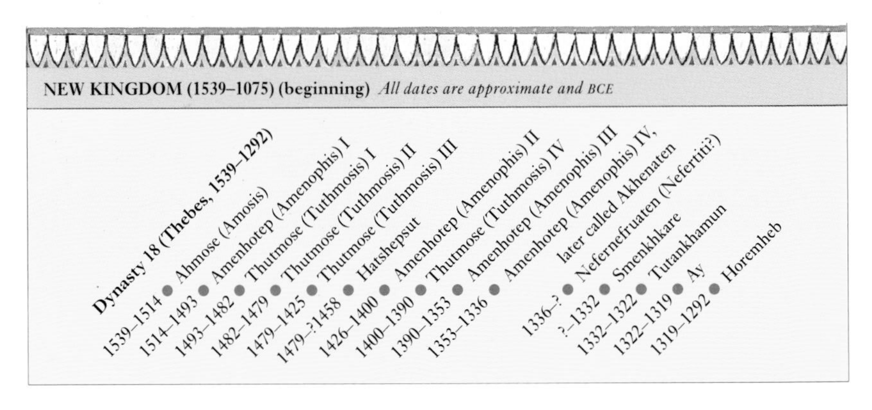

## SERVICE AND REWARDS: AMENHOTEP, SON OF HAPU

An appointment to high office in the pharaonic administration would ordinarily reflect the connections of the man, and the service shown by his family in the past, as well as his own merits. As a result, "dynasties" of officials might fill the same post for several generations and build alliances with the families of other powerful functionaries.

Every so often, however, a gifted outsider would break into the tight-knit upper ranks of government and leave an impression lasting long beyond his death. One such man was Horemheb, the general who became the Eighteenth Dynasty's last ruler (see p.35). Another arriviste, Amenhotep, son of Hapu, did not rise so high, but came to enjoy a posthumous fame among his compatriots that was equalled by few, even among the pharaohs.

Amenhotep (or Huy, the nickname that he and his contemporaries often used) came from Athribis in the Delta. The son of a petty official named Hapu, he trained as a scribe and spent most of his career in obscurity before being singled out by his royal namesake Amenhotep III (ca. 1390–1353BCE) for his organizational ability and wide knowledge of ancient lore. His promotion was rapid, and he was soon entrusted with nothing less than the reorganization of his country's finances and manpower. Under his supervision fell the collection of revenues for the crown and the organization and rotation of personnel engaged in the armed forces and on public works. While formally holding only a middling position in the civil service, Amenhotep effectively controlled Egypt's defences and virtually the entire civil administration.

The zenith of his career came during his master's first *sed* or jubilee festival, when, among other honours, he took the role of the crown prince in a re-enactment of the king's accession. Alone among his peers, Amenhotep was also granted a full-scale mortuary temple beside that of his king in Western Thebes. The civil servant of humble origins became the focus of a cult that flourished down to later antiquity, when he was deified, alongside another royal servant, Imhotep (see p.179), as a god of healing.

*Amenhotep was granted the right to erect statues of himself at the temple of Karnak, where this example was found. The great administrator is portrayed as a scribe, with a scroll on his lap and a writing palette over his left shoulder.*

*A statue (half life-size) of Thutmose III, the son of Thutmose II and one of his minor wives, Isis. For the first 20 years or so of his 54-year reign, he stood in the shadow of his stepmother, Hatshepsut. On assuming sole power, he launched military campaigns into Syria and Palestine, establishing Egyptian sovereignty in the region. (See also illustration, p.108.)*

Following the establishment of the empire by Thutmose III and his descendants, Egypt in the reign of Amenhotep III (ca. 1390–1353BCE) enjoyed a level of prosperity, stability, artistic creativity and international prestige unsurpassed in its history. These were undermined during the reign of Amenhotep III's son, Akhenaten, when the empire faced the external challenge of the Hittites and the internal disruption caused by Akhenaten's religious revolution (see pp.128–9). The religious experiment was reversed by Tutankhamun (ca. 1332–1322BCE), the last of the Thutmoside line.

A former general, Horemheb, emerged as pharaoh ca. 1319BCE. However, he produced no heir and was succeeded by his chief minister, Ramesses I, the founder of the Nineteenth Dynasty. Sety I and, especially, Ramesses II (see sidebar, right) restored Egypt's place among the great powers, but within a quarter of a century of the latter's death, the country was beset by invasions and civil wars. Under Merneptah, a horde of Libyans and "Peoples of the Sea" invaded the Delta (see p.45), and Nubia rose in the first general revolt against its Egyptian overlords since the early Eighteenth Dynasty. Subsequently, there was trouble between Merneptah's son Sety II and a usurper, Amenmesse. After the short reigns of Sety II's son Siptah and his queen, Tewosret, the dynasty ended amid a civil war between supporters of the chancellor, Bay, a Syrian who had risen from the position of royal cupbearer to become one of the powers behind the throne, and one Sethnakhte, who was backed by a number of senior figures in the administration.

Sethnakhte's victory (ca. 1190BCE) inaugurated the Twentieth Dynasty. His son, Ramesses III, the last of the New Kingdom's great warrior-pharaohs, repulsed two Libyan invasions and a new onslaught from "Peoples of the Sea". His successors, Ramesses IV to XI, were generally undistinguished, although they presided over a nation whose peaceful condition seems implicit in the orderly succession of kings and the absence of civil war. Yet the Twentieth Dynasty was an autumn, both for the empire and for the internal unity that the country had taken for granted for some five centuries. During the millennium that followed, Egyptians would come to live in a quite different world.

**RAMESSES THE GREAT**

**Of all the pharaohs who continued to live in legend long after their deaths, none was more celebrated than Ramesses II, otherwise known as Ramesses the Great. Reigning for close to sixty-seven years, he outlasted almost every other king in Egyptian history, with the possible exception of Pepy II. Ramesses IV later asked the god Osiris to "double for me the long duration of the prolonged reign of Ramesses II".**

**The general prosperity of Ramesses' reign, which was marked by diplomatic and military triumphs and monumental building programmes, combined to establish this period as a standard for all subsequent rulers. Many pharaohs adopted either his own personal name, Ramesses ("Re has fashioned him"), or a variant of his throne name, Userma'atre ("The justice of Re is powerful"), the "Ozymandias" of Classical writers. Ramesses III, the only one of his successors who came close to achieving a comparable record as a warrior and builder, conspicuously modelled his personal style and monuments on those of his illustrious predecessor. A number of later works of literature, too, harked back to the glorious days of Ramesses the Great.**

OPPOSITE: *The magnificent life-size funerary mask of King Tutankhamun, discovered when his almost intact tomb was excavated in 1922 (see p.196).*

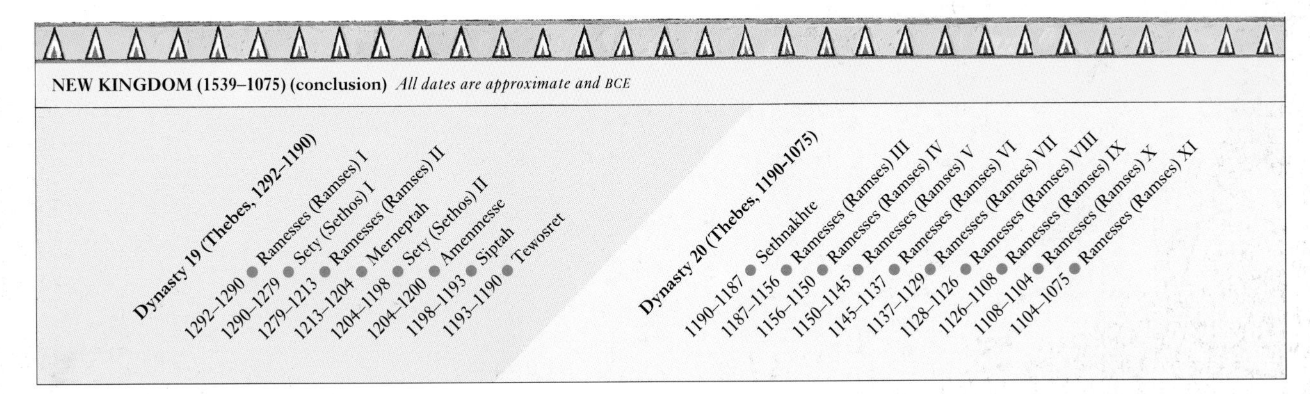

*King Taharqa, the most notable of the Nubian pharaohs who governed Egypt during the 25th Dynasty. His domains stretched from Nubia to the Mediterranean, and he was responsible for temples, pyramids and numerous other monumental projects throughout Egypt, such as the colonnade at the entrance to the temple of Amun at Karnak. Taharqa eventually withdrew from Egypt in the face of the threat from Assyria, leaving the country in the hands of his successor, Tantamani. This granite head portrays Taharqa wearing a distinctly Nubian style of cap.*

# DISUNITY AND FOREIGN RULE

From the twelfth to the eighth centuries BCE, Egypt experienced a reversal of fortune unprecedented in its long history. Its conquests in Western Asia, which Ramesses III (ca. 1187–1156BCE) had defended against the "Peoples of the Sea" ca. 1180BCE, were first to go. Early in the twelfth century BCE, the Hittite empire had disintegrated, and perhaps the pharaohs felt that what had served as a buffer zone between Egypt and its fellow superpower to the north now seemed expendable. In any case, Egyptian power in Asia fades from view ca. 1130BCE, so quietly that its disappearance might imply Egypt's loss of interest as much as any pressure from the Philistines and other groups that filled the vacuum.

At home, the seemingly peaceful procession of the last Ramesside pharaohs was bedevilled by strikes, inflation and rampant criminality at all levels of society. Worse was to come, for by the time of Ramesses XI (ca. 1104–1075BCE) there was widespread dissidence in Upper Egypt, first involving the high priests of Amun at Karnak, and then a number of figures who, ironically, had been called in to restore the king's authority. Order was finally re-established by senior military officers, who founded a new line of high priests. The price of this unrest was the loss of Egypt's provinces to the south, which were wrested from imperial control by the last viceroy of Nubia.

Ramesses XI's position was hardly secure, because power effectively passed to the Theban high priests long before the end of his reign. They controlled Upper Egypt while their relative, Smendes, ruled the area north of Memphis. Smendes became king on Ramesses XI's death, thereby founding the Twenty-First Dynasty, which continued to rule alongside the independent line of high priests of the south. There was a noticeable Libyan element in both ruling families, reflecting intermarriage with the descendants of Libyan prisoners who had settled in the Delta under Ramesses III and had since developed into an influential military caste governed by their own chieftains. On the death of Psusennes

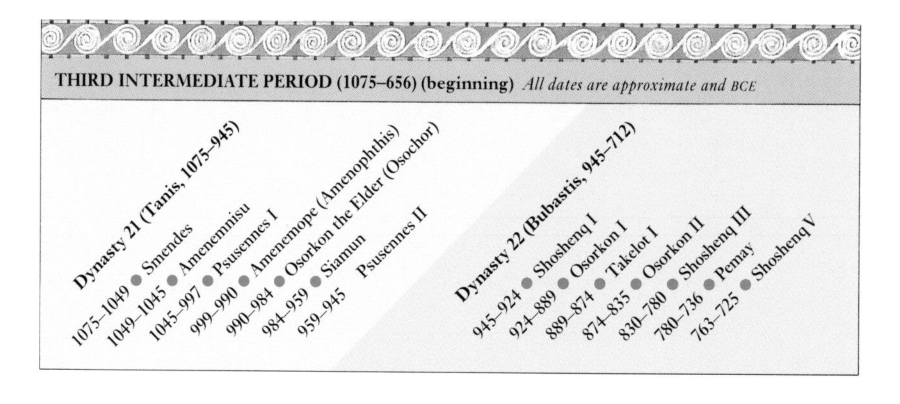

II, the last king of the Twenty-First Dynasty, the throne passed to one of his Libyan relations, Shoshenq, who bore the title "Great Chief of the Meshwesh Libyans". As Shoshenq I he became the founder of the Twenty-Second Dynasty.

Shoshenq is best known for leading a campaign into Palestine that is mentioned in the Bible (1 Kings 14.25–6), where he is called "Shishak". At home, he managed to reintegrate southern Egypt into the kingdom for almost a century. However, the forces of separatism proved stronger in the end, and by the reign of Shoshenq III they affected not only Thebes but also the Delta, which saw the rise of an increasing number of autonomous princes. By the middle of the eighth century BCE, at the height of this so-called "Libyan anarchy", no fewer than nine major kingdoms and principalities (most of them known collectively as the Twenty-Third Dynasty) coexisted in Egypt.

In this fragmented condition, Egypt began to fall under the influence of a state that had once been its colony. The Nubian kingdom of Kush, which had become a prosperous and united power, was an Egyptianized state whose rulers coveted their spiritual motherland. Under Piye, the Kushites swept north to Memphis. The Delta remained independent, although its assorted kings and princes submitted to nominal Nubian overlordship. A few years later, however, they accepted the sovereignty of Tefnakhte, the organizer of an anti-Nubian coalition. His capital of Sais, in the western Delta, became the seat of a new dynasty, the Twenty-Fourth, which lasted just fifteen years. Tefnakhte's successor, Bakenrenef, had barely begun to reign when he was deposed and killed by Piye's brother and successor, Shabaka, who went on to complete the Nubian conquest of Egypt.

The Kushite kings (the Twenty-Fifth Dynasty) liked to portray themselves as pharaohs in the grand tradition of their native predecessors, and Shabaka's successor, Taharqa (see illustration, opposite), perhaps came close to living up to such claims. But in the end they proved hollow, because soon after Taharqa's death, the Nubian regime succumbed to pressures from inside the Nile Valley and, especially, from the new superpower in the Near East: Assyria.

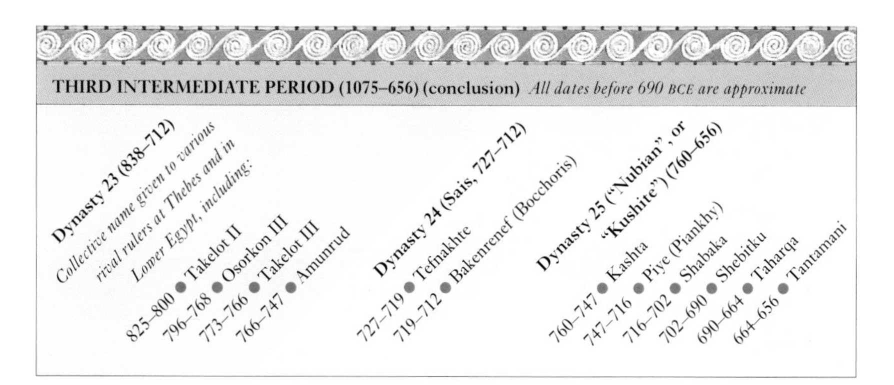


**A LOCAL STRONGMAN**

**For most of the "Third Intermediate Period" (Twenty-First to Twenty-Fifth dynasties), Upper Egyptian separatism was barely suppressed, even at times when the "Two Lands" were forcibly reunited. For example, during the Twenty-Fifth Dynasty and later, pharaonic power at Thebes reposed in the "God's Wife of Amun", a royal woman symbolically married to the Theban god Amun (see p.87). However, real power was wielded by subordinates such as Montuemhat (fl. ca. 655BCE).**

**Officially, Montuemhat ranked below the God's Wife and the High Priest of Amun. Yet, as mayor of Thebes and governor of Upper Egypt, he ruled a domain extending from the Nubian border into Middle Egypt. His career, which witnessed the end of the Nubian domination of Upper Egypt, illustrates how successive pharaohs accepted the power of the Theban aristocracy as the price of peace in the south.**

*A black granite bust of Montuemhat, effective ruler of Upper Egypt under the Nubian pharaohs and their immediate successors. The bust was discovered at Karnak, where he also occupied the position of Fourth Prophet of Amun.*

# A CONQUERED KINGDOM

The Nubian pharaohs of the late eighth and early seventh centuries BCE, the first foreigners to rule in Egypt since the Asiatic Hyksos in the sixteenth century BCE (see pp.30–31), were sufficiently Egyptianized to maintain classical Egyptian cultural and religious policies. Subsequent conquerors were decidedly more alien, beginning with the Assyrians, whose growing power in Western Asia the Nubians rashly challenged. This precipitated a duel lasting roughly half a century, in which Memphis repeatedly changed hands and even Thebes was sacked (656BCE).

The victor, however, was one of the Egyptianized Libyan magnates of the Delta (see pp.36–7), whom the Nubians had suppressed but not eliminated. This was Psamtik, ruler of the western Delta city of Sais, who outmanoeuvred his former masters, the Assyrians and Nubians, and his rivals among the other princes in northern Egypt, to found the Twenty-Sixth Dynasty. As Psamtik I, he used a combination of force and diplomacy to reunite the "Two Lands". Apparently for the first time, Greek mercenaries were hired to serve in the Egyptian forces, and Psamtik negotiated the transfer of Upper Egypt from Nubian control by having his daughter adopted as "God's Wife of Amun" at Thebes (see p.37).

The "Saite revival" under Psamtik and his descendants marked Egypt's last appearance as a great power under native rule. The Twenty-Sixth Dynasty kept at bay first Assyria and then its successor, the Chaldean, or "Neo-Babylonian", empire (612–539BCE). The dynasty oversaw the development of Egyptian naval power and increased diplomatic and commercial contacts with Greek states and other Mediterranean powers.

However, despite the shrewd diplomacy of Amasis, the Egyptians saw most of their allies swallowed up by the Persian empire of Cyrus the Great (550–529BCE). The same fate befell Egypt itself, when Cyrus's son,

**AN EGYPTIAN COLLABORATOR**
**Cambyses, the Persian conqueror of Egypt (525–522BCE), established his rule through a number of loyal Egyptian servants. One such was Udjahorresne, who had previously commanded the Egyptian navy under the Twenty-Sixth Dynasty. On his statue, Udjahorresne states that he was "chief physician", "king's friend" and personal advisor to Cambyses, who showed great piety and generosity toward Egypt's gods and temples. While tainted by self-interest, Udjahorresne's testimony suggests that Cambyses may not have been quite the monster that a hostile tradition later made him.**

*A coin issued in 30BCE to commemorate the incorporation of Egypt – symbolized by the Nile crocodile – into the Roman empire. The legend reads "Captive Egypt".*

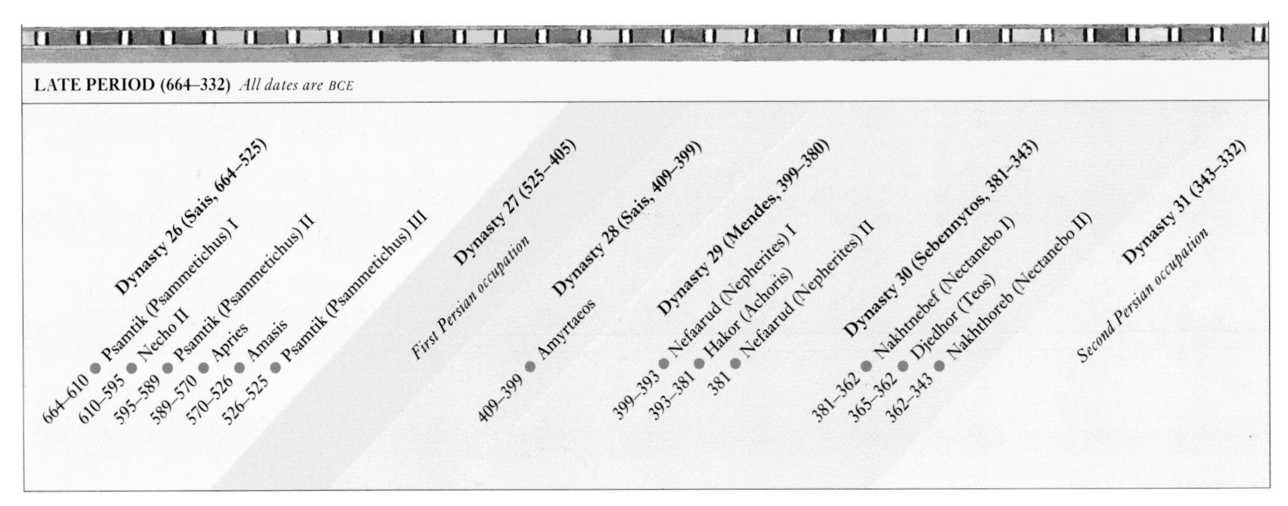

Cambyses II (529–522BCE), made short work of Egypt's last Saite ruler, Psamtik III. For all the moderation of the first Persian occupiers (known as the Twenty-Seventh Dynasty), Egypt adjusted poorly to its new status as an imperial province. A long insurgency (beginning ca. 463BCE in the northwest of the Delta) exploited Greek hostility toward Persia to keep this corner of Egypt detached from the empire. The area became a base for a wider rebellion which regained independence for the entire country (404BCE).

For nearly sixty years (Twenty-Eighth to Thirtieth dynasties) Egypt remained free by playing off factions among the Greeks against one another and against the Persians. Internally, however, Egypt was weakened by a persistent divisiveness and the result was a brief and troubled second Persian occupation (the Thirty-First Dynasty). This came to an end when the Macedonian forces of Alexander the Great destroyed the Persian empire and occupied Egypt.

During the second century BCE, the country remained firmly in the hands of its new Greek masters and their successors, the Ptolemaic Dynasty (the Thirty-Third). This, the last canonical dynasty of Egyptian rulers, was founded by Ptolemy I and ended with the famous Cleopatra VII. Under the Ptolemies, who ruled from a new capital, the Hellenistic coastal city of Alexandria, native Egyptian interests and institutions were overshadowed by those of Greek immigrants, who reinforced the Nile Valley's links with the wider Mediterranean world. These links became even stronger when Egypt became a province of the Roman empire in 30BCE, until the Arab general Amr ibn el-As brought Egypt into the world of Islam (641–2CE).

*Depicted in conventional Egyptian style and in traditional pharaonic dress, the Greek ruler Ptolemy V makes offerings to the Buchis bull, a sacred animal revered as the living incarnation of the god Montu, who also appears in the form of a falcon. Ptolemy V was responsible for the bilingual decree in Egyptian and Greek that appears on the celebrated Rosetta Stone* *(see p.233)**.*

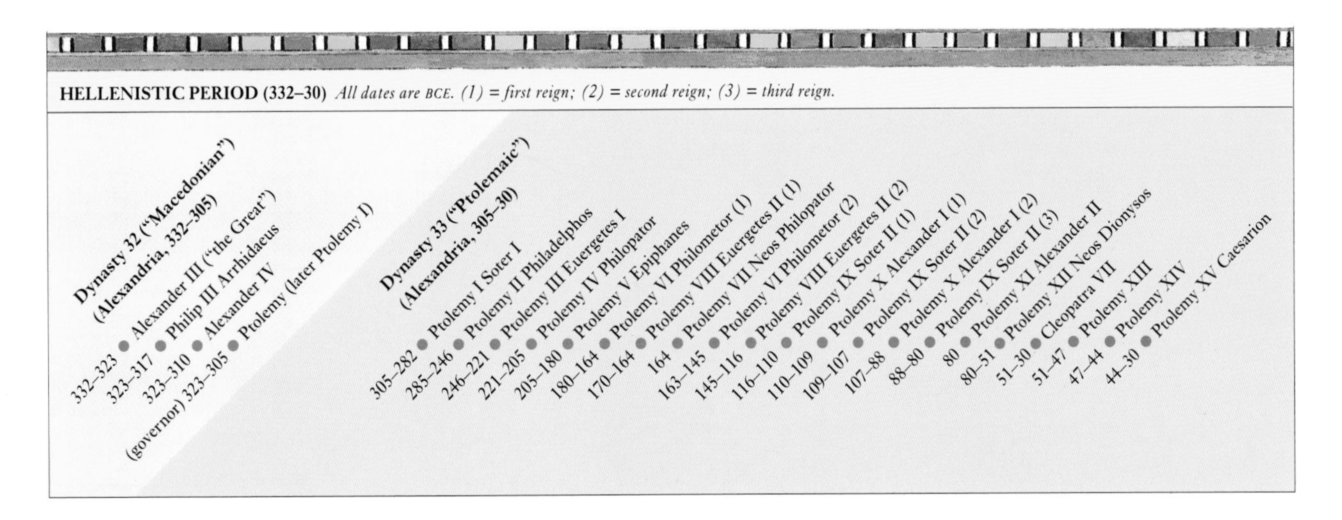

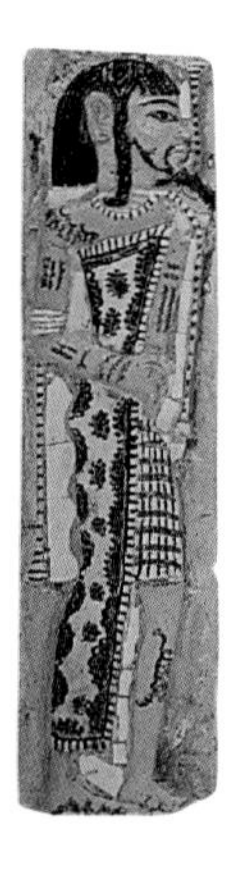

● CHAPTER 3

# EGYPT AND THE WORLD BEYOND

With virtually impassable deserts to east and west, coasts treacherous to seafarers, and the Nile blocked at the south by cataracts, Egypt was well protected from alien incursions. There were few points at which foreigners could enter the country, and for most of Egypt's history the pharaonic government was able to control their movements. At the same time, Egypt could freely choose the opportune time and place for expansion in northeast Africa and Western Asia. It was only with the rise of the great empires of the Near East and Mediterranean in the later centuries BCE that Egypt was forced to defend – and eventually relinquish – its isolationist outlook and, finally, its independence.

▲



ABOVE: *Coloured ceramic tiles from a palace of King Ramesses III (ca. 1187–1156BCE) depicting foreigners. Left to right: a Libyan, a Canaanite and a member of the "Peoples of the Sea" coalition (see pp.45–6).*

## THE FEAR OF HORUS IN THE SOUTH

For the ancient Egyptians, the south was "up"; the word for west was cognate with the West Semitic word for "right hand"; and places south were at the "forefront". This southward orientation betokened a cultural and political fact: for most of its history until the New Kingdom, Egypt was far more interested in its African roots than any relationship with Asia.

During the Old and Middle Kingdoms, Egypt saw Nubia, to the immediate south, as legitimate prey. From the end of the fourth millennium BCE on, the country's ability to amass manpower on a vast scale gave it the edge in dealing with its neighbours. A host of twenty thousand men sent to Nubia overwhelmed villages, enabling Pharaoh's plenipotentiaries to seize booty. If necessary, the expedition leader used force: "I slaughtered the Nubians on several occasions," says the vizier Antefoker, who lived in the twentieth century BCE. "I came north uprooting the harvest and cutting the remainder of the trees and torching their houses." Such bellicosity was termed "setting the fear of Horus [that is, the king] among the southern foreign lands, to pacify [them]".

Nubia possessed rich resources, as did the sub-Saharan region to which it had access; this goes a long way toward explaining Egypt's aggression. Egyptian expeditions to the south returned with ivory, ebony, incense, myrrh, throw-sticks, aromatic wood, leopard skins, giraffe skins, cereals and cattle. From the Middle Kingdom onward, efforts were made to tap mineral resources, especially Nubian gold deposits east of the Nile. The gold mines may even be commemorated in the region's name: "Nubia" may derive from the Egyptian *nbw*, "gold" and mean "gold-land". Strangely, the Nile was not often used as a transit corridor: expedition leaders (even if they were the royal agents resident in Aswan) preferred the "Oasis Road" which left the Nile in lower Middle Egypt, traversed a chain of oases including Farafra, Dakhleh and Dush, and regained the valley at Toshke in Nubia. Rather than going on foot, such expeditions

*Nubian tribute-bearers bring gifts of gold for Pharaoh. The gold has been cast into hoops so that it can be transported more easily. A fragment of a wall painting from the tomb chapel of Sobekhotep; 18th Dynasty, ca. 1400BCE.*

might employ hundreds of donkeys, the chief means of overland transort. Sometimes, as a result, the journey would last only seven months.

The Egyptians had long been in contact with the chiefdoms to the south, some of which were graduating to statehood. They often called the natives by names that had – for the Egyptians – slightly pejorative overtones: "bow people", "kilt wearers", "blacks". Many early expeditions in the Old Kingdom were little more than slave raids: texts record expeditions that brought back two thousand, seven thousand and, on one occasion (according to a Fourth-Dynasty graffito), even seventeen thousand Nubians. Once in Egypt, the foreigners were pressed into service on the land and in construction, in paramilitary forces abroad, or as police.

The Nubians did not accept the rape of their land with total quiescence. Chieftains did occasionally show obstinacy in the face of Egyptian pressure. A text from Thutmose II's reign records cattle thefts and an attack on Egyptian colonists. If punitive campaigns did not work, the Egyptians could always resort to magical curses: the names of Nubian chiefs were written on bowls or figurines which were ritually smashed.

Toward the close of the Old Kingdom there was serious unrest among the Nubian tribes – a defensive reflex that led to the rise of the first Nubian state, called Yam, centred upon the town of Kerma, south of the Second Cataract. This evoked strong reactions from the Middle Kingdom pharaohs. The Valley was occupied down to the Second Cataract and fortresses built. Frontier stelae were set up, limiting the entry of Nubians to merchants, and forbidding immigration on any scale. Senwosret III derides the foreigners: "They are certainly not a people to be respected: they are craven wretches! My Majesty has seen them – it is no lie!"

*Egyptians attack Nubian tribesmen on this panel of a painted wooden chest from the tomb of Tutankhamun (ca.1332–1322BCE).*

# EGYPT AND ASIA

**THE ASIAN CROCODILE**

**The attitude to Asiatics among the Egyptian population was that they were "vile", "wild", "doomed", "the abomination of Re". The following text of the twenty-first century BCE paints a vivid picture: "Speak now of the bowman! Lo, the vile Asiatic! ... He has been fighting since the time of Horus, never conquering nor yet being conquered. He never announces a day for fighting, like an outlaw thief of a [criminal] gang ... Don't give them a thought! The Asiatic is a crocodile on the river bank: he snatches on the lonely road, [but] he will never seize at the harbour of a populous city."**

**Late Period texts from Ptolemaic temples include Asiatics, along with the unwashed, the insane and the bearded, among those forbidden to enter temple sanctuaries.**

Oriented in their geopolitics toward the south and Africa, the Egyptians had turned their back on the north. Across the Suez frontier lay Asia, known as "Setjet" ("East") or the "northern lands", less densely populated than the Nile Valley and separated from it by a hundred miles (160km) of desert. If this frontier was well-fortified, Egypt had little to fear from invasion, and could exploit Canaan at will either commercially or militarily. Although Palestine possessed few resources that the Egyptians wanted, the area was important for its transit corridors providing communication with people farther to the north in Syria and Mesopotamia. Various routes were used by Egyptian prospectors, traders and couriers, who had to commit to memory a lengthy list of stopping places.

The Egyptians showed little inclination to settle in Western Asia, but the wealth of the Nile Valley and the Delta was always attractive to Asiatics. Artefacts found on the eastern side of the Delta at Ma'adi show evidence of trade with the Negeb Desert and Palestine in the late Predynastic Period (ca. 3300–3000BCE). There is evidence that the Delta had actually witnessed a demographic shift of peoples speaking a West Semitic language into the northeast of Lower Egypt. Immigrants continued to flow in during the Old Kingdom, attracted by prospects of work. "Hail to thee, thou perfect god Sahure!" shouted the new Canaanite arrivals from aboard ship as they approached the dock; "may we behold thy beauty!" Bedu nomads, too, were drawn to the fertile lands of the Nile to pasture flocks

*A family of Asiatic Bedu who have come to trade eye paint in Upper Egypt. A copy of a wall painting in the tomb of the nobleman Khnumhotep II at Beni Hasan; Middle Kingdom, 19th century BCE.*

## THE "FOREIGN RULERS"

The myth of universal dominion prevented Pharaoh from considering any foreigner of whatever rank on an equal footing with himself. Even the most influential foreign potentate would receive from the Egyptians only the banal designation "big man", never "king". All alien heads of state, whether formal kings or tribal headmen, were lumped together under the rubric "foreign chiefs"; and what was expected of them was loyalty and tribute. The theory consistently distorted the reality: innocent embassies are portrayed on the monuments as "obeisance through the power of His Majesty", and customary gifts become "tributes", even though Pharaoh may have reciprocated.

In reality, of course, "foreign chiefs" were often unwilling to consider Egypt's interest their own. Their recalcitrance might be punished by military retaliation, or by magical cursing rituals, involving the smashing of pots or figurines on which the names of rebels were inscribed.

The term "foreign rulers", or "Hyksos", is applied to an Asiatic regime, the Fifteenth Dynasty, which imposed itself on northern Egypt (see pp.30–31). Scholars differ as to whether the Hyksos came to power through invasion or peaceful settlement. But for just over a century, from shortly before 1630BCE, they held sway over the Delta and Middle Egypt from Avaris in the eastern Delta. Linguistically, they belonged with the West Semitic-speaking peoples of the Levant. Expelled by the Theban king Ahmose (ca. 1539–1514BCE), they continued to be seen as a threat to Egyptian security. Kings from Thutmose III to Ramesses II used their presence to justify pre-emptive strikes into Asia which eventually developed into wars of conquest.

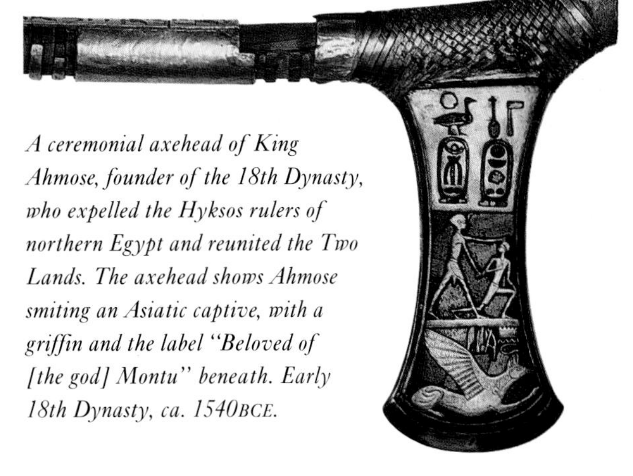

*A ceremonial axehead of King Ahmose, founder of the 18th Dynasty, who expelled the Hyksos rulers of northern Egypt and reunited the Two Lands. The axehead shows Ahmose smiting an Asiatic captive, with a griffin and the label "Beloved of [the god] Montu" beneath. Early 18th Dynasty, ca. 1540BCE.*

and herds in the dry season. A scene from the tomb of Khnumhotep II, nomarch of the sixteenth township of Upper Egypt (ca. 1870BCE), shows a group of thirty-seven Asiatic Bedu who have come to trade eye paint. Under normal conditions, however, Bedu were unwanted guests, opposed by farmers whose fields were laid waste by passing flocks, and hence deterred by border forts.

The average Egyptian reacted to the Asiatic incomers with contempt. But once settled inside the country, Asiatics could marry Egyptians, acquire important jobs and rise in society. Yanammu became a high state official under Akhenaten, 'Aper-el became prime minister (ca. 1400BCE) under Amenhotep III, and Bay became chancellor under Siptah and virtual king-maker (ca. 1210BCE). Actual acquaintance with a northern alien often banished the stereotype: Thutmose III's barber, who went on campaign with the king, received an Asiatic captive as a slave. In his will, the barber stipulates: "he is not to be beaten, nor is he to be turned away from any door of the palace. I have given my sister's daughter ... to him as wife. She shall have a share in [my] inheritance just like my wife and sister."

# EGYPT AND THE MEDITERRANEAN

No one knows when contact was first made between Egypt and the Aegean. But early Egyptian texts refer to the "Hau-nebu", which may mean the Greeks, because it was certainly used in later contexts as a vague term for Greece. From an early period, also, the Egyptians spoke of "Keftiu", a name identical with the Canaanite Kaptara, or Crete. The famous Cretan palace at Knossos was known around the Levant in the seventeenth century BCE. Palaces of the sixteenth century BCE in the Hyksos capital, Avaris, in the Nile Delta contain "bull-jumping" scenes in the Cretan manner. The name "Alashiya", Cyprus, turns up in a fragment of text from the reign of Amenemhet II (ca. 1876–1842BCE).

From the Eighteenth Dynasty on, the evidence for contact with the Aegean world becomes more plentiful and at the same time more specific. As the unrivalled major power of the ancient world, Egypt attracted the diplomatic suits of many foreign ambassadors from regions that were independent of the pharaohs, but under their political and commercial thrall. Fifteenth-century BCE tombs at Thebes in Upper Egypt depict the presentation of diplomatic gifts by strangers clad in costumes that are familiar from scenes discovered at Knossos. Merchant ships from the

*Captives brought back from the victory, ca. 1176BCE, of Ramesses III (ca. 1187–1156BCE) over the "Peoples of the Sea", a coalition of groups of Aegean origin that attacked Egypt (see map, opposite, and p.46). Among the prisoners are Peleset (Philistine) warriors, distinguished by their characteristic plumed headdresses. From the temple of Ramesses III at Medinet Habu in Western Thebes.*

Aegean plied the triangle of sea-lanes between Crete and Greece, Syria and the Nile Delta; and an inscription from the reign of Amenhotep III (ca. 1390–1353BCE) confirms that Egyptian merchants and seafarers were acquainted with the Aegean Sea. Until quite recent times, Egypt was regarded as an almost inexhaustble source of grain for the entire Mediterranean, and the peoples of the Aegean, who eked out a marginal agricultural existence, had a pressing need for Egyptian emmer-wheat and barley. For these commodities they exchanged spices, unguents, oil, opium and exotic manufactured goods.

During the New Kingdom, Libyan enclaves in northern Egypt enjoyed commercial contact with seafarers from the Aegean, Anatolia and Cyprus, probably on account of the suitable anchorages that were available along the Mediterranean coast. While trade dominated in these exchanges, piracy was also rife. The Lukka people,who gave their name to Lycia, a region on the southwest coast of Asia Minor, earned a reputation as raiders all over the eastern Mediterranean; and the Shardana, probably from the Sardonian plain around Mount Sardene, south of Troy, so impressed Ramesses II when they attacked the Delta that he recruited them as his bodyguards.

From the reigns of Merneptah to Ramesses III (ca. 1213–1156BCE) pirate raids began to assume the aspects of organized movements of "Peoples of the Sea" and other groups. Whether prompted in their trav-

*The routes of invasion and piracy by the Labu and "Peoples of the Sea" during the New Kingdom.*

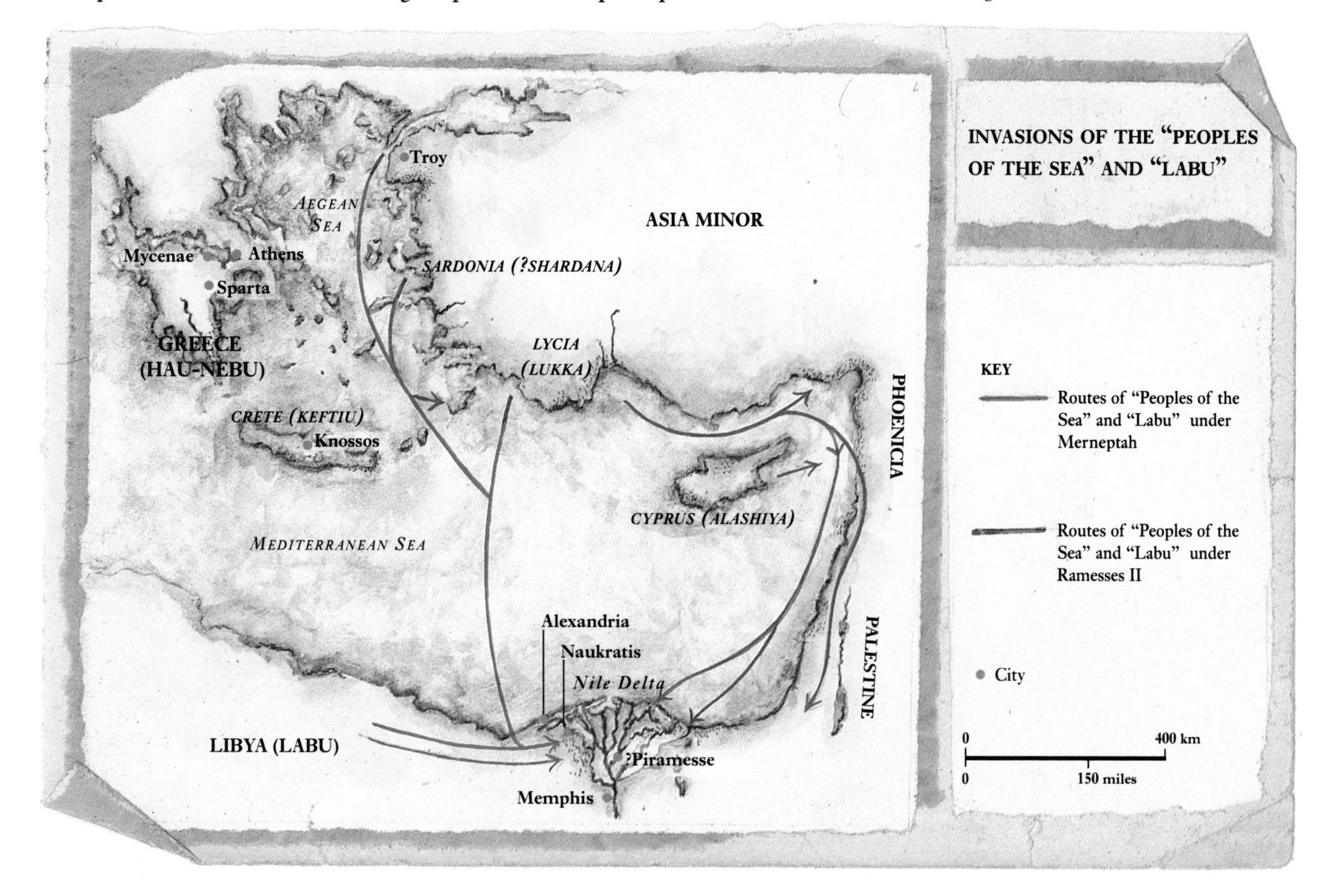

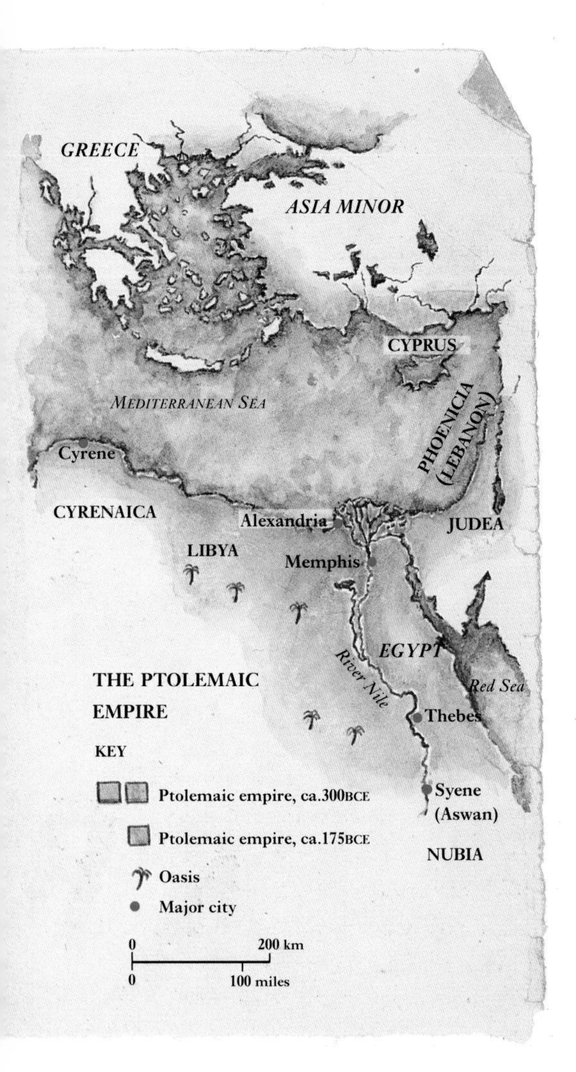


*The empire of the Ptolemaic rulers of Egypt in the 3rd and 2nd centuries* BCE.

els by famine or plague, or displaced by newcomers, or for whatever other reason, these ethnic groups showed a singleness of purpose that had hitherto been absent. No longer acting singly, the sea-rovers made common cause as coalitions whose purpose was conquest and settlement in Egypt. On three recorded occasions (the fifth year of Merneptah, and the fifth and eleventh years of Ramesses III) the Meshwesh, the Labu and the sundry "Peoples of the Sea" invaded the western Delta and the northern oases, only to be beaten off by a desperate Egyptian army. In the eighth year of Ramesses III's reign a major coalition of five peoples from the Aegean (Peleset, Tjeker, Shekelesh, Denyen, Washosh) set off along the south coast of Anatolia in ships and carts making for the Nile. So destructive was their passage that city after city in northern Syria and Cyprus was wiped off the map, never to be re-inhabited. (The contemporary archaeological record finds an echo in Greek traditions six centuries later concerning the aftermath of the Trojan War: such heroes as Teuker brother of Ajax, Agapenor king of the Arcadians, Mopsos the seer and Tlepolemos from Epirus are said to have led migrations eastward into Syria, Cyprus and Palestine in the years following the great siege.)

As the fleet and land forces of the Sea Peoples, with wives and children in train, descended the Levantine coast, Ramesses III mustered his army and took up position on the Delta coast. The Egyptians fought the invaders to a standstill (possibly in southern Palestine), while their warships devastated the enemy flotilla. The coalition was broken up. The Peleset settled on the southern coast of a region, Palestine, that takes its name from them; they appear in the Bible as the Philistines. Similarly, Sicily and Sardinia may be named after groups of the Shekelesh and Shardana who may have migrated to the western Mediterranean.

After the Aegean "Dark Ages" – the period between the collapse of Bronze Age Greece in the twelfth century BCE and the beginnings of Classical Greek culture – the Greeks and Egyptians experienced a mutual rediscovery. As in the New Kingdom, Egypt retained a fascination for peripheral Mediterranean communities as a fabulous source of material and cultural riches. It is small wonder that archaic Greek statuary and architecture should show more than a passing resemblance to Nilotic prototypes, or that it should have become fashionable for Greek savants and nouveaux riches to make the journey to Egypt for instruction and sightseeing. Lycurgus, the Spartan law-giver, and the philosophers Pythagoras and Thales are reputed to have visited Egypt in the seventh century BCE, and the Athenian political leader Solon certainly did so in the following century. Once again, Egypt proved irresistibly attractive to merchants, and ca. 630BCE a consortium of Greek merchant adventurers established a trading post called Naukratis in the Nile Delta with permission from the Egyptian king Psamtik (Psammetichus) I (664–610BCE), founder of the Twenty-Sixth Dynasty, which governed from the Delta city of Sais.

In its struggles with such foreign invaders as the Assyrians and Babylonians (671–600BCE), Egypt came to appreciate military assistance. Through international recruiting, at first by treaty with King Gyges of Lydia (another anti-Assyrian monarch), Psamtik bolstered his forces with thousands of Greek auxiliaries. These were posted mainly to the frontier forts where, although life was rigorous, pay was good, and a successful commander might end up with the royal gift of a city.

During the final sixty years of Egypt's independence under native rule, 404–343BCE, the Persian empire – which had already occupied Egypt twice – was successfully kept at bay largely owing to the presence in the Egyptian army of Athenian and Spartan hoplites (footsoldiers), who manned the borders in their tens of thousands for the pharaohs against the numerically superior Persian forces.

## THE TRIUMPH OF HELLENISM

Asiatics and Greeks had long since come to Egypt as mercenaries and merchants: from 525BCE, when the Persians invaded the country, they came as servants of alien empires and as settlers. With the arrival of Alexander the Great in 332BCE, Greco-Egyptian relations changed. For the next three centuries, a Greek line of kings – the Ptolemaic Dynasty (see p.39) – ruled Egypt, and Greek settlers came to dominate its administration and commerce. Alexander founded a Hellenistic new capital, Alexandria.

It was upon Alexandria that the intellectual life of Egypt under the Ptolemies was focused. Its Library and Research Institute (the "House of the Muses", or "Museum") was founded by Ptolemy I and Ptolemy II in the third century BCE. Although Ptolemaic Egypt was a major international power (see map), native Egyptian society receded increasingly into a backwater, symbolized by the temples to the ancient gods which maintained themselves as an unchanging but progressively weakening bastion against Classical modernity. The new Classicism would take even stronger hold in the eastern Mediterranean with the advent of the Roman empire, which overthrew Ptolemaic Egypt in 30BCE.

Although Egypt contributed the Isis cult to the list of popular "mystery" religions of the empire (see p.57), the advent of Christianity as the moral force behind Hellenistic culture spelled the end of the ancient Egyptian civilization. As elsewhere in the empire, the new faith grew in popularity, until finally the emperor Theodosius issued an edict in 392CE closing all pagan temples in his domains. The old Egyptian religion clung on for a while at places such as Philae in the south, but Hellenism had effectively triumphed.

*A burial chamber in an Alexandrian catacomb of the Roman Period (2nd century CE), displays Hellenistic architectural features but also Egyptian decorative devices such as the winged sun disc.*

**ROYAL GIFTS**
**Fragments from the international correspondence of pharaohs show the political importance of gift exchange. The king of Cyprus sent presents of copper to the king of Egypt, the latter sent three thousand talents of gold to the king of Babylon, while the king of Mitanni sent vast quantities of gifts with a daughter who was destined to become Pharaoh's wife. Egyptian statuary, *alabastra* (containers for unguents) and gold-encrusted furniture have been unearthed at Ebla, Ugarit, Byblos (to name but a few foreign states). While this cannot be considered trade of itself, it was clearly designed to facilitate commerce at lower levels of society by creating a congenial atmosphere at the head-of-state level.**

*Courtiers of the king of Punt, a northeast African land in the region of modern Somaliland, bear gifts for Panehsy, leader of an Egyptian trade mission to Punt in the 9th year of the reign of Hatshepsut (ca. 1479–?1458BCE). The expedition sought to exchange Egyptian products for African goods such as gold, ebony, myrrh, malachite and ivory. From Hatshepsut's mortuary temple at Deir el-Bahari, Western Thebes.*

# INTERNATIONAL TRADE AND TRAVEL

Modern connotations of "trade" (importing, exporting, middlemen, markets) have only limited application to ancient Egypt. The acquisition of foreign goods and services was essentially a royal monopoly, and if Pharaoh at times allowed individuals to trade abroad, he could easily limit or revoke that right. Moreover, the small foreign communities flanking Egypt on the south, west and north had little choice about entering a relationship with their powerful neighbour. Nubians and Canaanites who resisted such pressures were inevitably subjected to punitive campaigns.

However, beyond the immediate sphere of empire the exchange of commodities did in fact assume true commercial dimensions. Egypt was well endowed with resources, and its needs were clearly defined. It possessed foodstuffs, gold, copper, malachite, gemstones, natron (soda) and sundry minerals in abundance; but it lacked timber, iron, silver, tin and lead. Timber could be procured from the Lebanese coast (Egypt cultivated friendship with the Phoenician city of Byblos from earliest times), but the absence of metals had economic impact only when bronze and, much later, iron came into common usage elsewhere in the ancient world.

The merchants who conducted Egypt's trade comprised both natives and foreigners. At the highest level Pharaoh had his plenipotentiaries, who were sent abroad for diplomacy, with trade as a subsidiary agenda. Much more numerous were the "agents" of an institution, often a temple, who scoured the Middle East on specific missions, including the quest for slaves. At the height of the empire, the temple of Amun possessed innumerable agents and a commercial fleet of eighty-three ships on the Mediterranean. Occasionally, Egyptians would take up residence in foreign ports, where they acted as commercial intermediaries.

Texts from Ugarit tell us of entrepreneurs on the Phoenician coast pooling their resources to fund a commercial voyage to Egypt. The profits were enormous, despite import duties (at one time ten percent on Phoenician products, twenty percent on Greek). From all over the eastern Mediterranean came copper, oil, aromatic wood, lumber, resin, unguents, wine, opium, and manufactured items. In the other direction went commodities that Egypt could export in abundance: grain, natron and precious metals (see sidebar, left).

Land travel by messengers was on foot, although king's emissaries of high rank often had use (in the

New Kingdom) of a chariot and a retainer. For most of Egypt's history, large commercial ventures on land employed donkey caravans (camels were introduced only in the seventh century BCE). There were few "road" maps in our sense: travellers had to carry with them, or memorize, itineraries of stopping-places. Distances between points en route might vary slightly, but one day's march was usually ten to twelve miles (16–19km). Travel between Egypt and the Phoenician coast, Cyprus and the Aegean was by sea. The coastlands of the Red Sea were also reached by boat. Journeys to the Sudan were by the Nile or by the route of "Forty Days" (overland via the Saharan oases). The most common ocean-going boat was the "Byblos-ship", a vessel named, like the English "East Indiaman", after its overseas destination. Sailors had to memorize a formal itinerary (periplus). A fast trip from the Aegean to Egypt took five days, and the Nile between the Delta and the Nubian frontier could be traversed in three weeks. Caravans between Memphis and Nubia using the oasis route would take seven months there and back. Often, however, ships and traders would linger on their travels, vastly extending normal durations.

*Ancient Egypt and its international trade network. Among the most important routes were the Egypt–Phoenicia (Lebanon) axis, by means of which Egyptian goods reached the central and western Mediterranean and Mesopotamia; the Red Sea route, bringing produce from Africa and Arabia to Egypt; and the oasis routes, another artery of African trade.*

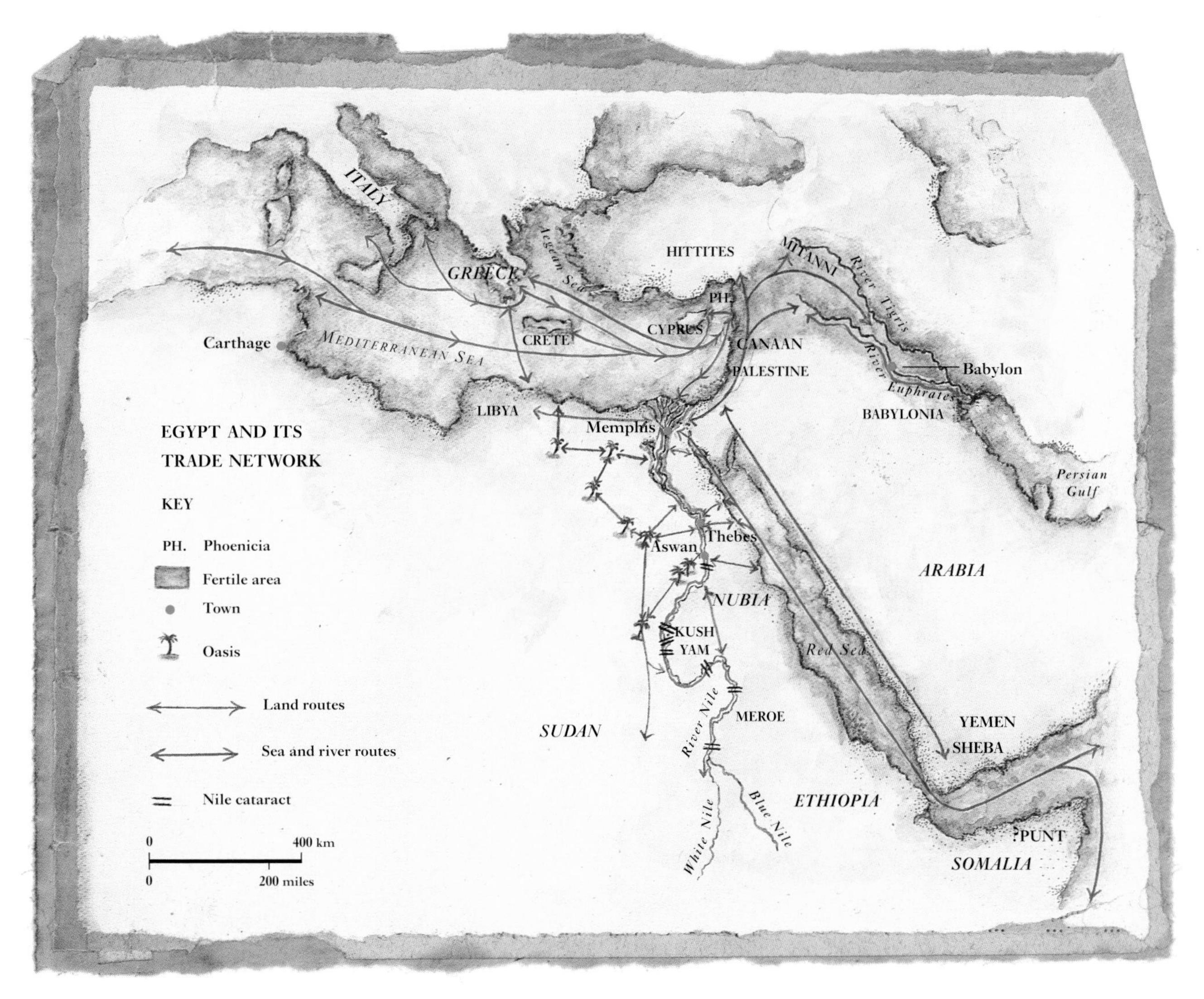

# THE GRIP OF EMPIRE

When Thutmose III and Amenhotep II launched imperial wars in Western Asia, their aim was to pre-empt any attempt at invasion, but in time the economic and social benefits of empire (in particular, foreign wealth and forced labour) also began to impress themselves on the pharaohs. Amun-Re, king of the gods, granted Pharaoh "title" to the dominion over foreign lands as part of his royal inheritance. The imperial frontiers came under the god's protection, and were his frontiers as well as Egypt's. It was in the presence of this supreme deity that enemy chiefs were despatched by the king's mace and sword. Pharaoh ruled "all that the sun disc encircled," and "every foreign land was under the feet of the Perfect God".

**THE TRIBUTE-BEARERS**

**Egypt imposed its own internal tax structure on its empire. The harvest was requisitioned from Canaan and Nubia: fine wheat came from the Esdraelon plain in central Palestine; fruit and vegetables from Lebanon and Syria; emmer-wheat and barley from the Sudan. A quota was levied on boxwood from Syria, cedar from Lebanon, gold from Nubia, natural glass from the Negeb. At Pharaoh's command, his governors also sent him human resources, such as maid-servants and unskilled labourers, shipped to Egypt in great numbers during the Nineteenth Dynasty.**

**At regular intervals Egyptian army captains would appear to collect taxes and settle disputes. Taxes from the Canaanites could also be despatched by ship at their own expense. Both in Asia and Africa refusal to co-operate or pay taxes was deemed rebellion. Especially recalcitrant enclaves might suffer deportation: Nubians to Canaan, and Canaanites to Nubia.**

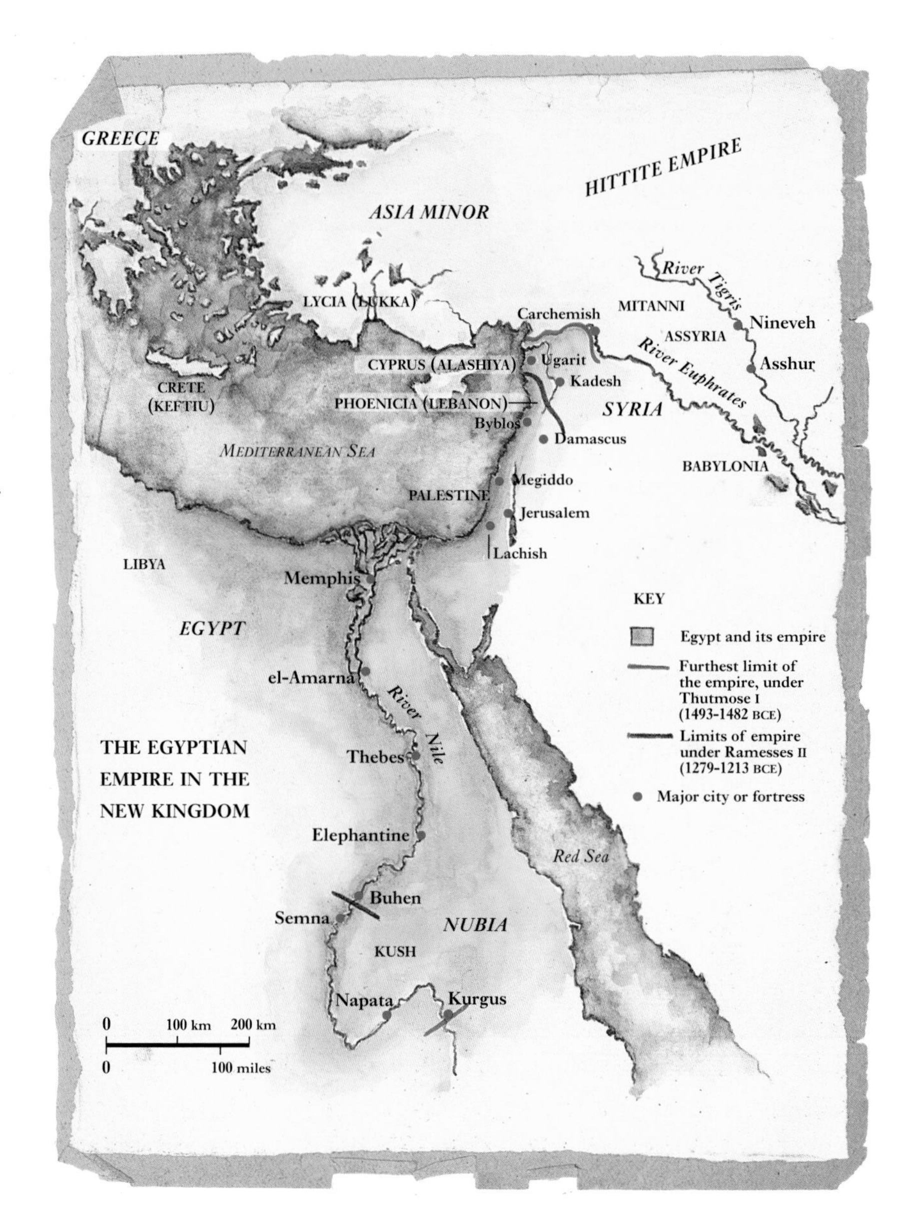


*The extent of the Egyptian empire in the New Kingdom (14th–13th centuries BCE). Egypt's sway stretched from deep inside Nubia (modern Sudan) to the Eleutheros valley in Lebanon. For a brief period under Thutmose III, the northern imperial frontier was on the river Euphrates.*

## THE EGYPTIAN ARMY

The Egyptian New Kingdom army had a core of full-time soldiers distributed in peacetime among garrisons in Upper and Lower Egypt, Nubia and Asia (see pp.78–9). For large-scale expeditions into Asia to fight Mitanni or the Hittites or to put down revolts, the pharaoh would conscript one in ten from the able-bodied of the temple communities to augment the standing army.

Footsoldiers were grouped into companies of two hundred (consisting of twenty platoons), and companies into "divisions" (in Egypt loosely called an "army") of around five thousand men under the banner of their local god. Companies were commanded by captains ("standard-bearers") who carried a staff topped with the company insignia; divisions were led by a general or lieutenant-general. The infantry used javelins, stabbing daggers and short scimitars; but their armour was restricted to padded caps, elliptical raw-hide shields and triangular sporrans. Chariots, of light wicker-work, were valued more for speed than armament. The driver wore a helmet of leather or bronze and body armour; his companion was armed with a bow and arrows and javelins. Chariots were grouped in bodies of fifty under a major, with larger groupings under colonels or lieutenant-generals. The most feared Egyptian units were the archers, who used the powerful composite bow. They were grouped in battalions under their own commanders, or seconded to infantry units. We know little of tactics, except that armies were deployed in "centres" and "wings", and that charges were common. The Egyptians proved inept at siege warfare, usually having to sit things out until the besieged submitted to hunger. Only in the Late Period did they borrow siege-towers battering rams from Asia.

*Forty Egyptian soldiers, equipped with spears and rawhide shields. From the tomb of the nomarch Mesehti at Assyut; 11th Dynasty, ca. 2000BCE.*

The Egyptian experience of empire differed markedly in the two major theatres, north (Western Asia) and south (Nubia). In the north there were autonomous metropolitan states. These were not strong enough to stand up to Egypt alone, but in coalition, or backed by a major power, they could hope to hold their own. Thus, we find, for example, the empire of Mitanni using Syrian city-states as a buffer against Egypt. In the south few major powers or urban centres existed. Once the kingdom of Kerma had been reduced, shortly after 1500BCE, Egypt with relative ease raised her frontiers within three hundred miles (482km) of modern Khartoum. An imperial administration developed more quickly in the African acquisitions than in the Canaanite province. Nubia and Kush were governed by a viceroy heading a provincial bureaucracy on the Egyptian model which milked the country of its resources, especially gold. In Asia, Pharaoh was content to allow the towns to retain their mayors and local social structure, while obliging Canaanite headmen to take an oath in his name not to rebel.

# THE COSMOPOLITAN REALM

Egypt's extended frontier in Western Asia survived for four centuries, and involved the country deeply in Levantine affairs. At the same time, however, a reverse flow of influence brought an Asiatic presence to the banks of the Nile and exposed the Egyptians to the culture of their subject peoples. Foreign merchants established a settlement on the north side of Memphis and their ships frequented harbours as far south as Thebes. Occasional concentrations of Bedu tribesmen and mercenaries dotted the landscape of both Middle and Upper Egypt. The international exchange of goods is signalled dramatically in the archaeological record by the extensive presence of Canaanite amphoras and store jars in Egypt and, conversely, by Egyptian vessels and artefacts discovered at sites in Canaan, within the reach of Egypt's imperial grasp.

Egypt in the New Kingdom was also interested in the manpower that its northern empire had to offer. Canaanites appear in Egypt, mostly from

## THE INFLUENCE OF ASIA

The empire opened Egypt's doors not only to foreign goods but also to ideas and languages of which Egypt previously had been only vaguely aware. Soon the expression "to do business in the Syrian tongue" became virtually synonymous with "to haggle" – that is, in the marketplace. By the thirteenth century BCE, the Egyptian language had become permeated with Canaanite words and expressions to designate foreign products, techniques, foods and patterns of behaviour, all of which were making there presence felt in the country.

A similar impact was felt in the realm of religion, cult and mythology. Northerners entering Egypt, whether in bondage or at liberty, brought their gods and their beliefs with them. Soon, strangely-garbed Western Asian deities began to make their appearance in Egyptian art, notably the storm-god Ba'al ("Lord") and his female consorts Astarte and Anath; the war-god Reshef; and the voluptuous goddess Qudshu, the "Holy One", who was normally depicted unclothed (see illustration, left).

*A stela of ca. 1250BCE depicting (left to right) the Egyptian god Min and two Asiatic deities, Qudshu and Reshef. In the lower register, the craftsman Qeh and his family worship the Asiatic goddess Anath.*

In some cases, the myths surrounding these figures were translated into Egyptian (as in the story of *Astarte and the Sea*, or *The Tale of Truth and Falsehood*), or else strongly influenced native mythology in the form of plot motifs. Heroes such as the protagonist in the tale *The Doomed Prince* take refuge in Syria, and Anpu, or Anubis, one of the protagonists of a popular story called *The Two Brothers*, lives in Lebanon. This story also has resonances with the biblical account of Joseph in Egypt.

*Tribute-bearers from Palestine and the Orontes valley present Pharaoh with some of the ornate metalware for which Syria was famous. One man in the upper register has brought his son – depicted naked and wearing a "sidelock of youth" in the Egyptian manner – for "safe-keeping" and education at the Egyptian court. Part of a wall painting in the tomb chapel of Sobekhotep; 18th Dynasty, ca. 1400BCE.*

the later years of Thutmose III, as prisoners of war from campaigns of conquest; but when their numbers proved insufficient, Pharaoh had recourse to forced deportation: Amenhotep II transported more than eighty-six thousand men from Palestine and southern Syria. To discourage insurrection Pharaoh required that the Canaanite mayors send their children to the royal court on the Nile where they were kept against the good behaviour of their parents. While there, they were brought up as Egyptian pages, grooms or guardsmen, and after their fathers died were sent back to their patrimonies to succeed to the mayoralty. The rank and file of the deported Canaanites were put to work on menial tasks, but skill and intelligence could bring rewards. Most served as farmhands on royal, temple or private estates, or as weavers in the workhouses. The 'Apiru, a social class of mafia-like brigands, were used on construction sites and quarries, cutting and hauling stone. The better-educated could hope for higher status: construction engineers, middle-ranking bureaucrats, doctors, scribes and soldiers. By the twelfth century BCE, the upper ranks of the royal catering service – the very butlers who waited on Pharaoh – were frequently filled by Canaanites. Sometimes foreign princesses would be brought to Egypt for diplomatic marriages, firing the popular imagination and giving birth to folklore: "Pharaoh wins the fair princess."

The Egyptian disdain for foreigners meant that Asiatic immigrants in general remained culturally discrete. However, some intermarriage did occur, and there can be little doubt that the greater contact with outsiders that was a direct result of imperial power served to undermine traditional attitudes toward the "vile Asiatic".

# THE MINELANDS OF THE SOVEREIGN

**A TURQUOISE TRAIL**

**Trekking to the mines in the Sinai and 'Arabah involved travel by sea and land. One route to the turquoise mines of Sinai led through the broad wadi (dry river bed) running east from the latitude of the Faiyum to a point on the Gulf of Suez near its northern end. There ships would be constructed and launched to take the miners to the west coast of Sinai near modern Abu Rudeis. Because they had to transport ore back to Egypt, the vessels were usually large cargo ships, carrying a crew of one hundred and fifty sailors. From the Sinai coast, a further journey by land brought the expedition and its donkeys to the site of the turquoise deposits, where a seasonal camp was already prepared. Exploited from the Predynastic Period into the early Middle Kingdom, the region is today pock-marked with mine shafts to the extent that modern Arabs refer to it as the "Valley of Caves".**

The pharaonic state celebrated itself in ways typical of any ancient complex society – through monumental architecture, art and sculpture, costume and adornment. To provide such things it was necessary to look beyond the state's borders. For temples and statues Pharaoh needed stone such as diorite from the south, greywacke from the Eastern Desert, and granite from the cataracts; for jewelry, Sinai turquoise and Nubian gold; for countless implements and fixtures, copper from the 'Arabah.

Craftsmen, miners, masons, prospectors and unskilled labourers would be seconded to a specially appointed plenipotentiary, rations and draft animals assigned, and the mining expedition despatched. Numbers of workers varied widely, from a few hundred to more than ten thousand. Taking with him the official papyrus copy of the "work-order", sealed in the king's presence, the expedition commander would often have the document copied and inscribed on the rock-wall adjacent to the mine or quarry. Thus we know, for example, that in the thirty-eighth year of the reign of Senwosret I (ca. 1919–1875BCE), a certain Amenemhet went to Hatnub for eighty blocks of stone, to be dragged by two thousand men, and that the consignment reached the Nile two weeks after quarrying.

Mineral-working often brought Egyptians and locals into confrontation, often to the disadvantage of the latter. Nubians worked the gold-mines in the Wadi Allaki, east of the Nile, under Egyptian supervision in conditions amounting to abject servitude. Ores were transported to the Nile and smelted at Egyptian outposts such as Buhen and Kubban, then sent to Pharaoh's court as part of the "impost of Wawat and Kush". Galena (lead sulphide), used for eye paint, was extracted from deposits at Gebel Zet on the Red Sea. It was sometimes transported to Middle Egypt and traded through passing Bedu. When, as in the Sinai, Egypt found it difficult to control the local population during the long periods between expeditions, tableaux designed to ward off evil might be prominently carved on rock faces showing Pharaoh overpowering a nomad. (See also pp.64–5.)

*Asiatic hunchback cattle haul stone from a quarry at el-Ma'asara near Tura in Lower Egypt, which was famous for the fine quality of its limestone. The two overseers are foreigners: an Asiatic (left, with goatee) and a Libyan (with sidelock). Early New Kingdom, ca. 1530BCE.*

# EGYPT'S LEGACY

*Linen-clad priests and devotees of the goddess perform morning sacrifices in this mural from the 1st-century CE Iseum (sanctuary for the mystery cult of Isis) at Pompeii, Italy. Particular Egyptian elements include the two sacred ibises and the horned altar (centre); the flute played by one worshipper (right); and the* sistra, *or sacred rattles, wielded by most of the other figures (compare illustration, p.148).*

Egyptian culture did not transplant easily; and Egyptians themselves were disinclined to proselytize on their own behalf. Some Nubian communities on the middle and upper Nile did, it is true, adopt Egyptian ways. The result may be seen in such lingering "pharaonic" vestiges as the kingdom of Meroe in the Sudan (third century BCE to early fourth century CE), with its pyramids, derivative art style and "Egyptian" temples. On the Phoenician coast, Byblos and other neighbouring communities borrowed what suited them (art motifs, some creation myth material, occasionally the hieroglyphic script), but to a lesser extent. By and large they retained their own indigenous languages and culture.

Importantly, Egyptian cultural influences can be detected in two regions – ancient Israel and the Aegean – that were to provide a conduit for ideas into post-Classical Europe. In Israel, archaeology has revealed the type of practical borrowings that commerce would naturally impose on a neighbouring state. The weights and measures and number designations used in Israel from the tenth to the sixth centuries BCE were in part derived from Egypt. Trade items and other commodities for which

*The Roman emperors removed numerous obelisks, statues and other artefacts from Egypt to decorate the city of Rome. This obelisk of Sety I (ca. 1290–1279BCE), which now stands in the Piazza del Popolo, was originally brought to the city in 10BCE to adorn the Circus Maximus.*

*A mosaic from Praeneste (Palestrina) in Italy, showing an Egyptian landscape of the 1st century CE, with Greco-Roman and Egyptian temples and boats, and Nile animals including hippos and crocodiles. All things Egyptian were very fashionable during the early Roman empire.*

Hebrew had no terms arrived from Egypt with their original designations intact, and these duly entered the Hebrew lexicon as loan words. Examples include the words for reeds, lotus, ebony, apes, linen and alabaster. In the realm of literature, the Hebrew Scriptures also show Egyptian influences; for example, it is well known that Proverbs 22.17–24.22 derive from *The Instruction of Amenemope*, as does the First Psalm. Some Hebrew imagery, especially in poetry, seems to recall Egyptian idiom. But it would be unwise to make too much of this: Hebrew culture in antiquity remained a distinctively Western Asiatic phenomenon.

From the seventh to the fourth centuries BCE, Egypt experienced increasing contact with Greeks from the Aegean. Merchants, adventurers and mercenaries flocked to Egypt for trade and employment; and soon it became fashionable for the Greek-speaking intelligentsia to make the five-day voyage for pleasure and study. Solon (early sixth century BCE), Hecataeus and Herodotus (both fifth century BCE) certainly spent some time on the Nile. By the time that Alexander the Great arrived in Egypt in 332BCE, a period had begun in which Greeks, and later Romans, would dominate the country for a thousand years, living alongside the natives.

The love-hate relationship that developed between Greeks and Egyptians was hardly conducive to any large-scale exchange of ideas. On the one hand, the Greeks stood in awe of Egyptian antiquity and "the wisdom of Egypt" (which soon became a cliché); on the other hand, they had nothing but contempt for Egyptian animal worship. Early in Greece's Archaic Period (seventh to sixth centuries BCE) Egyptian architecture and sculpture exerted a limited influence on the development of the Doric column and the standing statue; but in general Greek craftsmen developed their art along a wholly different path from that of their Egyptian counterparts. Certain Egyptian cosmogonies that highlight the elemental importance of water, air, earth and light (or flame), and introduce the agency of "Mind" and "Word" into creation (see pp.124–5), resonate around the eastern Mediterranean; but the question of dependency or influence is much debated. In the realm of mathematics, Egypt's contribution proved limited (see pp.94–5), but in medicine the Egyptians enjoyed a glowing reputation (see pp.96–7). Echoes of their methods of diagnosis and treatment, preserved in papyri, may be heard in the writings of Roman physicians. Egyptian astronomers bequeathed to the world the zodiac (see illustration, p.115) and also the 365-day, twelve-month calendar (see pp.92–3).

## EGYPT'S RELIGIOUS IMPACT

Although ancient Egypt exerted some cultural influence beyond its frontiers, when it suffered its ultimate transformation into the Coptic Egypt of late antiquity, it left no "successor" community among the nations of the world. However, certain aspects of Egyptian religion constitute a legacy, and consciousness of this adds a new dimension to our understanding of European Judeo-Christian culture.

The cult of Isis (and Osiris), offering personal salvation for the soul, spread widely throughout the Roman empire. The major themes of this "mystery religion" have come to be expressed in forms that subsequently influenced Christian literature and iconography: the Holy Mother with the divine Child in her arms; the judgment of the soul after death; for the saved the city of Heaven; and for the damned the underworld "Hell" with its tortures (see p.135).

*A terracotta ewer from the Roman temple of the goddess Isis in London. It is inscribed, in Latin, Londini Ad Fanum Isidis ("London, at the temple of Isis").*

In addition, the "Wisdom of Egypt" was a (misinterpreted) point of reference for Hellenized Egyptian texts in Greek known as the Hermetic Corpus, as well as for the Neo-Platonism promoted by Plotinus (third century CE). These and other belief systems, such as Gnosticism, that are based on esoteric learning and meditation as prerequisites for the soul's salvation, pointed to the Egyptian hieroglyphic inscriptions (which no Greek or Roman could read) as the alleged repository of all wisdom.

● CHAPTER 4

# THE WEALTH OF THE LAND

To the ancients, the bounty of Egypt was legendary. An abundance of crops grew in the rich, fertile soil of the floodplain. The Nile (see pp.10–16) was the fount of the country's agricultural wealth, but this also depended on the Egyptians' expert management of the land. Similarly efficient systems enabled the pharaonic government to exploit to the full the country's many other sources of economic prosperity – including the seemingly inexhaustible supply of minerals that were highly prized both in Egypt and beyond its frontiers.

▲



ABOVE: *A faience tile of ca. 1350BCE depicting a cow among the papyrus beds of the Nile. Cattle herders first appeared on the Nile in the late 6th millennium BCE.*

## THE FIRST FARMERS

Before the development of agriculture, the inhabitants of the Nile Valley lived by fishing, hunting, gathering and fowling. Toward the end of the last major ice age (ca. 11,000–8000BCE), the main source of food was probably fish, as the increased rainfall caused by postglacial warming led to high floods and the appearance of ephemeral lakes (*playas*) in the Sahara. Wetter conditions were frequently interrupted by shorter cool and dry intervals, when game became scarce and unpredictable. The bones of domesticated African cattle from ca. 8000–6000BCE have been found in the Egyptian Sahara. Cattle were probably domesticated at this time to help them withstand the effects of the droughts.

By ca. 5000BCE people in the region were gathering wild sorghum for food, as indicated by finds in Nabta *playa* and Farafra oasis. Those who dwelt in the valley relied on the aquatic resources of the Nile, and hunted and foraged in the low desert areas adjacent to the floodplain. The effect of the climatic instability and high floods of the postglacial period was an increased emphasis on fishing.

There is no evidence for food production on the Nile floodplain before the fifth millennium BCE. Traces of the earliest undisputed farming community in Egypt have been discovered at Merimde Beni Salama, a site on the western fringe of the Delta dating to ca. 4750BCE (see p.68). The possibility of whether there were earlier farming settlements along the Nile has often been linked to the discovery of ceramics, on the grounds that pottery, like other "urban" skills, could not conceivably have arisen before the establishment of settled communities. Pottery dating to ca. 5500BCE has been found in the Faiyum, where ceramics and stone tools display stylistic similarities with those of Merimde, suggesting some kind of cultural affiliation between the inhabitants of these sites. Pottery dating to ca. 5200BCE has also turned up at el-Tarif near Luxor in Upper Egypt. But there is no accompanying evidence of food production, and it is now known in any case that the development of ceramic techniques may predate the arrival of farming.

*Gathering the harvest, part of a relief from the mastaba tomb of Ipy at Saqqara. On the right, a worker is cutting a sheaf of grain (either emmer-wheat or barley) with a sickle, while another man talks to the harvest supervisor, who holds a long staff. Old Kingdom, 6th Dynasty, reign of Meryre Pepy I (ca. 2338–2298BCE).*

The appearance of Saharan pottery, cattle, sheep and goats on the Nile floodplain between ca. 5300BCE and ca. 4000BCE, together with wheat and barley from southwestern Asia, coincides with a transition to drier conditions throughout the region and in particular with the second of two spells of extreme drought that occurred ca. 6000BCE and ca. 5000BCE. Evidence from Farafra oasis in the Western Desert suggests that between these two dates the climate fluctuated severely, before the advent of cold and dry conditions ca. 5000BCE. During this millennium the Saharan lakes dried up and the human presence in the desert became sparse and ephemeral. It is highly probable that many inhabitants of the Western Desert, the Sinai and the Negev Desert began to trickle toward the Nile during this time of unstable climatic conditions. Some of these former desert-dwellers remained outside the Nile floodplain, becoming nomadic herders who moved between sources of water, or settling at spring-fed oases such as Kharga and Dakhla.

However, other migrants moved southward up the Nile Valley or settled near the Mediterranean coast. They established themselves along the edge of the floodplain, where they began to raise animals and to cultivate cereal crops. These "colonists" lived alongside the indigenous hunters, fishers and foragers of the Nile Valley, and the newcomers' cultural traditions – such as their use of cattle symbolism and their techniques of ceramic and crop production – became amalgamated with the hunting and fishing traditions of the existing inhabitants, a process that can be witnessed in the development of communities such as Hierakonpolis (see p.69). By ca. 4000BCE, farming villages had sprung up all along the banks of the river. The practice of fishing and fowling continued, but hunting as a primary means of obtaining meat was replaced by the raising of sheep, goats, cattle and pigs.

*This wall painting from the 18th-Dynasty tomb of Menna at Western Thebes includes a detailed depiction of Egyptian farm workers on the deceased's estate. In the bottom register they rake out the grain on the threshing floor, under the eyes of a supervisor. In the middle register, the harvest is threshed by having it trampled by a team of oxen, while other men winnow the grain in baskets (see also illustration, p.61).*

# LIVING OFF THE LAND

*A Roman Period Nilometer at Elephantine. A Nilometer consisted basically of a flight of steps on the bank of the river, and was used to gauge the rise in water level. The rate of rise enabled the extent of the inundation within a given period to be estimated. Measurements were recorded over many centuries (see p.13), and to some degree helped farmers to predict periods of scarcity or plenty.*

Ancient Egypt was known as a land of abundance, and kings sometimes boasted of the good harvests during their reigns. For example, it was said of Amenemhet III (ca. 1818–1772BCE) that "He makes the Two Lands verdant green more than a great Nile ... He is Life ... The King is food and his mouth is plenty." In an inscription at the temple of Abu Simbel, Ramesses II put the following words into the mouth of the god Ptah: "I give to you [Ramesses II] constant harvests ... the sheaves are like sand, the granaries approach heaven, and the grain heaps are like mountains."

This agricultural prosperity relied on the river Nile, on good land management and, above all, on hard work. The rich silt from the Nile's annual flood regularly renewed the fertility of Egyptian farmland (see pp.10–11). The floodwater irrigated the fields and the depth of the inundation determined how much land could be cultivated. To measure how much the river rose, the Egyptians built flood gauges known as "Nilometers" at various places along the Nile (see illustration, left).

The Egyptians built embankments and dykes in order to protect buildings and land during the inundation and to control the flow of water into the fields. They took advantage of natural depressions in the floodplain, which formed flood basins. Water was allowed to flow from one basin to another following the slope of the land, while artificial channels carried water to the farthest areas if the flood was low. No tools were used for irrigation until the New Kingdom, when a method for lifting water was devised, known in Arabic as a *shaduf* (see illustration, p.10). A post acted as a pivot for a cross-pole, which could swing in all directions and had a container attached to one end and a counterweight on the other. The container was filled by dipping it into the channel, and the counterweight then raised it to the appropriate level so that the water could be emptied out. In post-pharaonic times the *shaduf* – which is still in use today in some parts of Egypt – was supplemented by the water-wheel and the Archimedean water-screw.

After the floodwaters receded, much work was required to repair dykes and canals, to re-establish land-markers and to prepare the soil for sowing. Lightweight wooden ploughs were often all that was needed to turn the earth, but sometimes a hoe was used to break up heavy soil. Ploughs were pulled by teams of cows or people, and seed was often scattered in front of the plough. Crops ripened and were harvested before the next flood. In some cases, the use of irrigation extended the cultivable area and enabled two crops per year to be grown.

The harvest was another time of great activity. Cereal crops – barley and wheat – were harvested using wooden sickles with flint teeth, and the grain was taken to the village in large baskets. Men used forks to break up the stalks on the threshing floor and then donkeys or oxen were driven around the floor to trample the grain (see illustration, p.59). After winnowing (see illustration, below) the harvest was taken to a granary, where it was stored. A scribe recorded the amount of the harvest.

The Egyptians grew a range of vegetables in irrigated plots (see p.12), but the staples of their diet – bread and beer – were made from cereals. The grain was first crushed in large mortars, and then ground to obtain flour using grinding-stones and a quern (hand-mill). Loaves were baked in many different shapes over an open fire, often in conical moulds. People also made cakes flavoured with honey from wild or domesticated bees. Beer was as much a nutritious food as a drink, being produced from fermented barley-bread and often sweetened with honey, dates or spices. This was the Egyptians' principal beverage, but wine was also produced. Vineyard workers picked grapes by hand; they were then trampled in huge vats by up to six men. The juice underwent primary fermentation in large, uncovered pottery jars, and was then left to ferment a second time in stoppered jars on racks. These would be labelled with information such as the year, the place of origin and the winemaker.

Farming also included the rearing of animals, most commonly cattle. Large herds grazed on the rich grass of the Delta. Egyptians generally ate beef only on special occasions or if they belonged to the élite; however, cattle were also kept for their dairy produce and as beasts of burden. The value of an estate was calculated every couple of years according to the size of its herd. People also raised sheep, goats and pigs for meat, while geese and ducks were often reared at home for meat and eggs.

**LAND USE**

**The productivity of the various parts of the floodplain depended on the proximity of the fields to water. Land close to the Nile was generally too waterlogged to be suitable for growing cereals. The uplands, adjacent to the desert, were often very dry, and could be cultivated only during years of high flood discharge or by artificial irrigation. The best land was in the central area of the floodplain. Following a long period of cultivation, the fields of this fertile midland were occasionally left to lie fallow. Animals were allowed to graze in the fields after the harvest.**

*After threshing, the grain was winnowed – lifted up in wooden scoops and tossed into the air so that the chaff blew away in the wind while the wheat fell to the ground. This winnowing scene comes from the tomb of Nakht in Western Thebes; New Kingdom, 18th Dynasty, ca. 1400BCE.*

# HUNTING, FISHING, FOWLING AND FORAGING

The hunting of animals for their meat was common in Predynastic Egypt. Wild cattle (*Bos primigenius*) and hartebeest were hunted in the Nile Valley, while deer, wild ass and hares were found in the desert margins. Wild cattle were well adapted to the northern marshlands of the Delta, and tolerant of other wet areas; hartebeest preferred drier conditions; nonetheless, the bones of both are often found together in archaeological sites. The main prey for hunters in the desert was the dorcas gazelle, which was abundant in wadis along the edge of floodplain. Addax and oryx antelope, jackals and desert cats were also hunted.

Wild cattle were easy to hunt, especially during the dry season, when they congregated around sources of water and food; similarly, hunters found it easy to stalk gazelle that would seek the shade of trees and shrubs. Egyptians hunted on foot and used bows and arrows, spears and lassos, and dogs; they caught smaller animals in nets and traps. Their arrowheads were of small sharpened flints, occasionally attached to the lower jaw and teeth of a catfish. Animals were slaughtered and skinned with flint knives and the hides were prepared using flint scrapers.

After the development of agriculture, the hunting of large animals as a subsistence activity declined, but fishing and fowling were still pursued in dynastic times. The Nile and its surrounding marshes provided ample resources. Turtles and mussels were collected from the river. Hippopotami and crocodiles were occasionally hunted for food, although the

*Depictions of the tomb owner hunting birds in the Nile marshes were among the most common scenes in Egyptian tombs, as the examples on this and the opposite page illustrate. In the finely detailed wall painting from the tomb of Nebamun in Western Thebes (right), the deceased is fowling on a papyrus skiff. His hunting weapon is a wooden throw-stick shaped like a snake, and he uses a cat as a retriever. New Kingdom, 18th Dynasty, reign of Amenhotep III (ca. 1390–1353BCE).*

*In this fowling scene from the tomb of Nakht in Western Thebes, the deceased is shown twice, hurling his throw-stick and preparing to use it. Fowling with these weapons was a popular sport among wealthy Egyptians; those who hunted birds for a living would have employed nets and traps. As in the scene on the opposite page, the tomb owner is depicted in the company of his family; here, two servants are also present. New Kingdom, 18th Dynasty, ca. 1400BCE.*

former trampled crops and were regarded as pests, so that people often culled them with spears in hunting expeditions organized either for sport or ritual purposes. A major source of protein was fish from the river and irrigation canals. Especially abundant after the floods, fish were easy to catch from floodpools using nets or fishhooks. The most common were catfish and the Nile perch; fishermen harvested catfish during the inundation, when they migrated in large numbers to their spawning grounds in the shallow ponds of the floodplain. After mating had taken place, they could easily be trapped by hand. Fish were preserved either by being smoked or by being skinned and then dried over a hearth or in the sun. Hunters would often snare wildfowl, including coots, geese and ducks, in nets and traps on or near the banks of the river.

Wild foods included many seeds, fruits and vegetables. Egyptians would simply use a stick to dig up the tubers of wild nut-grass, a type of sedge still abundant today, which was a staple food. It was rich in carbohydrates, but when mature it contained a high level of toxins that meant it had to be ground and leached before being eaten. The tubers and seeds of the club rush were also harvested, and a number of other seeds were collected and probably ground into flour to make bread. The margins of the Nile provided catstail, bulrushes, papyrus and common reed, whose starchy rhizomes would be baked, steamed or roasted. People would also consume the tender leaves and young shoots of several other plants, as well as wild fruits, including palm nuts, melons and figs.

# EGYPT'S MINERAL WEALTH

**THE DIVINE METALS**
**The Egyptians regarded gold as a divine substance. The flesh of the gods was thought to be made of gold, their bones of silver, a rare metal not available locally in economical amounts and thus usually imported. The sun god Re was sometimes referred to as the "mountain of gold", while his daughter, the goddess Hathor, was often called the "golden one". The use of gold in the death masks of pharaohs, as in the spectacular example of Tutankhamun (ca. 1332–1322BCE; see p.34), reflected the belief that they became gods after death. Similarly, the tips of obelisks and pyramids (see pp.170–71) were often gilded, emphasizing their link to the cult of the sun. Great reserves of gold were built up by the pharaonic government, and gold was used for barter and the payment of mercenaries during the Late Period.**

Egypt was fortunate in possessing large deposits of many different types of stone and metal, which were extensively exploited. Certain minerals, including gold, copper, malachite and alabaster, were particularly highly prized, as were limestone, granite and other rocks used in the construction of temples and other monumental architecture. Some minerals had religious and symbolic significance in addition to their practical value (see sidebar, left).

The Egyptians mined both gold and copper from the earliest times; the name of one of the statelets of Predynastic Egypt, Nubt ("Gold town"), suggests that its prosperity was based at least partly on the exploitation of this metal. Mining operations were highly organized: mines had their own wells and were operated by a labour force supervised by the military. Especially during the New Kingdom, gold was mined in the Eastern Desert and Nubia (see box, below). Copper, mined in the Eastern Desert, Sinai and Nubia, was first used during the early Badarian Period (ca. 5000–4000BCE) for needles and fishhooks, but by the late Predynastic Period it was also employed in order to produce larger objects such

## WADI HAMMAMAT

The world's earliest geological map is the "Turin Mining Papyrus" of the mid-12th century BCE, which depicts the Fawakhir gold mines in the Wadi Hammamat in the Eastern Desert. It may have been produced in connection with an expedition in the reign of Ramesses IV (ca. 1156–1150BCE).

The Wadi Hammamat was exploited for its greywacke continuously from the Old Kingdom into Roman times. The quarries produced stone mainly for statues and sarcophagi. The rock was dislodged using chisels of bronze, and later iron, or by pounding with harder stones such as diorite. Blocks were transported on sleds either drawn by oxen or hauled by gangs of men.

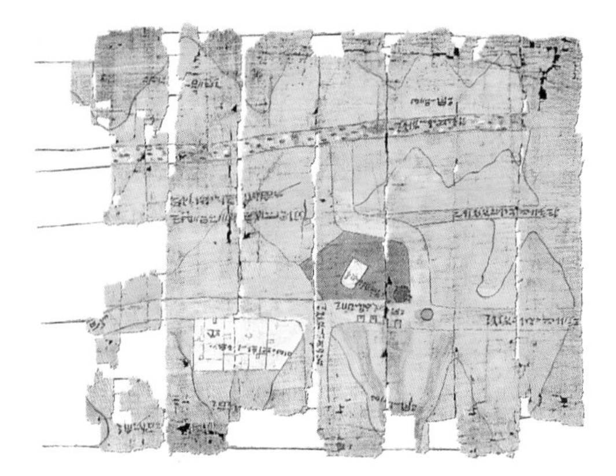

*The 20th-Dynasty map, now in Turin, Italy, of the Fawakhir mines in the Wadi Hammamat.*

The number of people involved in quarrying expeditions varied from fewer than one hundred to over ten thousand. Besides the expedition commander and his deputies, the workforce included stone-cutters, masons, scribes, animal-drivers, labourers, prospectors, guides and caterers. The men camped in rough stone huts, and worshipped in the neighbouring grotto dedicated to Min. Usually, expeditions would be commemorated by an official hieroglyphic inscription on the south side of the wadi.

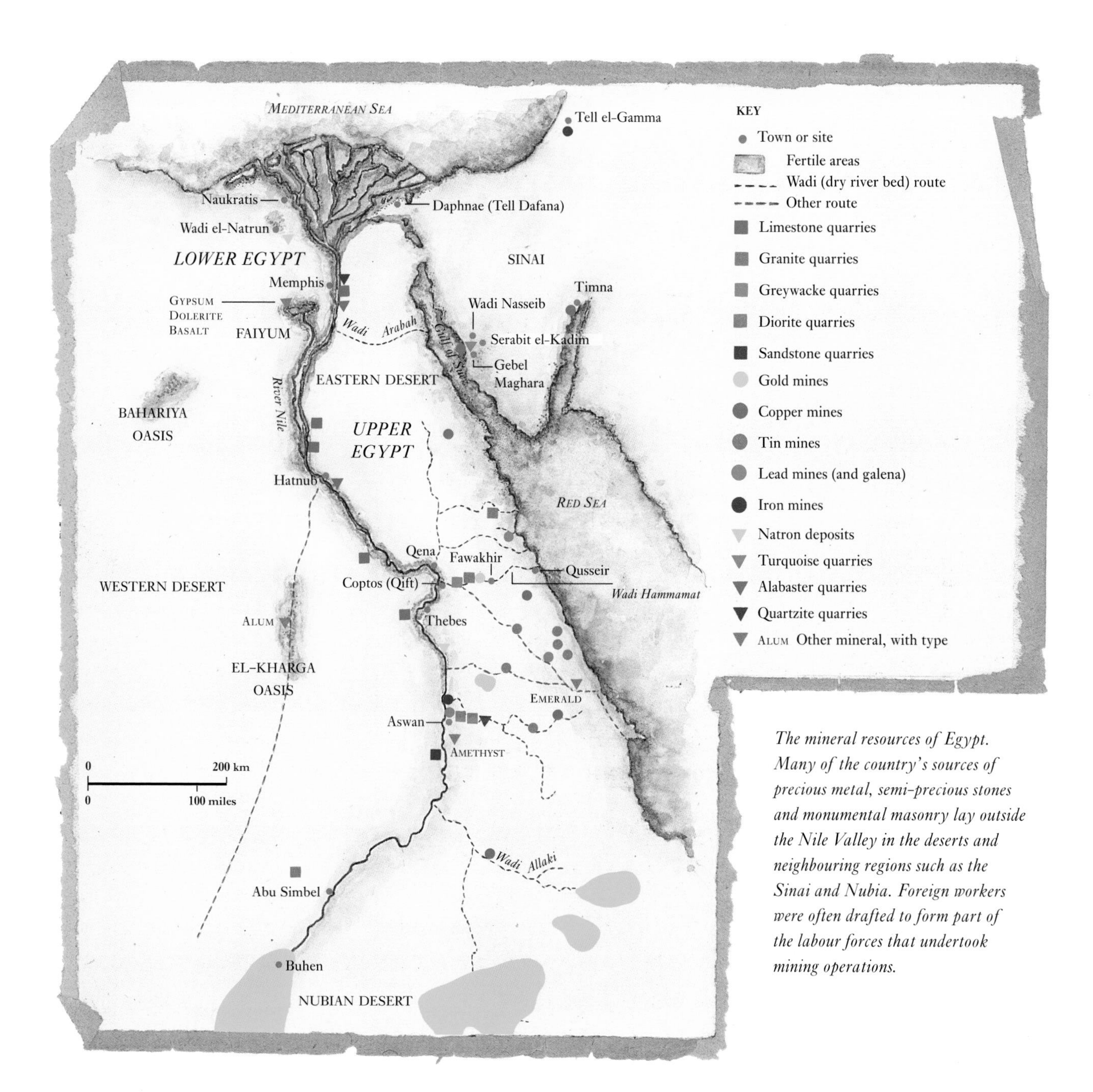


*The mineral resources of Egypt. Many of the country's sources of precious metal, semi-precious stones and monumental masonry lay outside the Nile Valley in the deserts and neighbouring regions such as the Sinai and Nubia. Foreign workers were often drafted to form part of the labour forces that undertook mining operations.*

as harpoons and daggers, and decorative items such as rings and beads.

Of the country's other resources, iron ore seams in Aswan and the Baharia oasis do not seem to have been exploited until as late as the Twenty-Sixth Dynasty (ca. 644–525BCE). Malachite, a copper-green mineral, was used as a cosmetic pigment in Predynastic times, when it represented vegetation and life. Egyptians adorned themselves with beads, amulets and ornaments made from a variety of semi-precious stones found in gravel deposits in Upper Egypt. These included amethyst, red garnet, jasper, galena, red feldspar and carnelian. Galena was also used for eye-paint. Natron, a desiccating agent, was extracted from Wadi el-Natrun and used in the mummification process.

# MANAGING THE ECONOMY

The continuity and resilience of Egyptian civilization was primarily a result of its agricultural economy. The majority of the population were subsistence farmers, and independent local economies, based on bartering, flourished in the villages. However, the economy was also organized at a national level by the pharaonic government and at a regional level by the administrators of Egypt's forty-two nomes (provinces). The nome capitals were the key centres for the management of the agrarian system, responsible in particular for the collection of taxes – in the form of grain, meat, leather, textiles and minerals – on behalf of the central government.

*Junior officials, such as scribes, soldiers and tax collectors, were responsible for the day-to-day collection, recording, storage and delivery of the annual revenues. In this wooden model of a granary from a tomb at Beni Hasan, a scribe records the amounts of grain being stored by the four granary workers. Middle Kingdom, 12th Dynasty, ca. 1850BCE.*

The Egyptian taxation service came under the supervision of a royal vizier. Agricultural yield in any one year depended on the extent of the Nile flood, which determined the land available for cultivation. It was in the government's interest to keep the irrigation systems in good condition, and responsibility for this was delegated to the nome authorities, which kept a register of landowners and tenants.

Away from the countryside, economic activity was largely generated by an urban élite that supported and benefited from the ideology of divine kingship (see pp.112–13). This ideology was manifested in the construction of great royal mortuary monuments such as pyramids or temples. These were a major economic undertaking, and a proportion of the royal revenues was earmarked to pay for the required workforce of skilled artisans and unskilled labourers. The latter consisted largely of rural workers

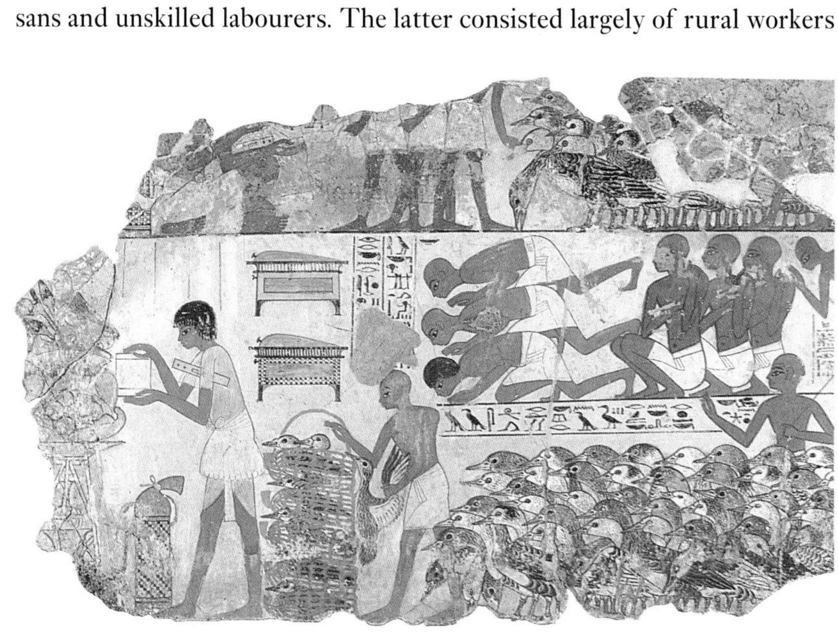

*In this fragment of a wall painting from the tomb of Nebamun in Western Thebes, a scribe presents the deceased (who would have appeared to the left of the scene) with the records of his goose flock, which is tended by gooseherds. Eighteenth Dynasty, reign of Amenhotep III (ca. 1390–1353BCE).*

conscripted for a few months at a time when they were not otherwise engaged in farming (for example during the inundation). They were paid in grain and other staples. In one case, the pay of an overseer is recorded as being twenty-eight times that of his lowest-paid underling.

The population of Egypt was small, probably no more than two million during the Old Kingdom and three million during the New Kingdom. The urban population was a fraction of this, perhaps no more than five percent. A rise in the urban population, greater demands by the élite, or an increased need for defence, put greater pressure on Egypt's farmers. For example, during the New Kingdom, when Egypt possessed a large empire, food producers had to help maintain an almost permanent imperial army of up to forty thousand troops, many of them drawn from the rural workforce. The problem was partly solved through agricultural reclamation projects in the Delta and the use of slaves for labour. In the Late Period, costly wars and the use of Greek mercenaries – who were paid in gold at a time when the gold mines were almost exhausted – contributed to Egypt's economic decline.

## THE ECONOMY OF EGYPTIAN TEMPLES

Egypt's two thousand or so temples owned large estates and constituted an important sector of the economy. The richest and most powerful were those in the royal capitals, such as Memphis and Thebes (see pp.208–9). Temples received endowments from the king and also generated an independent income from their substantial land and livestock holdings, as well as from donations by private individuals. In the Late Period, King Apries (589–570BCE) presented the temple of Ptah at Memphis with a perpetual, tax-free endowment of a district with all its arable land, inhabitants and cattle. According to the Great Harris Papyrus, which records the temple benefactions of Ramesses III (see illustration, right) about a third of all cultivable land in Egypt belonged to the temples.

*Part of the Great Harris Papyrus, a record in hieratic script of temple donations by Ramesses III (ca. 1187–1156BCE).*

Just as the temples benefited from royal patronage, they in turn benefited the king by legitimating his divine authority during his lifetime and perpetuating his cult after death. They also served as an occasional source of revenue for the royal coffers. The economic and political independence of the temples depended on that of the central government. When the royal authority was weak, as toward the end of the New Kingdom, the power of the high priests increased proportionately. Sometimes they were even able to challenge the royal authority, as was the case with the high priests of Amun at Thebes in the latter years of the Twentieth Dynasty (see pp.36–7).

● CHAPTER 5

# THE SETTLED WORLD

Egypt's magnificent temples and tombs, the houses of the gods and of the dead, have mostly withstood the ravages of time better than the dwellings of the living. Little remains of the mudbrick villages that were home to the great majority of the people – farmers on the rich agricultural floodplain of the Nile. Others lived in towns and cities, where labourers and unskilled workers dwelt alongside artists, scribes, priests and bureaucrats. Finally, there were the palaces of the king and his court, and the forts and garrison towns that kept watch over Egypt's frontiers.

▲



ABOVE: *A pottery model of an Egyptian single-story village house, which has one window, steps up to a roof terrace and "air-conditioning" – an aperture in the roof designed to catch and circulate breezes inside the house. Such models, known as "soul houses", were placed in graves by less wealthy Egyptians for use in the afterlife; various foodstuffs for the deceased are depicted in the enclosed courtyard. Middle Kingdom, ca. 1900*BCE.

## THE RISE OF URBAN LIFE

Ancient Egypt used to be thought of as a "civilization without cities". Until comparatively recently, only a few urban settlements of the pharaonic period had been seriously studied. In recent years, however, detailed archaeological investigations at sites such as Abydos, Elephantine and Buto have begun to reveal the existence of large and complex communities at a time when Egypt's distinctive civilization was beginning to emerge in the northern Nile Valley.

The first known settlement in Egypt, dating from ca. 4750BCE, was at Merimde Beni Salama, on the western edge of the Delta, about fifteen miles (25km) northwest of Cairo. Although it covers an area of 216,000 square yards (181,000$m^2$), and the total population has been put at sixteen thousand, not all the site may have been occupied at the same time. The earliest shelters here were simple wind-breaks and tiny pole-framed huts, but later houses were of mudbrick, perhaps with pitched roofs, and as much as ten feet (3m) in diameter. The level of organization in the village is indicated by an ordered street pattern and numerous public granaries.

Merimde Beni Salama seems to have been a simple rural community, but the late Predynastic site of Ma'adi, about three miles (5km) south of Cairo, was evidently a merchant town. To judge from the presence of imported pottery vessels and Palestinian products (such as copper pins, chisels, fishhooks, basalt vessels, bitumen and carnelian beads), it thrived on trade with the Near East. Among the wattle-and-daub huts and large semi-subterranean houses, covering an area of about forty-five acres (18ha), there are remains of agricultural activity and also of craft specialization – for example, the trading and processing of copper, which almost certainly came from the Sinai (see pp.64–5). There are also many fragments of Upper Egyptian pottery and stone artefacts, and it seems likely that the town prospered on trade with Upper Egypt and Palestine at some time in the mid- to late fourth millennium BCE.

Although most of the earliest villages have so far been found in or near the Delta, the first real towns, in the sense of substantial walled settle-

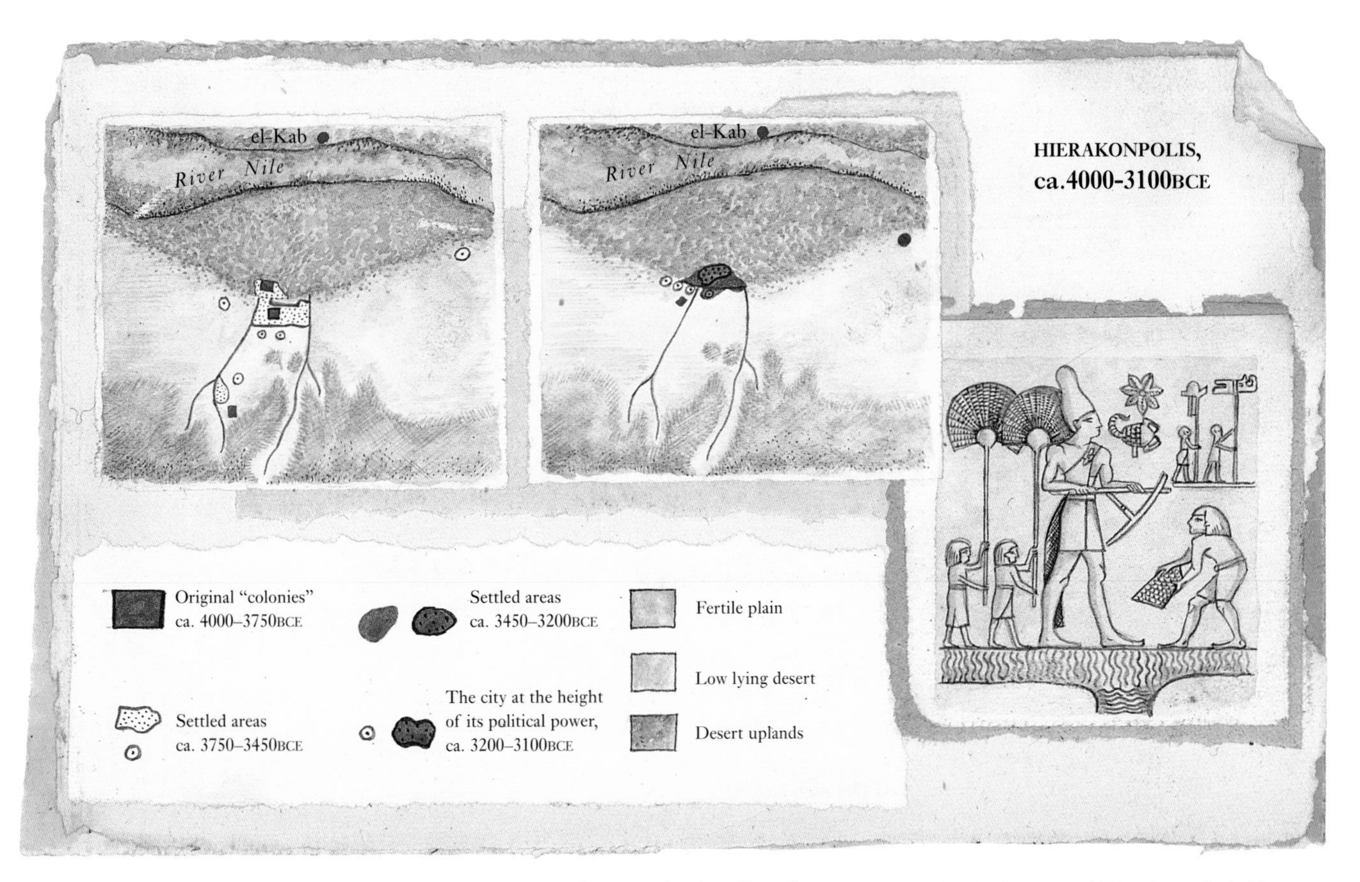


ments, were in Upper Egypt. Nekhen (better known by its Greek name, Hierakonpolis) and the South Town at Naqada emerged in the late Predynastic period as the centres of proto-states, which eventually amalgamated into a much larger Upper Egyptian state between 3200BCE and 3000BCE (see pp.22–3, 106–7). Hierakonpolis is a good example of how settlements developed in the Nile Valley. Its history begins ca. 4000BCE, when local hunter-gatherers were joined by farming and herding "colonists". By ca. 3500BCE, the population had risen probably to around seven and a half thousand and included such specialized workers as potters, masons and weavers. In the late fourth millennium, most people moved into the floodplain, perhaps as a result of low Nile floods (making it difficult to get irrigation water to the desert margins) and because the use of wood in potters' kilns had led to gradual deforestation. At this time, monumental architecture began to be built in and near the town, including a religious complex and brick-lined tombs. One of the latter is the earliest known decorated tomb in Egypt and may have belonged to a king. By ca. 3200BCE, a wall had been built around the settlement.

Naqada and Hierakonpolis (and probably also This, near Abydos) must initially have evolved through a combination of agricultural prosperity and trade. But as the emergent state developed, the towns were transformed into provincial capitals. They tended to develop around ancient religious centres, so the cults of Horus and Seth, at Hierakonpolis and Naqada respectively, may have played a key role in the cities' growth.

*The development of Hierakonpolis in Upper Egypt. The drawing alongside, based on a scene on an Early Dynastic ceremonial stone macehead discovered at the town, shows an Upper Egyptian king founding an irrigation channel. Hierakonpolis continued to expand in the 1st Dynasty (ca. 3000–2800BCE), but declined in political importance.*

*The so-called "Libyan Palette", a late Predynastic mudstone votive item found at Hierakonpolis, is decorated on one side with representations of fortified towns. They are probably in the process of being captured by an early Upper Egyptian ruler.*

# TOWNS AND HOUSES

There were several Egyptian words for different types of settlement. *Dmi* is sometimes translated as "town", and the terms *whyt* and *niwt* are often taken to refer to "village" and "city" respectively. However, although later Egyptian texts often appear to use these terms to denote different sizes of settlement, many Old and Middle Kingdom inscriptions seem to adopt a different usage, based on factors such as status and economics. For example, *niwt* may have been used in the Old Kingdom to mean an "ordinary town", whereas a *hwt* may have been a "centre of royal power", which was able to exact taxes from a *niwt*. On the other hand, *dmi* seems to have denoted a shrine or cult centre.

More information on Egyptian towns can be derived from archaeological and anthropological studies rather than linguistic ones. The uncertainties of archaeological survival are always an important factor in the reconstruction of the elements of an ancient culture, and this is particularly true of Egypt. Most towns were located on or near land subjected to the Nile's annual inundation, so that their preservation has tended to be relatively poor compared with that of cemeteries in the adjacent desert. Also, the Nile near modern Cairo probably shifted gradually eastward during the pharaonic period and later, thereby obliterating many urban sites in Middle Egypt. Finally, many ancient settlements have been pillaged for centuries by farmers for whom the mudbrick remains are a source of a compost called *sebakh*. Often, pharaonic settlements have survived simply because they were unusual communities located for some specific purpose in the desert rather than the valley.

Our current picture of Egyptian towns, therefore, is probably based on anomalous and unrepresentative sites. For example, the best-preserved

*Painted vignettes in the Book of the Dead sometimes provide glimpses of Egyptian daily life. This example shows the scribe Nakht and his wife standing in the garden of their house with hands raised in adoration of the god Osiris (at left). In the garden is a pool surrounded by trees, while the house itself (at right) has two roof vents. New Kingdom, 19th Dynasty, ca. 1300BCE.*

urban settlement of the pharaonic period is el-Amarna (ancient Akhetaten), the capital city built on a virgin site ca. 1350BCE by the pharaoh Akhenaten (see pp.128–9). Toward the end of Akhenaten's reign, most of the population returned to Thebes and the city was eventually abandoned. Such unusual factors mean that el-Amarna is a potentially unreliable source for the study of Egyptian towns.

Most Egyptian houses were built of mudbricks, which were formed in rectangular wooden moulds and left to bake in the sun. The changing dimensions of brick moulds can sometimes be used to date buildings. Some houses, particularly those built for short-lived communities in isolated desert locations, are remarkably well preserved, compared with more permanent towns in the floodplain, which have often suffered from being under modern fields or villages. Among the most meticulously excavated Egyptian houses are those in the planned workers' village that lies just to the east of Akhetaten. The village, which has often been compared with Deir el-Medina in Western Thebes (see p.73), lies in a flat, south-facing valley, adjacent to cliffs and surrounded by a network of ancient patrol roads. There were some seventy-two houses enclosed by a square wall; forty houses have so far been excavated. The enclosure wall

*This New Kingdom limestone model of a town house depicts a narrow two-story dwelling with a semi-enclosed roof terrace. The windows, with grilles or shutters over the bottom half, are set fairly high in the walls to prevent the interior from overheating in the Egyptian sun.*

## THE PYRAMID TOWN OF KAHUN

Several towns have been found in the vicinity of the pyramids of the Old and Middle Kingdoms. The largest and best-preserved is Kahun, also known as Lahun, a rectangular, planned settlement measuring about a third of a mile by half a mile (384m by 335m). It was discovered in 1889 by the British archaeologist Flinders Petrie at the eastern end of the pyramid complex of the Twelfth-Dynasty pharaoh Senwosret II, on the southeastern edge of the Faiyum oasis. The official papyri found at the site suggest that the town was known as Hetepsenwosret ("Senwosret is satisfied"), and it is thought to have housed first the workers who constructed the pyramid, then the officials responsible for maintaining Senwosret's mortuary cult.

By the late Middle Kingdom, Kahun had acquired its own mayor and had probably grown into a town in its own right. If an average of six persons per house is assumed, the population of Kahun might have been about three thousand. However, according to one estimate its granaries could support a considerably greater population, in the region of five thousand to nine thousand. It is possible that more grain was stored than was needed at any one time in case of prolonged shortages or famine.

Unlike the sprawling cities of Thebes or Memphis, Kahun was very much a planned entity, segregated by a thick wall. On the western side, there were more than two hundred small rectangular buildings, each covering an area of up to 1,100 square feet (100m$^2$) and consisting of three to seven rooms. The eastern side was clearly the residential quarter of the wealthy: there were about a dozen very large houses, each with up to seventy rooms and covering 11,000 to 26,000 square feet (1,000–2,400m$^2$).

It would be unwise to estimate the distribution of wealth during the Middle Kingdom from the housing at a single and probably atypical settlement. However, the sharp residential contrasts visible at Kahun, together with the evidence from contemporary tombs, are an indication that Egypt's resources at this time were concentrated in the hands of a small élite.

**A TOWN REGISTER**

**In a papyrus of the late twelfth century BCE is a list of the names and professions of 182 residents of Western Thebes. Such documents usually suggest that the urban population was dominated by priests, soldiers, scribes and administrative officials, but this papyrus includes more humble occupations, such as gardeners, herdsmen, fishermen, coppersmiths and sandalmakers.**

**The houses where these artisans lived probably lay between the village of Deir el-Medina (see box, opposite) and the mortuary temples of Sety I, Ramesses II and Ramesses III. On the whole, the householders are not grouped by job or status, and may therefore simply be listed in topographical order. If so, all the different professions and social classes lived together rather than in separated specialized zones. However, not surprisingly in a community that dwelt alongside so many temples (see map, p.195), the households of priests are often grouped together, particularly in the immediate vicinity of the three royal mortuary temples. There are also a few concentrations of other trades, in pairs or groups of up to five in a row (such as coppersmiths, scribes, brewers, fishermen, herdsmen and sandalmakers). Most of the "land workers" are concentrated at the southwestern tip of the settlement.**

and lower courses of the external walls are of bricks made of alluvial mud from the banks of the Nile. However, the upper courses and the internal walls are of clay which was quarried just outside the village. This arrangement suggests two stages of building: it is possible that the basic foundations of the village were laid down by the authorities, after which most of the construction was left to the villagers themselves. One house, in the southeastern corner of the village, was much larger than the others: it was probably the residence of the head of the community. The other houses each consisted of four rooms: a front room, a central room, and a back room divided into two. In most cases, a staircase led up from one of the back rooms. There has been much debate as to whether such staircases simply led to the roof, but there is growing evidence that houses here and in Akhetaten itself – and presumably elsewhere – often had upper floors.

Akhetaten has probably contributed the strongest image of the houses of wealthy Eighteenth-Dynasty nobles. The typical "Amarna villa", housing the families of high officials, appears to have consisted of about twenty to twenty-eight rooms. In the centre of the building there was usually a somewhat taller, columned room with windows placed above the level of the smaller surrounding chambers. The largest of these villas were set in extensive walled courtyards that often included a well, a garden and granaries.

The distribution of grain and water supplies may provide the key to understanding the pattern of daily life in the city as a whole. The inhabitants of the many smaller houses, wedged in the gaps between the élite residences, were probably often employed by the villa owners, and no

*Some tomb paintings, such as this fragment from the tomb chapel of Nebamun, portray the villa gardens of wealthy nobles in some detail. This pool is well stocked with fish and ducks and is surrounded by a variety of trees and shrubs, including date palms, sycamores and mandrakes. The sycamore in the top right-hand corner of the pool is inhabited by a tree goddess who proffers trays of food and drink, symbolizing an idyllic afterlife for the deceased. New Kingdom, 18th Dynasty, ca. 1400BCE.*

doubt received basic rations from their employer's central stores. House sizes at Akhetaten differ less markedly than at Kahun (see p.71), suggesting a less stark contrast between rich and poor and the presence of a substantial "middle class".

The excavations at el-Amarna also provide a good sense of the furniture and fittings of a typical New Kingdom house. In the small houses, the furnishings consisted simply of mudbrick benches, wall-niches and the occasional rough stone stool or wooden table. In the élite villas, on the other hand, there were wooden beds, mattresses, stone-grille windows, plastered stone washing rooms and even toilets (consisting of stone or wooden seats with a keyhole-shaped aperture placed above a tray of sand).

Much information about the occupations, diet and standard of living of Egyptian town-dwellers is still missing. Many activities would have taken place outdoors, but it is only in the last two or three decades that streets and yards have been excavated as thoroughly as domestic interiors.

## THE WORKERS' VILLAGE AT DEIR EL-MEDINA

In 1929, the Czech archaeologist Jaroslav Cerny identified a small site at Deir el-Medina in Western Thebes (see map, p.195) as the village inhabited by the labourers, scribes and craftsmen who created the tombs in the Valley of the Kings. The houses, chapels and tombs of these workers provide an unusually detailed and intimate picture of a small community of government employees from the time of Thutmose I (ca. 1493–1482BCE) to the end of the Twentieth Dynasty (ca. 1075BCE). At its peak, in the Twentieth Dynasty, the village – known as the "Place of Truth" – consisted of seventy mudbrick houses arranged in rows within an enclosure wall. Another forty houses, scattered about the immediate vicinity, were probably the homes of less skilled workers such as donkey-drivers and fruit-pickers. Each of the houses at Deir el-Medina had on average four to six rooms, plus small cellars for storage. The function of each room is uncertain, and they would not necessarily have conformed to modern ideas of single-purpose chambers such as "the kitchen" or "the bedroom". Animals may well have been kept in some rooms.

The village at Deir el-Medina is unique in having yielded many different types of texts written by the villagers themselves. Most take the form of *ostraca* (inscribed potsherds or flakes of stone), but there are also papyri and many inscribed objects from houses and funerary chapels. Although the government provided the village workers with wages in the form of emmer-wheat and barley, as well as such commodities as beer, honey, fish and oil, the many surviving records of personal transactions suggest that people found numerous opportunities to supplement their state rations.

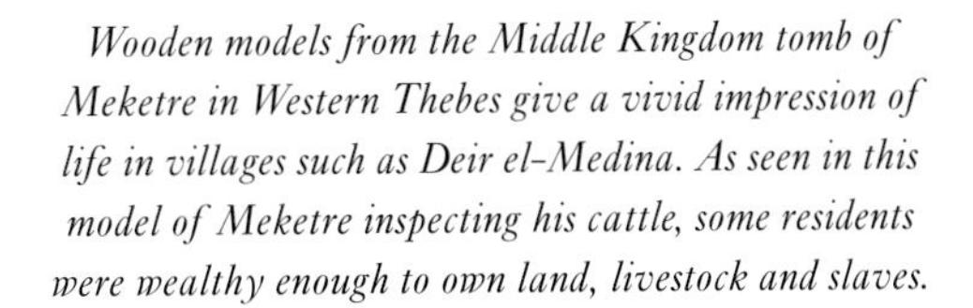

*Wooden models from the Middle Kingdom tomb of Meketre in Western Thebes give a vivid impression of life in villages such as Deir el-Medina. As seen in this model of Meketre inspecting his cattle, some residents were wealthy enough to own land, livestock and slaves.*

# CAPITAL CITIES

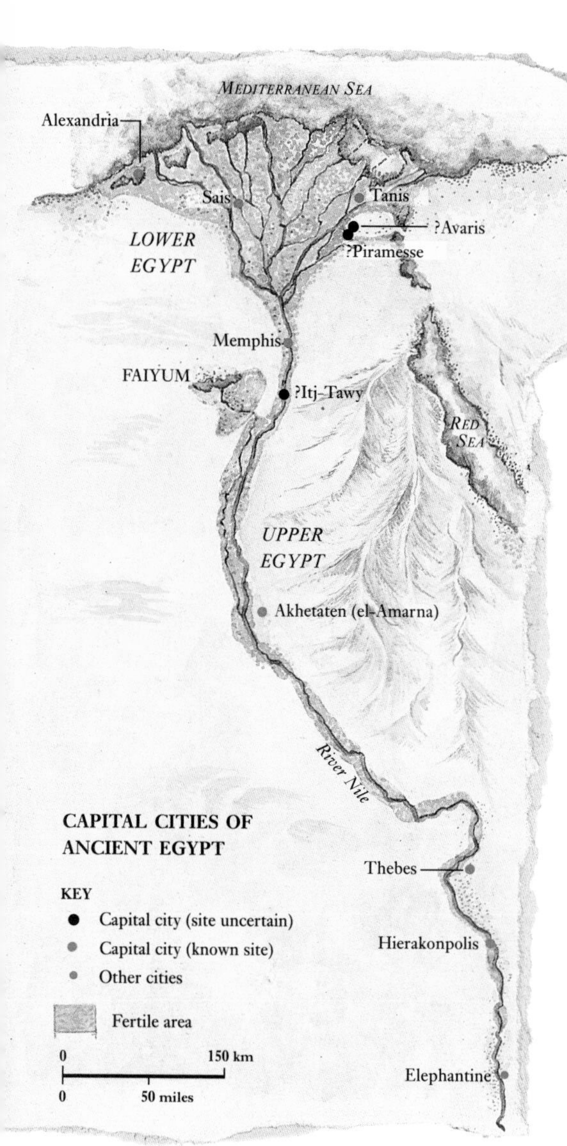


*The principal capitals of the pharaonic period. Excluded from this map are the chief towns of the more obscure or short-lived of Egypt's dynasties.*

The first and most enduring capital of Egypt was Memphis (see box, opposite), a few miles southwest of modern Cairo, near the junction of the Nile Valley and the Delta. It was founded at the beginning of the Dynastic Period or earlier, no doubt rapidly eclipsing the late Predynastic capital of Hierakonpolis (see p.69).

The city's earliest name was probably Ineb-hedj ("White Walls"), very likely a reference to a royal palace. It later became known as Men-nefer ("Established and Beautiful") after the nearby necropolis of Pepy I at Saqqara. "Memphis" is a Greek version of Men-nefer.

The emergence of a Theban family of pharaohs at the beginning of the Middle Kingdom saw Thebes, in Upper Egypt, transformed into a powerful religious and administrative centre which later rivalled Memphis during the Eighteenth Dynasty (ca. 1539–1292BCE) and after. However, the capital of the Twelfth Dynasty (ca. 1938–1759BCE) was a new city, Itj-tawy ("[Amenemhet I is] Seizer of the Two Lands"), to the east of the early Twelfth-Dynasty necropolis at el-Lisht. The history of the Twelfth and Thirteenth dynasties is very much one of political and economic centralization toward Itj-tawy.

At the end of the Thirteenth Dynasty (ca. 1759–after 1630BCE), the Middle Kingdom rulers were forced to retreat southward to Thebes with the emergence of the Asiatic Hyksos rulers in the north (see pp.30–31). They established their capital at Avaris (modern Tell el-Dab'a) in the eastern Delta, and transformed this small Egyptian town into an Asiatic city typical of contemporary Syria-Palestine.

In the sixteenth century BCE, the Hyksos were expelled and the incoming Eighteenth Dynasty restored Memphis to its former importance. The home of the dynasty, Thebes, became the burial place of New Kingdom pharaohs and took on administrative significance. However, Akhenaten (ca. 1353–1336BCE) founded his own capital, Akhetaten, in Middle Egypt. Akhetaten has furnished many fascinating details of everyday urban life, but the radical innovations of Akhenaten's reign (see pp.128–9) mean that it was probably not typical of Egyptian capital cities (see pp.71–3).

By the reign of Sety I (ca. 1290–1279BCE), Egypt's focus had shifted north to encompass its new sphere of influence in the Near East. Sety's successors built a new city, Piramesse, in the Delta, which became the power base of the Ramesside kings of the thirteenth and twelfth centuries BCE. However, Thebes and the traditional capital, Memphis, still both seem to have performed religious and administrative functions.

During the Third Intermediate Period, Egypt was split between Upper Egypt, ruled by the High Priests of Amun at Thebes, and the Delta, ruled by pharaohs of Libyan descent based at Tanis. Founded by Psusennes I (ca. 1045–997BCE), Tanis was built mainly of stone from Piramesse and Avaris, which were also the source of the new town's statuary. When the Kushite pharaoh Piye (ca. 747–716BCE) reconquered the north, he claimed to have subjugated as many as ten rulers, each with a different capital city.

The bias toward a northern power base continued in the Late Period, with Sais in the Delta (modern Sa el-Hagar) becoming the capital for much of the time. The seal was set upon Egypt's absorption into the Mediterranean world when Alexander the Great (332–323BCE) founded a new capital, Alexandria, on the Mediterranean coast. This Hellenistic city, facing toward Europe, was to be the last capital of ancient Egypt.

## MEMPHIS, THE CITY THAT DISAPPEARED

Remarkably little survives of Memphis, the capital of Egypt for most of its ancient history, largely because the ruins were quarried in the Middle Ages for stone to build Cairo's churches and mosques. The city's size and importance are suggested both by the extent of its necropolis at Saqqara and by the literary accounts of Greek authors such as Herodotus and Strabo. During the New Kingdom, Memphis was dominated by the sacred precinct of the god Ptah, probably as impressive a monument as the temple of Amun at Karnak (see pp.208–9).

The archaeological site of Memphis covers almost three square miles (7.5km$^2$). The oldest parts are thought to lie close to the cemeteries of northern Saqqara and Abusir, but modern excavations suggest that the city spread south and east as the Nile shifted eastward. Archaeological work has focused on large ceremonial buildings, such as the temple of Ptah, the palace of Apries (ca. 589–570BCE), and the temple and palace of Merenptah (ca. 1213–1204BCE; see p.77). However, excavations between 1984 and 1990 by the Egypt Exploration Society at Kom el-Rabi'a, a residential area, sought to gain a sense of Memphite daily life during the Middle and New Kingdoms. The thousands of discoveries included millstones, copper alloy fishhooks and limestone fishing weights. Further clues to the diet of New Kingdom Memphites were the remains of at least twenty types of Nile fish, and over three thousand animal bones, including those of sheep, cows, goats, pigs, ducks, geese and waders.

*This New Kingdom calcite sphinx is one of the few monuments to have survived the medieval pillaging of Memphite temples and palaces for masonry to build churches and mosques in Cairo.*

**ROYAL HOMES OF AKHENATEN**
**The North Riverside Palace, an extensive residence surrounded by a thick perimeter wall at the northernmost tip of el-Amarna, is thought to be the most likely residence of Akhenaten and his family. Unfortunately, most of this building is now lost under modern fields. Another royal home, described by its excavators as the "King's House", may have stood opposite the Great Palace in Akhetaten (see main text). The building was linked to the Great Palace by a massive bridge spanning the city's main thoroughfare.**

**The King's House was probably a temporary residence for the royal family when they were visiting the two principal temples in the centre of the city. In a room of this palace the British archaeologist Sir Flinders Petrie found a large fragment of a wall painting (see illustration, below) depicting two of the king's youngest daughters seated at the feet of their parents during a feast. They are portrayed with the informality that characterized the official art of Akhenaten's reign.**

# ROYAL PALACES

The palace of the king was the hub of the Egyptian administration. Itjtawy, the capital city of the Twelfth and Thirteenth dynasties, was known simply as "The Residence", a name that indicates the importance of the pharaoh's palace to the running of the country. In practice, most rulers seem to have had several different residences, ranging from the principal home of the royal family at the capital (Memphis for most of the pharaonic period) to small palaces for ritual use attached to Theban mortuary temples. During the Twelfth Dynasty there were also a number of ephemeral "campaign palaces", two of which were built near Egyptian fortresses in Nubia (see pp.78–9), perhaps to provide temporary accommodation for the pharaoh during his military campaigns.

Most surviving royal palaces date to the New Kingdom or later. They include the ceremonial buildings of the Ramesside kings at Piramesse in the northwestern Delta and the palace of Merneptah at Memphis. The design of the palaces varied, but they usually included a throne room, a columned hall and a "Window of Appearances", an opening in the wall where the king stood to observe rituals or bestow largesse on his courtiers. The best evidence for the architecture and decoration of royal buildings derives from the Upper Egyptian sites of el-Amarna, ancient Akhetaten, in Middle Egypt, and Malkata in Western Thebes. Both date to the fourteenth century BCE. Malkata (first excavated 1888–1918 and re-examined

*A wall painting showing Akhenaten's daughters (see sidebar, above) from the "King's House" at el-Amarna; ca. 1335BCE.*

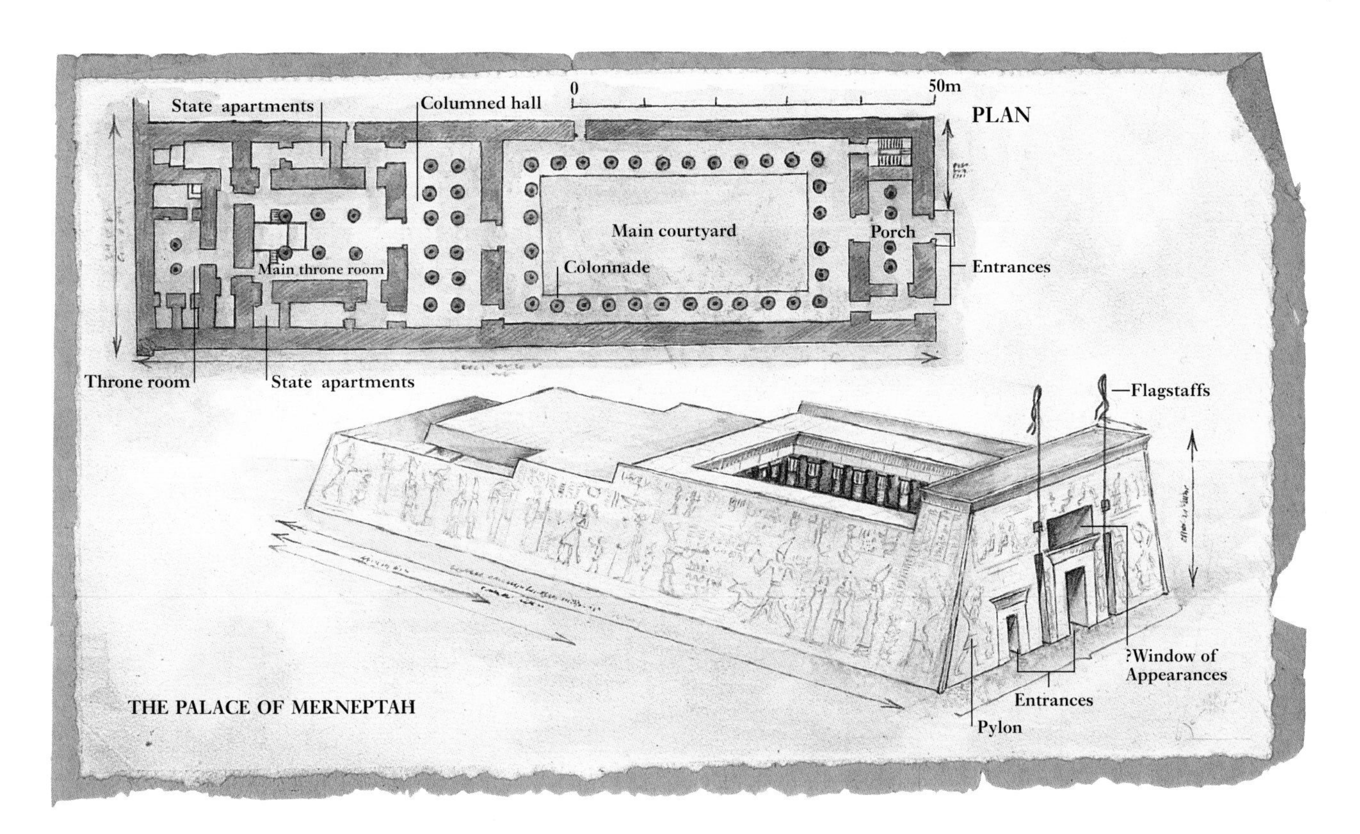


in the 1970s) is essentially the remains of a community that grew up around the Theban residence of Amenhotep III (ca. 1390–1353BCE). The excavations have revealed several large official buildings, including four probable palaces, as well as kitchens, store-rooms, residential areas and a temple dedicated to the god Amun. Just to the east of the royal residence are the remains of a large artificial lake, now known as Birket Habu. This seems to have been created at the same time as Amenhotep III's palaces and was probably used in conjunction with festivities to mark the *sed*, the jubilee festival held to commemorate the thirtieth anniversary of the pharaoh's accession.

*Most surviving royal palaces date to the New Kingdom or later, including the ceremonial buildings of the Ramesside kings at Piramesse and the palace of Merneptah at Memphis (plan and artist's reconstruction, above). The palace's well-preserved ground plan (top) reveals that the exterior of the palace may have resembled a typical Egyptian temple. Certain details, such as the position of the "Window of Appearances", must be considered as conjectural in this drawing.*

The best-preserved royal structure at Akhetaten, the capital city of the "heretic" pharaoh Akhenaten (see pp.128–9), is the North Palace, a large ceremonial building incorporating pools, gardens and an aviary, located close to the northern end of the city. It appears to have belonged at first to one of the royal wives, probably the famous Queen Nefertiti (see pp.88–9), and later to the princess Meretaten, one of Akhenaten's daughters. The walls and floors of both this residence and the "Great Palace", which was situated in the city centre, were decorated with remarkably beautiful paintings, many depicting typical Nile scenes, with birds and animals surrounded by papyrus plants and palm trees. In the Great Palace, some of the pavements around one of the pools were painted with representations of the "Nine Bows", figures symbolizing the king's traditional foreign enemies, on whom Akhenaten was able to trample as he walked around his pool (see also sidebar, opposite).

# FORTRESSES

**A LIFE ON THE FRONTIER**
**Insights into the lives of soldiers manning Egyptian garrisons and fortresses in Nubia are found in the "Semna Papyri", administrative reports sent by the garrison commander at Semna to the military headquarters at Thebes in the reign of Amenemhet III (ca. 1818–1772BCE). The reports (see illustration, below) suggest that soldiers in Nubia had to carry out regular and painstaking surveillance and reconnaissance of the deserts surrounding the fortresses. A typical report states that "the patrol that went out to survey the desert-edge near the fortress of Khesef-Medjau on the last day of the third month of spring in the third year has returned to report to me, saying 'We have found the track of thirty-two men and three donkeys ...' ". The dispatches convey something of the tedium (and perhaps paranoia) of life on Egypt's frontiers.**

Egypt's vulnerable northern and southern frontiers were policed from fortified garrison towns. These might house anything from a score to a few hundred troops, who would serve for up to six years. To man the fortresses, Egypt was willing to recruit Nubians, Philistines or Libyans.

One crucial frontier zone was Lower (northern) Nubia. During the Old Kingdom, Egypt was able to exploit the area's mineral resources relatively peacefully. However, in the Middle Kingdom, the growth of Nubian power led to the construction of at least seventeen fortresses, mainly in the period from Senwosret I to Senwosret III (ca. 1919–1818BCE). They were evidently intended to control the king's monopoly on the lucrative trade routes from sub-Saharan Africa. The enormous scale of the Nubian fortresses has led some archaeologists to regard them mainly as an expensive exercise in propaganda. They incorporated crenellations, bastions, ditches, triple loopholes and other features perhaps more readily associated with the castles of medieval Europe. There were two basic types of fortress, probably corresponding to two main phases of Middle Kingdom colonization in Nubia. "Plains fortresses", such as Buhen (see box, opposite) and Iken near the Second Cataract (the largest of eleven fortresses built by Senwosret III), stood on flat land and had a rectangular ground plan. Such forts no doubt looked impressive but would actually have been quite difficult to defend. The second type, more genuinely functional, was irregular and sprawling and designed to withstand siege warfare.

*One of the Semna Papyri (see sidebar, above). They were discovered in 1896 in the Theban tomb of a Middle Kingdom lector-priest. Among other things, the tomb contained a box of magical papyri, three of which were written on the back of old military dispatches.*

Not all forts were used exclusively by the military. Iken possessed granaries and a mud-lined slipway, along which traders could drag their boats to avoid the rapids at this point. Askut, midway between the Second and Semna cataracts, may have been the main grain store for all the garrisons in the area: half of it was taken up by granaries. Gold, mined at nearby Khor Ahmed Sherif, was taken to Askut for processing.

In the New Kingdom (ca. 1539–1075BCE), the Nubian forts became more like normal towns, with temples outside the fortress walls. So firm was Egypt's grip on Nubia at this time that a number of new towns, often with relatively perfunctory defences, were established south of the older Middle Kingdom fortresses.

In the northeastern Delta, Egypt's other most vulnerable region, Amenemhet I (ca. 1938–1909BCE) built a row of forts known as the "Walls of the Prince" (*Inbw Heka*) to protect the country against invasion from Palestine. Six centuries later, Ramesses II built his own garrisons along the same frontier.

## A NUBIAN FORTRESS

One of the largest and most sophisticated of Egypt's frontier fortresses was Buhen, which lay about 156 miles (250km) upstream from Aswan near the Second Cataract of the Nile in Lower Nubia. The fortress, on the west bank of the Nile, originated in the Old Kingdom as a centre for the organization and supervision of Egyptian mining expeditions in the region. Later, during the Twelfth Dynasty (1938–1759BCE), impressive mudbrick ramparts were built around the settlement, transforming it into a garrison that effectively controlled the region down to the Second Cataract.

The outer western wall, some thirteen feet (4m) thick, incorporated five large towers as well as a massive central tower serving as the main entrance or "barbican". This included two baffle entrances, double wooden doors and a drawbridge. The inner fortress, built on a more regular, square plan, was protected by large towers at each corner of its walls and bastions at intervals of sixteen feet (5m). The garrison town that grew up within this inner fortress consisted of several rectangular blocks of housing separated by intersecting streets. One block comprised the residence of the fort commander, and there was also a temple. In the New Kingdom, Buhen undoubtedly became much more of a civilian settlement, as the pharaohs of the Eighteenth Dynasty pushed Egypt's frontier farther south, beyond the Fourth Cataract (see map, p.51).

The remains of Buhen were first examined in 1819 but mainly excavated from 1957 to 1964 before the construction of the Aswan High Dam: like all Egypt's Nubian fortresses, Buhen was to disappear under the waters of Lake Nasser.

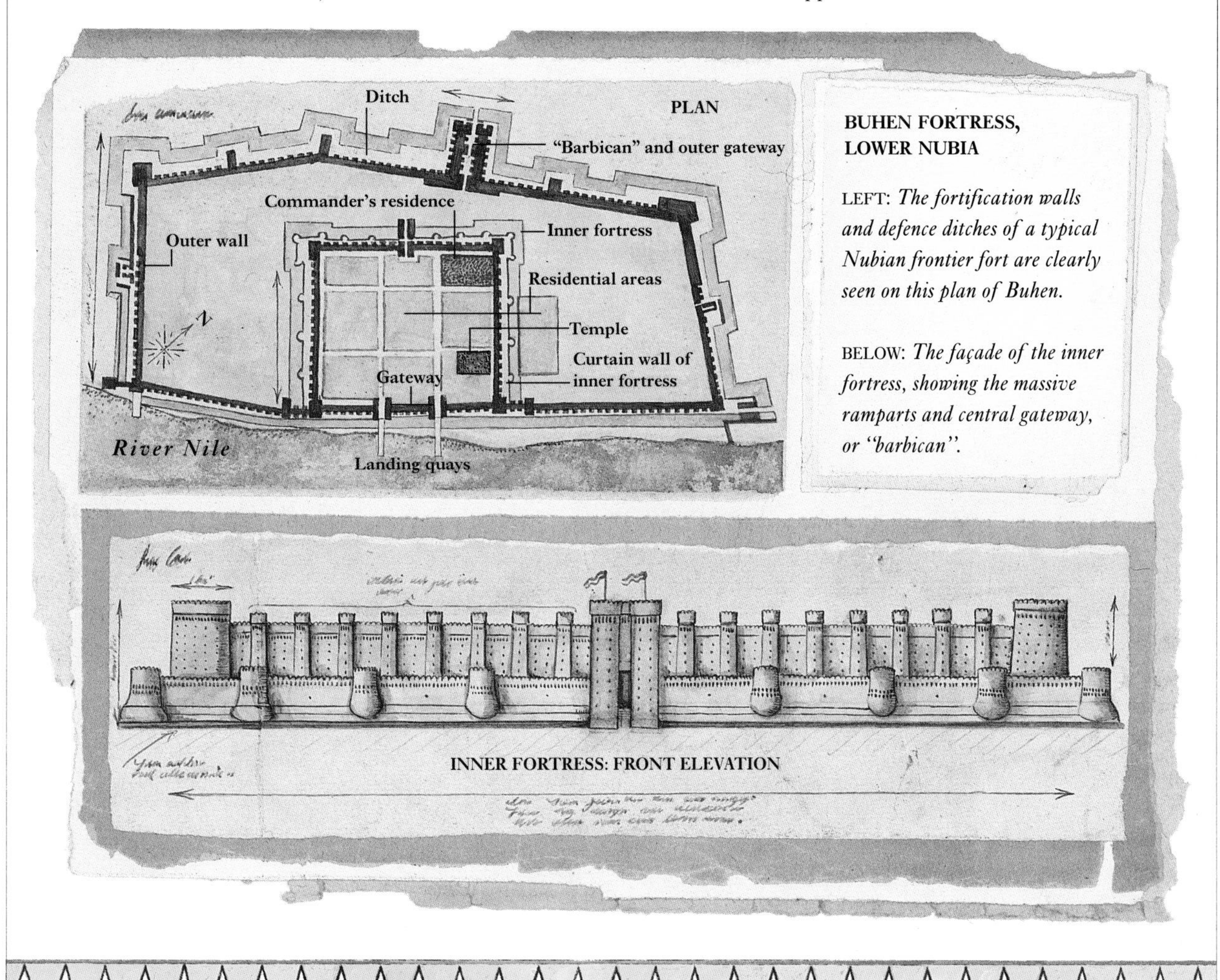


**BUHEN FORTRESS, LOWER NUBIA**

LEFT: *The fortification walls and defence ditches of a typical Nubian frontier fort are clearly seen on this plan of Buhen.*

BELOW: *The façade of the inner fortress, showing the massive ramparts and central gateway, or "barbican".*

● CHAPTER 6

# WOMEN IN EGYPT

Although their social position was determined principally by the status of their fathers and husbands, women enjoyed a higher profile in Egyptian society than was possible in many other civilizations of the ancient world. They were equal with men before the law and, although only men were permitted to hold government office, the wife, mother or daughter of a king or prominent official could wield considerable influence. In the event of a pharaoh's minority, his mother might rule as regent. On rare occasions, as in the case of Hatshepsut, the stepmother of Thutmose III, a woman might even assume the full mantle of royal power and rule as pharaoh.

▲



ABOVE: *The widow (standing, centre), with a female relative, bewails her husband, the scribe Hunefer, before his mummy and the god Anubis. From the Book of the Dead of Hunefer, Thebes, 19th Dynasty.*

RIGHT: *Elite couples attended by female servants, dancers and musicians, from the tomb of Nebamun at Thebes, ca. 1400BCE. On their heads, the guests and musicians wear wax cones that would melt as the evening wore on, releasing a strong perfume.*

## GENDER AND SOCIETY

Women did not form a homogeneous group within the hierarchy of Egyptian society, because their status depended on that of their father and, after marriage, of their husband. But this did not mean that they were without individual rights. In some other ancient societies, women were legally inferior to men and could not go to court without a man to act for them. Egyptian women, on the other hand, were not only equal with men before the law, but were also entitled to attend court, unaccompanied, as plaintiff, defendant or witness. They were responsible for their own actions and answerable for them to the court.

Because women could own or rent property in their own right, not all of them were economically dependent on their husbands. Unless a will specified otherwise, all children received an equal share of an inheritance, and a daughter could acquire a substantial income on her parents' deaths. Wives, too, could inherit wealth. One late Middle Kingdom papyrus, discovered at the pyramid town of Kahun (see p.71), contains the will of a priest called Wah, in which he bequeathes his property to his wife and family: "I am creating a transfer deed for my wife ... she herself shall pass

*A high-status woman at a funeral banquet for the royal vizier Ramose, depicted in the latter's tomb at Western Thebes, ca. 1375BCE. The woman's social position is indicated by her finery, which reflects courtly fashions of the time: she wears a broad necklace (probably of gold and semi-precious stones), linen clothing, a richly braided wig and a headband embellished with lotuses (compare illustration, p.82).*

it on to any children that she shall bear me, as she chooses. I give her the three Asiatics [slaves] given to me by my brother, Ankhreni ... She may give them to any of her children that she wishes."

A woman controlled a third of any property that she held in common with her husband, and could dispose of all her personal property as she wished. A certain Naunakhte, who lived during the Twentieth Dynasty at Deir el-Medina (see p.73), left a will in which she disinherited four of her eight children on the grounds that they had not looked after her properly in her old age. However, Naunakhte's will explicitly states that all eight would still share the property that was due from their father.

Egyptian women were able to engage in business, and they often traded surplus goods, such as cloth and vegetables, produced by their households. As Wah's will shows, they could also own slaves, whose services could be hired out for profit.

Papyri and monuments were produced for the élite and say little about less privileged Egyptians. Peasant women probably looked after their husbands' needs, raised children and worked on the land when necessary. Élite families also hired them as servants, musicians and dancers.

**FEMALE LITERACY**

**The extent to which Egyptian women were literate, if at all, has provoked considerable debate. The only people to receive a formal education were the élite men employed by the pharaonic bureaucracy, who were sent to school as boys to learn how to read and write. Women were barred from government office, so literacy was not a vital accomplishment, and no texts aimed specifically at a female readership have survived. But it is possible that girls may have been taught by their fathers or brothers, and then later handed on their skill to their daughters. It has been suggested that some extant short notes sent by women to female friends living nearby were written by the women themselves. Otherwise, the notes would have had to be dictated to a male relative or a scribe, and then read out to the recipient – it would have been simpler to transmit such brief messages orally.**

# ROLES AND IMAGES

*A wife embraces her husband at a banquet depicted in the tomb of Ramose, Western Thebes (compare illustration, p.81). On viewing this scene, an Egyptian would have understood that the couple were seated side by side, but the conventions of Egyptian art demanded that the figures be shown one behind the other in order that neither was obscured. The husband is placed in front of the wife in the more important position, reflecting the dominant role that élite men played in society as members of the governing bureaucracy.*

*A painted limestone figurine of a female brewer, from the tomb of Meresankh at Giza. The woman, who is bare-breasted and wears a wig and a bead necklace, is kneading moistened barley dough through a strainer into a large spouted pot, where it will ferment in water. Old Kingdom, end of the 5th Dynasty (ca. 2350BCE).*

In ancient Egyptian society, men and women fulfilled very different roles, and rarely did they overlap. Men of the élite held positions in the bureaucracy that administered the country. Élite women, on the other hand, were excluded from the bureaucracy, and were active in the domestic sphere. They were responsible for raising children, running the household and overseeing servants. In general, men enjoyed greater access to wealth than women because they were paid a government salary. This economic disparity is demonstrated by the far greater number of monuments – tomb chapels, statues, stelae and so on – erected by men than by women. It was rare for a woman to have the most expensive type of monument, a decorated tomb chapel; ownership of such chapels was mostly limited to high-ranking officials, and may have been one of the rewards of high office.

The different social roles of élite men and women are also reflected in the imagery used in ancient Egyptian art. People are not shown as individuals but made to conform to certain ideals. For women, the ideal was a youthful beauty with the emphasis on the hips and breasts, the areas of the body connected with childbearing. Pregnancy itself, and the fuller figure of the older woman who has borne a number of children, are seldom shown. Mature female images were probably avoided as negative, suggesting a stage of life too old for childbearing. As a result, there is little difference in the portrayal of a man's wife and his mother. For men, however, maturity had a different connotation, and male figures conformed to one of two ideals. The first was a man in the prime of youth, while the second was a mature, fuller figure with rolls of fat. This latter image represented the successful bureaucrat who had juniors to do the active work, and whose salary paid for him to eat well (see illustration, p.33).

Couples are frequently depicted in both two- and three-dimensional art. In many cases, the woman places her arm around her husband's shoulder or waist (see illustrations opposite and p.84), and in statues of the New Kingdom the gesture is often reciprocal, each partner embracing the other. Although ancient Egyptian society was male-dominated, the frequency with which the élite women were depicted alongside their husbands and sons shows that they were recognized as playing an important role within society.

The roles of non-élite men and women, and the division of labour between them, are less easily understood. Funerary monuments depict activities in the large households and estates of the élite, where poorer women are often shown carrying out household tasks such as grinding grain, baking bread and brewing beer. Although some women are depicted working in the fields, especially gathering the cut grain at harvest time, most of the outdoor labourers shown are men. It is not certain whether this reflects the actual division of labour or rather an ideal among the élite that women worked mostly indoors and men outdoors.

## MARRIAGE AND DIVORCE

Marriage would have been regarded as the natural state for both adult women and men, but we know little of how marriage partners were chosen. Most marriages seem to have been monogamous, although there is occasional evidence of a (non-royal) husband having more than one wife. There was no legal or religious ceremony by which marriage was formalized, and marriage occurred when a man and a woman established a household together. Divorce was not uncommon, and took place when couples who had been living together separated. Remarriage was possible for both men and women. Grounds for divorce probably included childlessness, and adultery on the part of the woman. In the latter event, a woman might forfeit the property that she was normally entitled to on divorce. In fact, adultery was taken very seriously by the community, and it was unacceptable for men to have affairs with married women. The reason, ultimately, was that men handed on their property to their children, and wanted to be sure that their heirs were indeed their biological offspring. Maternity, of course, was never in doubt.

A fascinating letter written by a woman of the Twentieth Dynasty who lived at Deir el-Medina, the artisans' village in Western Thebes, tells of a community's outrage at a married man who had been carrying on an affair with another woman for eight months without divorcing his wife. A government official only just prevented the wife's supporters from beating up the errant husband and his mistress, and the official makes it clear that the husband must regularize his situation regarding the two women one way or another. The letter indicates that, while marriage and divorce were not matters regulated by the state, they were of great interest to the community, and that social pressure was used to enforce conformity to accepted norms of behaviour in relationships between women and men.

# THE FAMILY

**CHILDREN**

**When a husband and wife failed to conceive, their marriage would probably end in divorce. However, there is also evidence that childless couples often adopted a child.**

**Infant mortality was high, with the greatest danger of death in the first five years of life. The longer a child lived, the better its chances of reaching adulthood. For this reason, children were only gradually incorporated into society, not becoming full members until they reached puberty.**

The importance of the family as a fundamental social unit is clear from numerous monuments where couples are depicted along with their offspring. Children provided for their parents in old age and were responsible for their parents' burials and funerary cults. In families of the élite scribal class, about whom we know most, the wife would have been in charge of running the household while the husband worked outside the home. This fact is reflected in the most common title for women, "mistress of the house". In large households, this would have meant overseeing the work of servants. In smaller households, the women of the family would have had the tasks of grinding grain, baking bread and preparing food. They would also have been responsible for spinning and weaving to make textiles for clothes and other purposes.

*The dwarf Seneb, his wife Senetites and their son and daughter: a painted limestone statue of ca. 2475BCE. The boy and girl are represented in a way typical of the Egyptian icongraphy of children: nude, with one finger to the mouth. The artist has sensitively positioned them in order to avoid leaving a conspicuous space that in other portraits would have been occupied by the sitter's legs.*

The importance of women's reproductive role is reflected in the Egyptians' concern for female health and hygiene. A number of papyri deal specifically with the welfare of women and their offspring, and cover such topics as infertility, conception, pregnancy, miscarriage, childbirth, milk supply and the well-being of newborn infants. Tests were devised to determine whether a woman was fertile, whether she was pregnant, and whether a newborn child would live or die. The Egyptians were acutely aware of the problems of infertility, the dangers of childbirth to both mother and infant, and the possibility of death in infancy. Women appear to have given birth in a crouching or kneeling position, supported on two blocks and assisted by two or more midwives, one holding her from the front, the other delivering the child.

Surviving texts say little about menstruation, although New Kingdom laundry-lists refer to what may be sanitary towels. According to one reading of a disputed passage in the Middle Kingdom *Satire of the Trades* (which denigrates every profession except that of the scribe), the launderer is unfortunate because he has to "clean the clothes of a menstruating woman".

The Egyptians believed that disease and misfortune derived from malign spirits and the hostile dead, and that maintaining health was a matter of providing protection against these forces. One remedy was to invoke the healing power of a deity by reciting a spell over whatever was to be used as a cure or means of protection. Such "medicines" might include brews to be swallowed or rubbed on the skin, vaginal tampons, or amulets representing suitable deities to be worn around the neck (see also pp.96–7).

Private houses contained an area devoted to the household cult, which involved the worship of the domestic deities Bes

## FERTILITY FIGURINES

The importance of female fertility to the ancient Egyptians is demonstrated by numerous images of nude women found throughout Egypt. Modelled from clay, faience, wood or stone, the figures are often shown wearing a necklace and hip girdle, with the pubic triangle clearly marked. Sometimes they lie on a bed, and they may have a child with them. These images were used in the household cult and may have been placed on the domestic altar. Their function was to ensure successful conception and childbirth, thereby guaranteeing the continuity of the family living in the house.

Fertility figurines were also presented as offerings at the shrines of Hathor, a goddess closely connected with sexuality, fertility and childbirth. Although no texts say why the figurines were offered, or whether the donor was a woman or a man, it is fair to guess that they were presented to the goddess with requests for children, or possibly in thanks for a safe delivery.

*This faience fertility figurine represents a nude woman – apart from elaborate jewelry that includes a hip girdle – who has tattoos or body paint on the skin. Middle Kingdom; of unknown provenance.*

Similar figurines have also been found at burial sites. As the dead were believed to be able to influence the condition of the living, some may have been offerings brought to the tomb to ask the dead for their aid in conception. One figurine, which includes the image of a child, carries the message: "May a birth be granted to your daughter Seh."

Other figurines may have been included in the original burial. At one time, scholars referred to such figures as "concubines of the dead", who would offer sexual gratification to the (male) owner of the tomb. However, they have been found in both male and female, adult and child burials, and it is now understood that they were in fact intended as a type of fertility figurine. For the ancient Egyptians, death represented a passage to a new life, and the transition was envisaged as a form of rebirth. So fertility figurines that were originally designed to aid conception and birth into this world also helped the dead to be reborn into the next.

and Taweret and a concern for the continuity of the family line through the fertility of its women. Bes and Taweret were connected with conception, pregnancy and safe childbirth and also helped to ward off malign spirits and the dead. They are often depicted with ferocious grimaces and carrying knives or the hieroglyphs for "protection" or "life". Their images appear frequently on household objects, such as chairs, beds and cosmetics jars. Numerous amulets of the two deities have been discovered; these were probably worn by women to protect them during pregnancy and labour. Houses also contained wall decoration relating to childbirth, including scenes showing mothers with their newborn infants.

Niches for stelae and busts of deceased family members were also found in houses. Because spirits of the dead were thought to influence – for good or ill – the affairs of the living, it was important that they should be honoured and given offerings (see pp.152–3).

*Three 18th-Dynasty necklaces made up of various amulets. The outermost includes tiny images of the goddess Taweret and would have been worn to guard the wearer against the dangers of pregnancy and childbirth.*

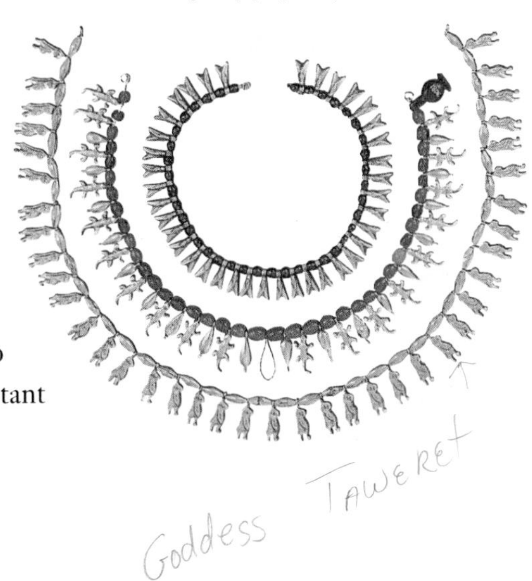

# WOMEN AND RELIGION

**VOTIVE STELAE**

**From the Middle Kingdom onward, the élite set up votive statues and stelae in temple precincts. Their purpose was to link for eternity the donor with the deity of the temple and the cult performed there.**

**Women rarely erected temple statues, but both men and women put up votive stelae, which usually show the donor offering to or adoring the deity to whom the stela was dedicated. There is no type of scene or text that was used exclusively by women. On the other hand, although women were permitted to dedicate votive stelae to either male or female deities, surviving examples suggest that women most often chose to dedicate their stelae to goddesses.**

In the Old Kingdom, the First Intermediate Period and the beginning of the Middle Kingdom, many élite women became priestesses of the goddess Hathor, although few women served in the cults of other deities. By the New Kingdom, the priesthood was all male, with the sole exception of the royal woman who held the position of the "God's Wife of Amun" at Thebes. Instead of being priestesses, many élite women of this period held the title of musician of a particular deity, and their role was to perform the musical accompaniment to temple ritual. They are often shown holding the sacred rattle (*sistrum*), which they shook rhythmically as they chanted. Scenes from a temple building of Queen Hatshepsut at Karnak depict groups of women shaking *sistra* as part of the musical ensemble that accompanied the sacred boat of the god Amun as it went in procession from the temple of Karnak to the temple of Luxor and back (see p.159). In charge of the musicians there would have been a woman of high social status who bore the title "The Superior of the Musical Troupe".

The ritual enacted in the hidden sanctuaries deep within state temples had little to offer on the level of personal religion. Nevertheless, both men and women visited the outer areas of temples to pray and to present votive offerings to the gods. Many such offerings have been recovered from shrines dedicated to the goddess Hathor at Deir el-Bahari. Although few reveal the identity of the dedicator, some include depictions of female donors. Because Hathor was closely connected with sexuality, fertility, pregnancy and childbirth, she was a particularly important deity for women, many of whom undoubtedly visited her shrines with gifts and petitions. One fragmentary statue of a man found at Deir el-Bahari con-

*The priesthood was almost entirely male during the New Kingdom, but women were still able to take part in religious rites. Here, the deceased's wife mourns before the upright coffin of her husband and the god Anubis; another group of women, perhaps family members but probably including hired mourners, bewails the deceased. From the Book of the Dead of the royal scribe Ani; 19th Dynasty, ca. 1125BCE.*

## THE GOD'S WIFE OF AMUN

In the Twenty-Fifth and Twenty-Sixth dynasties, one of the most important positions in the temple of Amun at Karnak, Eastern Thebes (see pp.208–9), was that of the "God's Wife of Amun". It was usually held by a daughter of the king, who never married, but would adopt her successor. Although we do not know the exact details of this princess's role, part of her function was to carry out ritual actions before the god, including shaking the sacred rattle or *sistrum*, in order to stimulate the deity constantly to re-enact creation and maintain the created world.

The role of God's Wife had its origins at the beginning of the New Kingdom, when the title was bestowed by King Ahmose on his queen, Ahmose Nefertari, together with a financial endowment to support the office. Later, Queen Hatshepsut was prominent as God's Wife before she became pharaoh (see p.89), and she may have used the authority of the position to help her rise to power. After her death, the God's Wife declined in importance for several hundred years.

The importance of the God's Wife is reflected in the decoration of a series of chapels in the precinct of Karnak. Here, she is depicted on an equal footing with the pharaoh in scenes which, before the Twenty-Fifth and Twenty-Sixth dynasties, showed only the king. She adores and makes offerings directly to deities, who embrace, crown and suckle her. The close relationship between the God's Wife and her divine partner, Amun, is seen in a faience statuette from Karnak that shows the God's Wife Amenirdis I on the lap of the deity, who holds her in his arms.

*This elaborately inlaid bronze statuette probably represents a God's Wife of the 3rd Intermediate Period, one of the forerunners of the powerful God's Wives of the 25th and 26th dynasties.*

tains a text addressed specifically to women who visit the temple. In it, the man promises that in return for offerings he will intercede with the goddess on the women's behalf for "happiness and a good [or perhaps 'potent'] husband" (see also p.85).

Women also played an important role as mourners in burial rites. Representations of the funeral procession to the tomb show the dead man's wife and groups of other women mourning the deceased. Their hair is dishevelled, their breasts bared and tears spill from their eyes. Two women, standing at either end of the bier that carried the deceased to the tomb, enacted the parts of the goddesses Isis and Nephthys mourning the death of their murdered brother Osiris (see pp.134–5). Although the funerary processions of women are rarely depicted, we know from excavated burials that women were afforded similar, if usually less rich, funerary equipment as men of equal status, and that they expected to share the same afterlife, with its attendant needs and dangers.

Women also participated in funerary cults. These were ideally carried out by the eldest son of the deceased, but depictions on funerary stelae suggest that female family members commonly played a part. They are shown burning incense, pouring libations and dedicating offerings.

# ROYAL WOMEN

*The famous painted bust of Queen Nefertiti found at Akhetaten in the workshop of a sculptor called Thutmose. It shows her as a strikingly beautiful woman, wearing a unique tall blue crown in which she is frequently depicted elsewhere. It has been suggested that the queen enjoyed the status almost of a co-ruler rather than simply a queen-consort. Some scholars believe that Nefernefruaten, an obscure pharaoh said to have ruled briefly following the death of Akhenaten, may have been Nefertiti herself. The bust appears to be unfinished and may have been used in the workshop for teaching trainee artists.*

Like the office of kingship, queenship, embodied by the king's mother and the king's principal wife, was divine. The two queens wore the same insignia, used the same titles and appeared in the same types of scenes, because they shared a single role. In Egyptian belief, the sun god perpetually renewed himself by impregnating the sky goddess each evening and being born of her again in the morning, so that the goddess was both his consort and his mother. The king was the earthly manifestation of the sun god, and the role of the sky goddess was split between his mother and principal wife. The divine aspect of their role was displayed through various items of insignia that these queens shared with goddesses.

How kings chose their principal wives is unknown. Some principal wives were the sisters of the kings they married, but others were of non-royal birth. It was once thought that kings married their sisters because the right to the throne was passed through the female line, so that the man who married the current "heiress" became king. However, for this to be true, there would have had to have been an unbroken line of "heiresses" in descent from one another, and such a line does not exist. It seems more likely that kings who married their sisters did so because such unions occurred between deities but not among ordinary people: in this way, the king stressed his divine aspect and separated himself from his subjects.

Surviving evidence tells us little about the personalities of individual queens, but the large amount of material associated with some in contrast

*Queen Nefertari (right), first principal wife of Ramesses II, makes an offering to the goddess Hathor in her tomb in the Valley of the Queens, Western Thebes. In addition to her beautifully decorated tomb, Ramesses II also built for his wife the smaller temple of Abu Simbel in Lower Nubia (see p.204), in which the queen is associated with Hathor and is shown making offerings to various deities. In one scene, she is herself depicted in the form of a goddess receiving offerings.*

## HATSHEPSUT, THE FEMALE PHARAOH

Few royal women became pharaohs, but one of these exceptional figures was Hatshepsut, the daughter of Thutmose I and principal wife of her half-brother, Thutmose II. Hatshepsut bore Thutmose II a daughter but no surviving son, so when Thutmose II died young, the title of king was inherited by the son of one of his secondary wives. Thutmose III was too young to rule alone, and so Hatshepsut became regent. At some time between the second and seventh years of Thutmose's reign, Hatshepsut took the titles of a king. At first, she was depicted in female dress, but soon she began to be represented in the traditional costume of a male king. To legitimize her claim to the throne, the queen set up inscriptions claiming a divine birth and stating that her father had proclaimed her his heir during his lifetime.

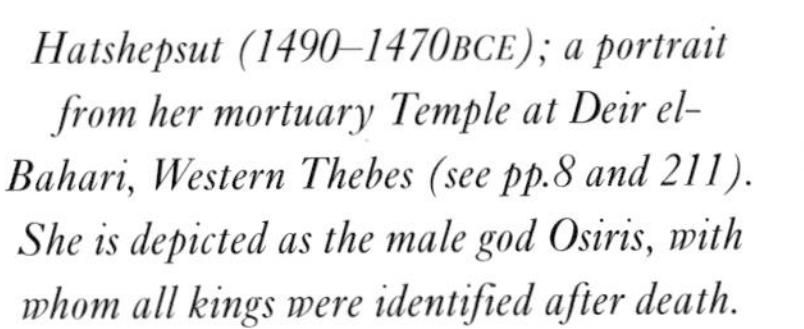

*Hatshepsut (1490–1470BCE); a portrait from her mortuary Temple at Deir el-Bahari, Western Thebes (see pp.8 and 211). She is depicted as the male god Osiris, with whom all kings were identified after death.*

Hatshepsut did not replace Thutmose III, who ruled as king alongside her. The period was a prosperous one for the Egyptians, and Hatshepsut was able to sponsor numerous building projects; she also mounted important trade expeditions and military campaigns.

By his twenty-second year as king, Thutmose was ruling alone, presumably after Hatshepsut's death. Late in his long reign – he ruled for fifty-four years – Thutmose instigated a campaign to mutilate Hatshepsut's monuments, apparently to erase all signs of her kingship. It was once thought that Thutmose's motivation was one of simple hatred. However, the reason may rather have been that Egyptians generally considered it unnatural for a woman to become king. As he neared the end of his reign, Thutmose may have wanted to prevent the succession of another female monarch by erasing the memory of Pharaoh Hatshepsut. Perhaps significantly, her name and image were not attacked on those monuments that represent her simply as the queen-consort of Thutmose II.

**THE HARIM**
**Most Egyptians were monogamous, but their kings practised polygamy. Important royal wives had their own establishments and staff, but others lived in one of several harims around Egypt run by male officials. Many harim wives probably came from élite families, while others were foreign princesses married in diplomatic alliances. A document from the reign of Ramesses III relates to a failed conspiracy hatched in the harim by a royal wife called Tiy, who plotted to kill Ramesses and put her son on the throne.**

to others suggests that these were of particular importance. Two of the best-known queens are Nefertiti, consort of the "heretic" king Akhenaten (ca. 1353–1336BCE; see pp.128–9), and Nefertari, the first principal wife of Ramesses II. No other queen attained the prominence of Nefertiti, nor was depicted more frequently on her husband's monuments. Akhenaten, who abandoned the worship of Egypt's traditional gods in favour of that of the Aten (sun disc), built a number of temples to his god (see p.203). Nefertiti appears everywhere in the decoration of these buildings, accompanying her husband in rituals and even performing them alone. However, the most famous portrait of the queen is the celebrated painted stone bust, now in Berlin (see illustration), discovered in a sculptor's workshop at el-Amarna, the site of Akhenaten's capital city of Akhetaten.

When Ramesses II built a temple to his divine self at Abu Simbel, he erected a smaller temple nearby dedicated to the goddess Hathor and Queen Nefertari. Today, Nefertari is most famous for her painted tomb in the Valley of the Queens in Western Thebes (see illustration).

• CHAPTER 7

# THE BOUNDARIES OF KNOWLEDGE

The "Wisdom of Egypt" has held a powerful grip on the Western imagination since antiquity, and the belief that the Egyptians prized knowledge, in both practical and esoteric fields, beyond all other earthly possessions is scarcely exaggerated. According to the author of *The Instruction of Amenemope* – a guide to wise living that influenced the biblical Book of Proverbs – to acquire wisdom was to gain "a treasure house for life", whereby "your being will prosper upon earth".

▲



ABOVE: *A detail from a late 4th- or early 5th-dynasty statue from Saqqara depicting a tomb owner as a scribe. The sagging breasts and rolls of fat are characteristic of this profession, and are commonly represented in Egyptian statuary.*

## THE TEACHING OF KNOWLEDGE

In ancient Egypt, literacy was the key to the whole range of knowledge. The literate "wise man" provided many services for his local community, from performing religious rituals and reciting literary texts, to magico-medical assistance and schooling. The ruling class was literate, yet even at the peak of Egyptian power this represented perhaps only one or two per-cent of the population. The ruling official was depicted as a scribe. Fat, well fed, elegantly dressed and exempt from physical labour, he was said merely to register the work of others, not to work himself. His palette and papyrus scroll were symbols of the authority of knowledge, and bureaucratic lists and registers were the tools of political and economic power.

Cultural knowledge, which was also conveyed through writing, brought a different sort of power. Literature was prized for its ability to influence men and because it brought enduring fame to the author. Writing was prestigious because it could acquire antiquity, as a lasting memorial. Medical recipes, rituals and magical spells were all given authority through claims that they were written by a god or an ancient king, and had been found effective over a long period.

The home of learning was the temples, where departments called the "House of Life" contained extensive libraries, and where learned men copied manuscripts. During the pharaonic period the palace and departments of state were also centres of learning.

The principal purpose of education was to prepare a child to enter the pharaonic bureaucracy. Some children of the highest class were educated at the palace, along with the royal princes. Most were taught at home, ideally by their own fathers, in schools associated with the temple or office for which their fathers worked. In the New Kingdom, the first school text was *The Book of Kemyt*, a compilation of formulae for letter-writing, funerary texts and autobiography that was copied from earlier manuscripts. The schoolboy practised reading and writing in the cursive

*The ibis-headed Thoth proffering the power of life, from the Book of the Dead of the scribe Hunefer (ca. 1285BCE). As the god associated with writing, Thoth was the patron of scribes and also the deity who gave writing its magical and ritual power. He was the scribe of the gods and recorded the vindication of the deceased at the "weighing of the heart" in the underworld* *(see p.137).*

hieratic script by rote learning, copying out first *The Book of Kemyt* and then other works of classic literature – texts that often he did not fully understand. Hieroglyphs were an advanced subject. Initial learning was tedious and slow. One text compares the process to training a horse or plough ox which learns through discipline and the stick until its understanding is sufficient for the task: "the ear of a boy comes to be on his back, and he listens to his beating." The word for "teaching" is etymologically the same as that for "punishment".

Once he had developed a good hand, the student progressed to copying and then composing administrative texts as well as literary works – model letters, administrative reports and variations on standard literary themes, such as the value of enduring a scribal education, or praise of the teacher. The student practised these exercises as an apprentice scribe, working as assistant to a professional. In this way he developed his literacy at the same time as he learned his trade and acted as secretary or clerk to his master. There is no evidence for free access to education, even at the elementary level; only the son of a scribe could expect to be educated.

**THE INVENTORY OF KNOWLEDGE**

**The Egyptian thinker, whether a theologian defining god or a medical practitioner discussing injuries and treatments, presented his knowledge in the bureaucratic style of a list and not through the exposition of abstract principles. The attempt to list everything that exists is characteristic of Egyptian learning. *The Onomasticon of Amenemope*, an advanced school text, captures this idea well:**

> **"Beginning of the teaching [for] enabling the mind, instructing the ignorant, and knowing all that exists; what Ptah has created, and Thoth has written out; the sky with all its constellations [?]; the earth and what is in it; what the mountains spew up; what is inundated by the flood; being all on which Re has shone; everything that grows on the earth, which the scribe of the god's book in the House of Life, Amenemope son of Amenemope has conceived ... "**

**What follows this opening is simply a list of words in thematic order, the nearest thing to a dictionary or encyclopedia that survives from ancient Egypt. As a word list, however, its practical use was to train the student in written vocabulary.**

# THE NATURAL WORLD

**THE VISION OF ORDER**
**In the Egyptians' view of their own topography, order is the norm: the Nile flood is never depicted, only the ordered landscape once the Nile has returned to its bed. From the early Middle Kingdom, inscriptions on a kiosk of the Twelfth-Dynasty king Senwosret I at Karnak include a geographical itinerary of Egypt, dividing it into its natural halves on the north and the south façades of the building. The administrative divisions of the country – the "nomes" – are tabulated. Their measurements are given, along with mention of their god and/or their capital town. These texts seem to be politically motivated, laying out regions of the country in detail to facilitate control by the central government, but they also define the kiosk as an image or model of Egypt, inhabited by its god. "Geographical texts" from the temples of the Late Period define the country in a similar way. The nomes are listed, and their sacred trees, their festivals, their taboos and even their snakes are tabulated.**

According to the Egyptians, their country lay at the centre of the cosmos, a living entity, with the Nile at its centre. The sky was sometimes personified as the sky goddess Nut, separated from the earth, her husband Geb (see pp.116–17). To travel out of the bountiful "black land" of Egypt was to enter the "red land" of the desert, an exotic and dangerous territory. Within the Nile Valley was a natural order of elements, which could be listed, and which defined the Egyptians' place in the physical world.

The Egyptian vision of the world moves seamlessly between the physical and the biological, the tangible and the metaphorical, the measurable and the mythological. The sky and earth worked according to daily, monthly or annual cycles of birth, growth and death, and the biological cycle is a constant theme in Egyptian interpretations of the world. For example, the pyramid temples of the Fifth and Sixth dynasties showed extensive details of the life-cycles of many plants and animals. Much later, Thutmose III included pictures of exotic animals and plants, probably brought home by foreign expeditions, in reliefs at Karnak.

Scientific geography as practical knowledge is not well represented in the textual record. The most important document is a map of the gold mines in the Wadi Hammamat, from the reign of Ramesses V (see p.64); although not accurate in detail, it seems a reasonably good sketch plan of an expeditionary route. The nearest parallels are sketch plans of a route into the underworld, on the floors of a group of Middle Kingdom coffins (see p.136). Foreign itineraries – lists of towns visited, conquered, claimed as tributary – become standard in New Kingdom royal inscriptions.

*An astronomical ceiling from the tomb of Sety I (ca. 1290–1279BCE) in the Valley of the Kings, Western Thebes. The constellations and divisions of the sky are represented as divinities.*

An extensive practical knowledge of the sky and the movements of the stars and planets is represented in a semi-mythological way on tomb ceilings and coffin lids. Ceilings in the tomb of Senmut, from the reign of Hatshepsut, and from the tomb of Sety I in the Valley of the Kings, depict the major constellations as mythological figures. The Egyptian calendar (see box) was based on the observation of the moon and stars. The cycles of the moon, and the seasonal movements of the stars, defined the calendar and regular religious festivals, based on observations made from the temple roof. The most significant of these observations was the reappearance of Sirius (Sopdet to the Egyptians), as a morning star in mid-July, after several months of invisibility, at the onset of the annual inundation of the Nile. This conjunction of natural phenomena marked the beginning of the Egyptian New Year. (See also pp.116–17.)

## THE EGYPTIAN CALENDAR

The Egyptian year was divided into twelve months, each thirty days long. Each month was divided into three weeks of ten days, or "decades". Astronomical texts divide the night sky into thirty-six decans, according to the rising of constellations at certain hours of the night. Each decan represented one decade (week) in the basic calendar. The months were grouped into three seasons: *akhet*, "inundation", roughly mid-July to mid-November; *peret*, "emergence" or "winter", roughly mid-November to mid-March, the main growing season; and *shemu*, "harvest", perhaps literally "low-water", roughly mid-March to mid-July, the time of grain harvest.

The Egyptian year was brought up to 365 days by the inclusion of five additional days following the end of *shemu*. These were referred to as the birthdays of the gods Osiris, Seth, Isis, Nephthys and Horus, and were days of ill omen. However, one quarter of a day a year was still unaccounted for, so the calendar progressively moved out of synchronization with the natural year by one day every four years. The rising of Sirius (Sopdet) only coincided with the calendrical New Year's Day every 1,460 years – the event is actually recorded for 139CE. Similarly, the thirty-day month did not coincide with the natural cycle of the moon. Many festivals were celebrated on the basis of observation of the cycles of the moon rather than on the civil calendar.

The day itself was divided into twenty-four hours, with twelve hours of day and twelve of night. The passage of time was measured by water-clocks in the form of a marked bowl from which water slowly dripped through a hole.

ABOVE: *The goddess Isis, whose birthday was marked by one of five intercalary days at the end of the Egyptian year. From the tomb of Princess Thuya, Western Thebes, ca. 1375BCE.*
LEFT: *A bronze figurine of the star-crowned goddess Sopdet, or Sothis, personification of Sirius. Late Period, after 600BCE.*

# MATHEMATICS

Our knowledge of Egyptian mathematics derives from a small number of papyri that were compiled as practical textbooks for training scribes who were already at an advanced stage of education. It is not clear how widespread serious mathematical training was, but basic accounting procedures – such as calculations of crop yields in relation to land areas, food commodities against consumption rates, or the input of labour and raw materials for work projects and the running of temples – were central to all administrative functions.

The hieroglyphic decimal system used by the Egyptians has separate symbols for the numbers 1, 10, 100, 1,000 and 10,000 (see box, opposite page), although it has no symbol for zero. Intermediate numbers were written as multiples of the single number.

All mathematical procedures seem to have been based on the underlying processes of addition and subtraction. Multiplication was carried out by adding a number to itself the requisite number of times, and division consisted of subtracting a number until an indivisible remainder was left. Multiplication tables do not seem to have been used, although multiplication or division by ten was a simple and standard process.

Procedures for dealing with fractions were based on the addition and subtraction of unit fractions. With the exception of the special cases of $^2/_3$ and very rarely $^3/_4$, the Egyptians did not use multiples of fractions, but only the single unit fraction. For example, $^1/_5$ was written as *r*5 "the fifth

*A page of calculations from Papyrus Rhind, the most famous Egyptian mathematical text. The problems mainly deal with calculating the volume of a granary. There is also a table for dividing a quantity of grain into fractions, as well as a formula for calculating the area when squaring a circle. Fifteenth Dynasty, ca. 1550BCE.*

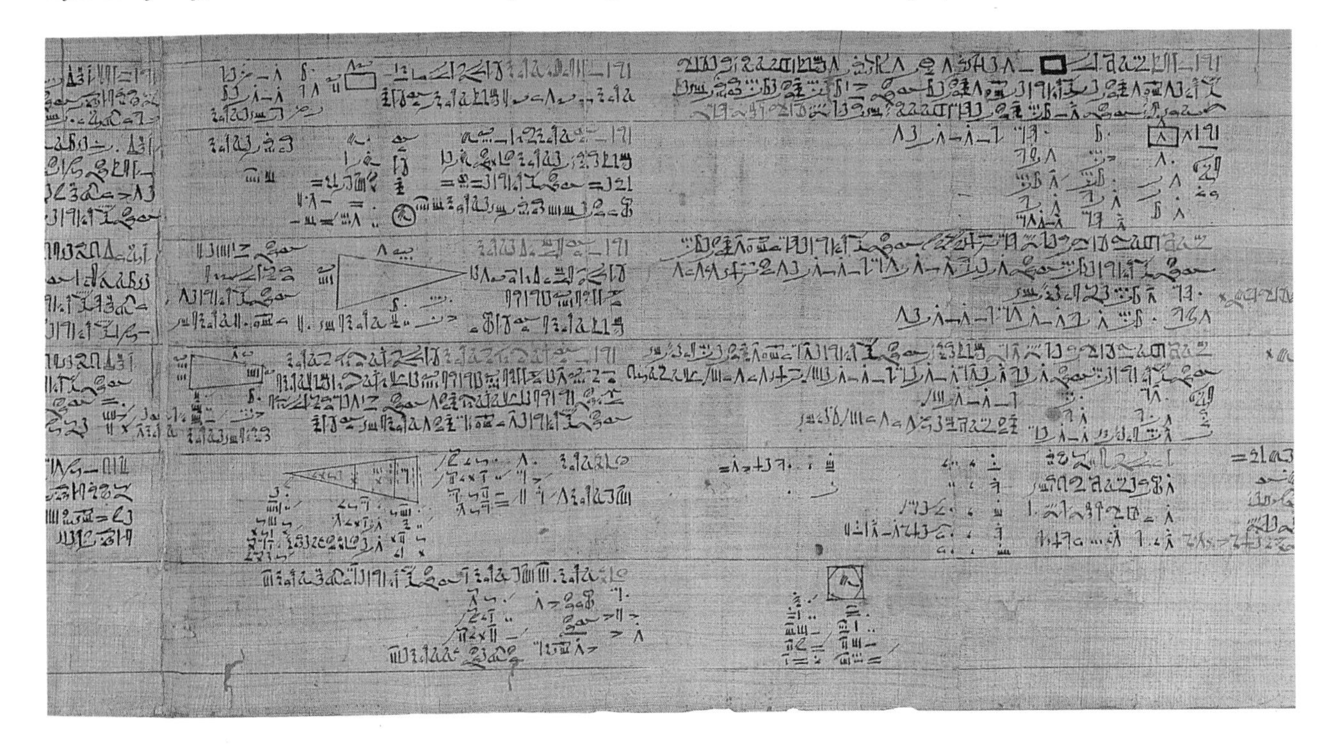

part", and $^{1}/_{6}$ as *r*6 "the sixth part". Where modern mathematics would write $^{11}/_{30}$, Egyptians would write *r*5 *r*6 – that is, $^{1}/_{5}$ + $^{1}/_{6}$.

The surviving mathematical papyri show only calculations that would have been used in practical applications – for example, methods for calculating the areas and volumes of a variety of shapes, including triangles and cylinders. (This is not to say that the Egyptians had no concept of numbers in the abstract, simply that there is no surviving evidence for this.) An Egyptian would calculate how long it would take so many men to build a brick ramp because he expected to supervise men building ramps. In one text, an army scribe, Amenemope, is challenged to calculate the number of men needed to transport an obelisk of given size from the quarries, to erect a colossus in a given time, calculate the rations necessary to feed men digging a lake, and to arrange stores for a major military expedition to Syria.

There is an evident concern in the papyri with formulae and ratios for practical application. A ratio *pesu* was used as a measure to define the quantities of bread or beer made from a single unit of grain. The term *seked* defines the slope of a pyramid as a ratio of lateral displacement against rise. There was a formula for calculating the area of a circle, on the basis of taking the diameter, subtracting $^{1}/_{9}$, and squaring the result. A circular granary nine cubits in diameter is thus calculated to have a base area of 64 square cubits, instead of the correct answer, 63.64 square cubits.

Unlike the Greeks, the Egyptians seem not to have been interested in producing proofs of mathematical formulae, but they do show a highly sophisticated manipulation of numbers. In the context of applied accounting – that is, the accurate measurement of commodities, areas and volumes absolutely – minor errors were not significant. It is, in fact, fairly typical of surviving accounts papyri dealing with consumables to include errors in arithmetic, with totals that show a minor shrinkage in the sum of the parts, to the advantage of the accountant.

## EGYPTIAN NUMBERS

| 1 | 8 | 40 | 362 | 10,000 | 1/2 |
|---|---|---|---|---|---|
| 2 | 10 | 100 | 1,000 | 100,000 | 1/3 |
| 4 | 24 | 200 | 4,281 | 1,000,000[1] | 1/6 |

[1] Also means "I can count no further!"

*In this relief from the 5th-Dynasty tomb of Ty at Saqqara, Ty and his wife inspect the produce of his estate. In the bottom two registers his scribes are keeping the accounts.*

### WEIGHTS AND MEASURES

**The basic Egyptian unit of length measurement was the "royal cubit" (about twenty inches or 50cm), consisting of seven "palm-widths" (three inches or 7.5cm), each of four "finger-widths" ($^{3}/_{4}$ inch or 1.9cm). (The standard cubit, based on the distance between elbow and fingertips, was six palm-widths.) One hundred cubits made one *khet*. The standard land measure was a *setjet* (usually referred to by the Greek equivalent, an *aroura*), consisting of one hundred square cubits (about $^{2}/_{3}$ of an English acre or just over 0.25ha).**

**Measures for grain were the *hekat* (just over a gallon or five litres), made up of ten *hin*; a quadruple *hekat*, often called an *oipe(t)*; and a *khar*, or sack of four *oipe(t)* (seventeen gallons or 80 l). Liquid was normally measured by the *hin* (approximately one pint or 0.5 l). The standard measure of weight was the *deben* (two pounds or 0.9kg), divided into ten *qite*, or at some periods divided into twelve *shat*.**

# MEDICINE

Medicine was a branch of advanced scribal learning. Egyptian doctors were of the highest social status and had a reputation throughout the ancient world. A knowledge of medicine was identified with a knowledge of ritual. However, although the physician acted as a specialized ritualist, this should not be taken to imply that recitation took priority over therapeutic practice.

Surviving medical papyri, which belong to a textual tradition stretching back to the Middle Kingdom, were intended for both teaching and reference. For example, in the Edwin Smith Surgical Papyrus dating to the late Middle Kingdom or slightly later, individual cases are described and their symptoms listed, together with specific findings from physical examinations. A diagnosis is given, together with a prognosis – curable, treatable, or untreatable – and treatment, if appropriate, is described.

Most medical texts consist of series of prescriptions – recipes for the treatment of every possible sort of ailment. Modern analysis of the efficacy of such prescriptions is at times complicated by difficulties in identifying both the ailments and the ingredients used in the cure. Some ingredients seem to rely more on sympathetic magic than pharmacology. Yet it is clear that medicine was also practised on a scientific basis.

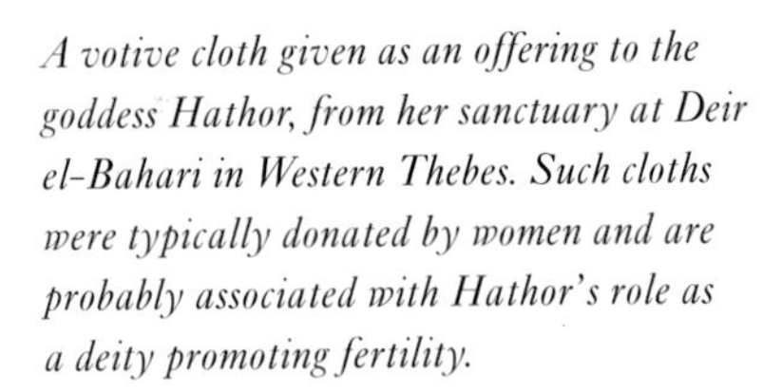

*A votive cloth given as an offering to the goddess Hathor, from her sanctuary at Deir el-Bahari in Western Thebes. Such cloths were typically donated by women and are probably associated with Hathor's role as a deity promoting fertility.*

Two documents (the Ebers and Berlin medical papyri) provide an account of the interconnection of the parts of the body by "vessels", a term applied to veins, arteries, ducts, muscles and tendons. The heart was at the centre, and the vessels were understood to transfer air, blood and other fluids, as well as disease, to different parts of the body. The papyri

also exhibit a good practical knowledge of the structure of the body, its organs and their functions, but there is no evidence for a more formalized study of anatomy. Nor is there evidence to suggest that medical practice was based on any knowledge of the human anatomy derived from the processes of mummification or butchery: internal surgery was not undertaken, and there was little practical motivation for such study.

To the modern observer, the use of "magic" appears in direct proportion to the limitations of other available treatments. Spells to charm snakes, and to prevent as well as cure or alleviate snake bites, are common. However, one papyrus identifies the snakes of Egypt, gives an analysis and prognosis of their bites, and lists detailed prescriptions and treatments, with only limited use of magical incantations.

Gynaecology and childbirth were an ancient Egyptian speciality. For conception, pregnancy, childbirth, and welfare of the new child, divine assistance was more useful than human, and while practical treatments for gynaecological problems are found in many of the papyri, there are also extensive collections of charms and incantations for birth, protection of babies and satisfactory provision of milk, as well as spells against childhood diseases and dangers. Several deities, such as Isis and Hathor, were invoked to help women to conceive and to bear children safely.

*A protective gilt and painted statue of the scorpion goddess Selqet, from the tomb of Tutankhamun (ca. 1332–1322BCE). Scorpions were a common hazard in Egypt and invocations to Selqet were among the many means employed to banish them or cure their effects.*

## DIAGNOSIS AND CURE

The Edwin Smith Surgical Papyrus provides a systematic analysis of injuries and their treatment. From the nature of the injuries, it was once thought to be the work of an army surgeon, but some scholars now attribute it to a doctor attached to a pyramid construction team. If this is so, the wounds listed are probably the results of on-site accidents. One case, "a man with a wound above his eyebrow, reaching to the bone", is to be treated in the following manner: "You probe his wound, and you bring together his gash for him with a stitch [?], and you say about him: '... An injury I will deal with.' After you stitch [?] it, you bandage it up with fresh meat the first day. If you find that this wound is loose with its stitches [?], then you bind it together with bandages, and anoint it with oil and honey every day until it gets better."

Not every injury was curable, however, as in the case of the unfortunate man partially paralysed following a fracture of the skull: "If ... you find that swellings protrude behind the break in his skull, and his eye is squinting under it, on the side that has the blow to his skull, and he walks shuffling with the sole of his foot on the side that has the blow to his skull, you diagnose him as one struck by what enters from outside ... An injury not treated." The phrase "what enters from outside" is explained: "It is the breath of god, or death."

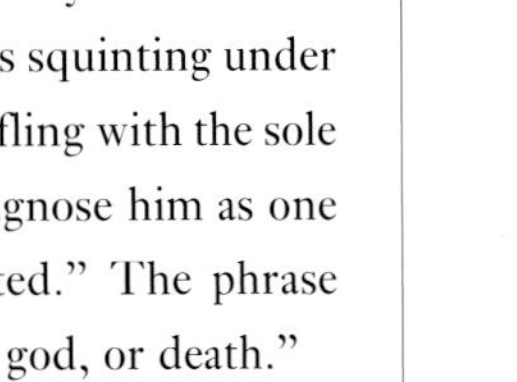

*A wooden model of a weavers' workshop from the tomb of Meketre in Western Thebes; 11th Dynasty, ca. 2000BCE. Our knowledge of Egyptian technology derives partly from depictions in tombs, but largely from the archaeological record. Written works describing or instructing apprentices in technical practice do not survive, and probably never existed.*

# TECHNOLOGY

All technology in the pharaonic period is marked by the absence of machinery or complex tools, and a reliance on simple methods, allied with slow and patient, extremely accurate, craftsmanship.

Perhaps the main limitation on technological advance lay in the relatively undeveloped state of metal technology. Copper and bronze were extremely costly commodities. Manufactured iron only began to appear in Egypt, as a foreign import, after the New Kingdom. Instead of deliberately manufactured bronze, natural alloys of copper, or copper hardened by working, predominated. This limited the practical application of metals for cutting and working tools. Flint provided the cutting edge for many ordinary tools all through the age of the pharaohs.

A high standard was reached at an early date in woodworking and carving (for luxury furniture), as well as boat-building. Metalwork attained its peak in the production of gold jewelry (some Old Kingdom tomb scenes associate goldsmithing with dwarves). Semi-precious stones such as turquoise, lapis lazuli, carnelian and amethyst were frequently used as inlays. True glass flourished only briefly, in the New Kingdom, probably under foreign influence, but glazes were important at all periods. Egyptian faience, widely used for decorative objects, was an artificial frit covered with a blue or green glaze.

In the Old Kingdom, stone vessels were manufactured for the tomb, as well as stone sarcophagi and carved reliefs. Metal tools that were hard enough to cut stone were not available, and the most important procedures were variations on pounding and abrasion. For the finest work, cop-

*The Nile was a vital communications artery (see pp.14–15) and, unsurprisingly, Egyptian boat-builders were famed for their skills. This royal barge, depicted in a wall painting from an 18th-Dynasty tomb in Western Thebes, displays the characteristic curved prow and stern of Nile vessels.*

per chisels and tubular drills were deployed, but wherever possible the work was carried out using hard stone pounders and polishing stones fed with sand as an abrasive.

Building stone was quarried and worked without cutting tools or machinery. In the quarries, channels were cut around the blocks where necessary using hard stone pounders; then wooden wedges were driven into the slots and dampened to split the stone away from the natural rock. Engineering of any sort was based on manpower, not machinery. Heavy stone blocks were transported on sledges, and because no lifting devices existed, they were raised into place via building ramps. Wheeled transport was not practical in Egyptian conditions, and probably not used in any significant way before the introduction of the war chariot from abroad in the New Kingdom. Efficient water-lifting devices are not known before the Ptolemaic Period, and even the simple *shaduf* (see p.10) for irrigation does not seem to have appeared before the New Kingdom. Agricultural technology was limited by the high cost of metal, which restricted the development of tools for ploughing and harvesting.

**CRAFTSMEN IN SOCIETY**

**Most craft production was for the élite, or for state and temple institutions, which could afford the necessary investment in long training and slow, labour-intensive work. The very highest value of production, such as statuary or gold working, seems to be entirely associated with the palace or temple workshops. The best craftsmen were schooled from boyhood at the palace, in parallel with the palace schooling of the children of high officials. Such master craftsmen might hold very high status. To a limited extent, craftsmen working for the élite, or in a temple context, formed a small but nevertheless significant middle class, better paid and provided for than the ordinary peasants.**

## "THE SATIRE OF THE TRADES"

A famous Middle Kingdom text known as *The Satire of the Trades* was intended to encourage the trainee scribe to regard his profession as the best of all. In pointing up the supposed failings of other trades, the Satire reveals fascinating details of the life of Egyptian skilled artisans. The following is an extract:

"[The scribe] is [merely] a child, [but] he is addressed respectfully.
He is sent to perform [official] missions, and before he returns he dresses himself in a gown.
I have never seen a sculptor as an [official] envoy, nor that a goldsmith is sent.
I have seen a coppersmith at his work, at the mouth of his furnace,
His fingers like the claws of a crocodile, and he stinks worse than fish roe.
Every carpenter who wields the adze, he is wearier than the labourer in the field.
His field is the wood, and his hoe the axe.
There is no end to his craft, and he does beyond what his arms are capable of [...].
The jeweller is boring carefully into every type of hard stone.
He completes the inlay of an eye; his arms are exhausted and he is weary.
He sits at sunset, with his knees and his back cramped [...].
The potter is under the soil, although he stands among the living.
He grubs in the mud more than a pig in order to bake his pots.
His clothes are stiff with clay, his loincloth in rags.
Breath enters his nose direct from his furnace.
He tramples [the clay] with his feet, and is himself crushed by it."

*This golden head of a statue of a hawk from Hierakonpolis ("Hawk town"; see p.69), a centre of the cult of Horus, illustrates Egyptian metalworking skills at their finest. The eyes are formed from a single rod of obsidian, rounded and polished at both ends. Sixth Dynasty, ca. 2350BCE.*

**DREAM INTERPRETATION**

**The following dream directory is taken from the Chester Beatty III Papyrus, which dates from the Ramesside Period (ca. 1292–1075BCE):**

"If a man sees himself in a dream slaughtering an ox with his [own] hand, good: it means killing his adversary.

"Eating crocodile [flesh], good: it means acting as an official among his people.

"Submerging in the river, good: it means purification from all evils.

"Burying an old man, good: it means flourishing.

"Working stone in his house, good: fixing a man in his house.

"Seeing his face in a mirror, bad: it means another wife.

"Shod with white sandals, bad: it means roaming the earth.

"Copulating with a woman, bad: it means mourning.

"Being bitten by a dog, bad: it means he will be touched by magic.

"His bed catching fire, bad: it means driving away his wife."

*A paste-glass scarab, symbol of the rising sun and hence of life and rebirth, is the centrepiece of this amulet of the Greco-Roman Period.*

# MAGIC

Egyptian magic involved an application of metaphysical knowledge for both practical and religious purposes. It was spoken of as a divine creation for the benefit of humanity. Personified as a deity, Heka, it was one of the manifestations of direct contact between the divine and human worlds. Knowledge of magic fell into the same category of advanced learning as ritual, myth, medicine and literature, and the magician was in practice indistinguishable from the ritualist or physician. Magic was simply one category of knowledge, to be used by the learned as an aid in all their relations with both the physical and the divine world.

The magical rituals that are most easily understood involved the deflection of enemies by cursing formulae (see pp.144–5) These are accompanied by the ritual destruction of wax or clay figures. Rituals devised for vanquishing or holding at bay the cosmic enemy Apep, or Apophis, the foreign and political enemies of king and country, and also private individuals were essentially similar in character. Some Greco-Egyptian spells invoke evil gods and demons to appear in a person's nightmares. Alternatively, magic could serve a benign purpose. A love potion might be presented in a way not significantly different from a medical prescription with its accompanying incantation. Recipes are likely to include ingredients whose power is understood to derive from principles of sympathetic magic, and from the punning force of names. Words and names were seen as particularly potent, as was verbal association, or association of ideas at a metaphorical or symbolic level (see pp.242–3).

The primary technique used in magic was to compel rather than request the assistance of divine powers. In spells, the magician or his subject was often identified by name with a deity in order to endow him with the power of that god; or the magician would threaten the deity with dire consequences if his demands were not met. The future could be ascertained by questioning the god's cult statue as an oracle or through the interpretation of dreams. Calendars of lucky and unlucky days provided guidance for behaviour, with mythological explanations for their recommendations. Evidence for the use of astrology or other forms of divination is, however, very limited in the pharaonic period. Dreams were a point of contact through which the gods could make themselves known to human beings. "Incubation", the practice of sleeping in a temple compound in order to receive a prophetic dream from a god, is not known before the Late Period, but it reflects earlier, less formalized practice.

Magical protection was provided by amulets, the carrying of which, by the living as well as the dead, was believed to make a real difference to a person's fortunes, far beyond the effects of what we might call "lucky charms". Such objects survive in huge quantities from burial sites, where little amulets representing gods and goddesses, parts of the body, animals, items believed to contain special power and magical symbols were included in the wrappings of mummies. The amulets are made from a wide variety of materials, because symbolic power was attributed to the substance of an object as well as the thing it represented. Magical spells, always written on a new sheet of papyrus, were also used as amulets. A group of amuletic texts of the Late Period contains decrees put into the mouth of a god, promising protection for the individual against a wide range of physical and supernatural dangers.

Magical practice involved the whole range of Egyptian gods and goddesses, depending on the type of divine assistance expected. Among the major deities, Isis was the most frequently addressed, owing to her role as the protector of her son, Horus, with whom the person seeking help would often be identified. Another benign maternal figure, Hathor, was also commonly invoked. Of the huge number of lesser protective deities, the most curious is Bes, the dancing dwarf, whose hideous features personify the supernatural world's mixture of frightfulness and beneficence.

*A Ptolemaic Period faience figurine of the dwarf god Bes, who represents the banishing of disruptive forces through the powers of dance, music and joy. A popular domestic deity, he was especially associated with procreation, childbirth and the protection of infants.*

## A SPELL AGAINST CROCODILES

The Nile crocodile was revered for its power and strength, but for boatmen and others, the creature was one of the main dangers of life on the river. The Harris Magical Papyrus, now in the British Museum, contains a vivid spell against the animal:

"The first spell of all water-songs, about which the magician has said: 'Do not reveal it to the common man! It is a secret of the House of Life [that is, the learned men of the temples].'

" 'O egg of water and spittle-of-the-earth – the eggshells of the Ogdoad of Gods – great one in heaven, great one in the underworld ... who is prominent in the Island-of-Knives, it is with you that I have escaped from the water. I will emerge with you from your nest. I am Min [a god associated with rejuvenation] of Coptos ... '

"This spell is said over a clay egg that is to be put in the hand of a man at the prow of a boat. If something surfaces on the water, it is thrown on the water."

Another way to ward off crocodiles was to point at them with the index and little finger of one hand (see p.156).

*King Senwosret I (ca. 1919–1875BCE) before the god Amun, who is depicted in the form of Min (left), the ithyphallic god invoked in the spell against crocodiles and associated with fertility and rejuvenation. From the pharaoh's* sed *(jubilee) pavilion at Karnak in Eastern Thebes.*

# THE ARTS

Egyptian literary forms belong to the realm of recitation and performance, and not individual reading. Classic literature is composed in short lines for recitation, typically couplets, which have a rhythmic cadence that varies between rhetorical narrative and self-evidently metrical styles. Despite the incomplete nature of the script, which does not represent vowels, it is clear that the sound of the words was important in the recognition of literary merit. The prosodic weight lay at the beginning of words and lines, rather than at the end, so that alliteration and homophony provide the rhythm; rhyme was probably not used, but the rhythm of the individual line or couplet was pointed by parallel or contrasting meaning in the paired clauses.

Most important, however, was the use of plays on the sounds of words to reinforce meaning. As a literary device, the pun or *double entendre* transcended superficial verbal cleverness and was taken to represent a deeper reality. Identity of sound was felt to represent a community of meaning. Plays on words were vital in the verbal performance of ritual, magic and medicine, where they provided the association that gave force to the action. For example, the dressing of a wound in a medical papyrus includes honey (see p.97). The wound recovers: the Egyptian word used is *ndm*, "get better", but also has the commoner meaning, "is sweetened". Such games of metaphorical association lie at the centre of the ancient Egyptians' understanding of their relationship to the world and their ability to control it. Words and names carried the power of meaning and made reality accessible; their use also gave pleasure. Both tomb autobiography and royal inscriptions are addressed to a public. Both have their roots in

**THE SONG OF KING INYOTEF**
**This lyric, recorded in the New Kingdom document known as the Harris Papyrus, was sung at the funerary feast of one King Inyotef, a name borne by a number of Eleventh- and Seventeenth-dynasty pharaohs. It is one of the few examples of something approaching religious skepticism in Egyptian literature:**

"Song which is in the chapel of Inyotef, justified, which is in front of the singer with the harp.
He is flourishing, this prince.
Good is fate; it is good to perish.
A generation passes, another endures, since the time of the ancestors.
The gods who lived before rest in their pyramids.
The transformed dead likewise are buried in their pyramids.
The builders of chapels, their places no longer exist.
What has been done with them?
I have heard the words of Imhotep and Hordjedef.
Their sayings are widely quoted.
Where are their places?
Their walls are abandoned.
Their places no longer exist, as if they had never been.
Nobody returns from there,
that he may tell of their state,
that he may tell of their things,
that he may calm our hearts,
until we go to the place that they have gone.

[Refrain:]
Have a good time. Do not grow tired of it.
Look, there is nobody allowed to take his property with him.
Look, there is nobody who is gone and returns back."

*Male and female singers, musicians and dancers depicted on a painted limestone relief from the tomb of Nenkhefetka at Saqqara. The musicians play the harp, flute and a reed instrument. Old Kingdom, 5th Dynasty, ca. 2400BCE.*

*Female dancers beat tambourines and clapsticks before an approaching funeral procession in this limestone tableau from Saqqara. New Kingdom, 19th Dynasty, ca. 1250BCE.*

formal recitations of praise for the individual or the king, to an audience on a public occasion. Narratives for entertainment survive in some number, the stories based on mythology or focused on a king or his court. "Wisdom texts", in which a father teaches his son, or a wise man bemoans the state of the country, present themes that can be classed as practical philosophy.

There is no basis for believing that narratives might have been performed to any degree of musical accompaniment: the absence of evidence probably implies that they were not. However, more lyrical styles of literature were clearly performed to music. These range widely in content, embracing love poetry, hymns to gods, praise poems and work songs, all of which could be sung with or without accompaniment.

Although our technical knowledge of Egyptian music is very limited, the depiction of musicians is common. Music of every sort was used to accompany ritual and for the adoration of a god, in festivals and processions. A harpist is commonly shown, performing for the tomb owner, or at the funerary feast, and his song typically addresses the relationship between this life and the next. The harp seems to be the normal instrument to accompany a limited recitational style of singing, but more extensive bands of musicians are often shown, associated also with dancing. The basic instruments are string and woodwind: harps of different sizes, lutes, flutes and oboes, and, for percussion, tambourines, drums and *sistra* (rattles). A simple trumpet is only known from New Kingdom military contexts, and brass instruments were not part of the normal ensemble.

Musical accomplishment did not form part of the expected syllabus for the educated or cultured Egyptian, although in the New Kingdom it was normal for high-status women to be associated with a divine cult through the position of "chantress" to the local god. The little that we know of musicians and dancers implies that the performers shown on tomb and temple scenes were professionals whose skill was unwritten: a craft that was related to, but of distinctly lower status than, the fine use of words.

*Part II*

# BELIEF AND RITUAL

*An avenue of ram-headed sphinxes at the temple of the god Amun at Karnak, Eastern Thebes. Amun, one of the greatest of Egyptian deities, was frequently depicted as a man with a ram's head or as the animal itself, which was associated with strength and fertility.*



OPPOSITE: *Osiris, with whom the pharaoh was believed to be assimilated after death, enthroned as god of the underworld. He wears a plumed crown (*atef*) and clasps the crook and flail, other emblems of royalty. A copy of a vignette from a New Kingdom papyrus.*

CHAPTER 8

# THE LORD OF THE TWO LANDS

Presiding over the land and people of Egypt was the king, or pharaoh, an absolute monarch who oversaw every function of the state, from the collection of taxes and the administration of justice to the fighting of wars. However, unlike the rulers of other lands, he – or, rarely, she – was not simply the chief of state, but also an essential component of the Egyptian cosmos, who was believed to share the divinity of the gods from whom his power derived. Without Pharaoh, the world would descend into a state of chaos that would threaten the very existence of the universe.



ABOVE: *A painted relief from Deir el-Bahari, Western Thebes, depicting King Thutmose III (ca. 1479–1425BCE) wearing the* atef *crown (see p.109) of Osiris, with whom the king was identified after death (see p.108).*

## THE ORIGINS OF KINGSHIP

Kingship in ancient Egypt was an essential element in the proper functioning of both the state and the cosmos. The reigning pharaoh was the link between the world of the gods and the world of humankind (see p.108), and in his position at the centre of the Egyptian state he embodied a divine power. While concepts of kingship such as this are detailed extensively during Egypt's historical period in texts and in the programmatic decorations of tombs and temples, certain elements are discernible in much earlier times. It is possible to trace the development of the royal iconography that was characteristic of the later pharaonic tradition in Egypt and neighbouring regions right back to the Predynastic Period and even earlier, when Egypt was ruled by local kings or chieftains, well before the unification of the country ca. 3100BCE (see pp.22–3).

In Upper Egypt at least, based on the current evidence, there appears to have been some continuity of development from the early (Badarian) Predynastic settlements into the later (Naqada) phases – when royal iconography becomes more apparent – and into the Dynastic Period. Although there were settlements in both the north and the south, the culture in each was distinct, and perhaps in these early differences lie the roots of the later two kingdoms that comprised pharaonic Egypt.

However, early scholars tended to see the Protodynastic culture as emerging in a fairly developed form just prior to 3100BCE, and to attribute this to the direct or indirect impact of cultures beyond Egypt's borders. Traditionally, Mesopotamia and, most recently, Nubia have been seen as influences or sources: the former because of certain decorative motifs and its early and strong tradition of monumental architecture and writing; the latter because of the presence of what appears to be Egyptian royal iconography on artefacts discovered in Nubia (although such material may have come originally from Egypt). Recent archaeological evidence indicates that more well-developed settlements existed earlier and in more places in Egypt than was previously thought, and such information is reshaping present theories of the origins of kingship.

*The head of a colossal statue of Ramesses II (ca. 1279–1213BCE) at the temple of Luxor, Eastern Thebes. The king promoted the cult of the divine pharaoh by erecting many statues of himself as a god throughout the country (see p.113).*

*A painted limestone head of Nebhepetre Mentuhotep II (ca. 2008–1957BCE), the Theban ruler who reunited Egypt at the end of the First Intermediate Period. He wears the White Crown of Upper Egypt. The Egyptians believed that their country consisted of two kingdoms – Upper and Lower Egypt – which had been unified by the legendary King Menes (see p.23). Kings were customarily depicted wearing either the Lower Egyptian Red Crown (compare the above illustration with that on p.26); or the White Crown; or the* Pschent *(see p.108), a combination of the two. (See also p.109.)*

Cultural and political unity appears to have existed quite early in the fourth millennium BCE. A typical element of dynastic iconography, the Red Crown, first appears during the early Predynastic Period (Naqada I, ca. 4000–3500BCE): an image in raised relief on a jar from Naqada clearly resembles the later crown. Other characteristic imagery developed during the ensuing Predynastic phases (Naqada II and III). At Hierakonpolis (see p.69), what is probably the tomb of a Predynastic ruler contained a scene of a man smiting three smaller figures. The "Smiting Scene" was a symbol of royal subjugation throughout Egypt's history (see pp.23 and 155).

The Naqada phases also witnessed the first steps toward an organized state in Egypt. In various parts of the country during Naqada I there were small dominions, each of which was ruled by a chief. From these early political units emerged the concepts of both kingship and government that evolved during Naqada II and III and would become fully developed when Egypt became united under a single ruler just before the recorded First Dynasty.

# THE GOD-KING ON EARTH

*King Thutmose III (ruled ca. 1479–1425 BCE) is portrayed in this granite statue wearing the* Pschent *or Double Crown, symbolizing his rule over the "Two Lands" (see illustration, opposite page). The crown is adorned with a rearing female cobra, the* uraeus, *which represented the pharaoh as the protector of his realm. Another item of regalia is the tightly-plaited false beard. Such beards were considered an attribute of the gods and therefore indicated the king's divine status. The few women who ruled as pharaoh – most notably Thutmose III's predecessor, Hatshepsut (see p.89) – also wore the false beard.*

Like most monarchs, Pharaoh occupied a unique position at the top of the social and political hierarchy. However, the king in Egypt was more than just a head of state: he was a discrete and essential element within the cosmos. His unique position alongside the gods, humankind and the spirits of the deceased (*akhs*), was necessary to maintain *ma'at,* the divine order. Without Pharaoh, the cosmos would be in disarray and the world would descend into chaos. The king was an active participant in the mythology associated with kingship, fulfilling on earth the role of the god Horus, son of Osiris. His title *Sa-Re* ("Son of Re"), first attested in the Fourth Dynasty (see p.173), clearly indicates that the monarch was also regarded as the living descendant of the sun god. A pharaoh acquired this deified status upon his coronation. Myths associated with kingship, particularly those surrounding Osiris and his family – parts of which occur in the popular New Kingdom story known as *The Contendings of Horus and Seth* (see p.22) – derive from the earliest periods and persisted throughout ancient Egyptian history.

As a secular ruler, Pharaoh oversaw the bureaucracy of the state and all its judicial actions. As a religious figure, he was the theological leader of the country, and inscriptions and decorations project the image of the king as the chief officiant to the gods at all rituals. However, in reality, it was the temple priests who actually performed this role on a day-to-day basis (see pp.150–51).

Pharaoh's military role as the commander-in-chief and controller of all military operations is well documented in texts and images from all periods of Egyptian history. According to such records, the king's frequent battles were always successful. The gods sanctioned Pharaoh to fight bat-

*King Ramesses III (ca. 1187–1156BCE) in the white linen robes of a high priest (right) before the deities associated with Heliopolis – Re-Harakhte (a form of the sun god), Atum, Iusaas and Hathor Nebethetepet. The scene is taken from the Great Harris Papyrus – the longest papyrus known to exist – which records temple donations by Ramesses III (see also p.67).*

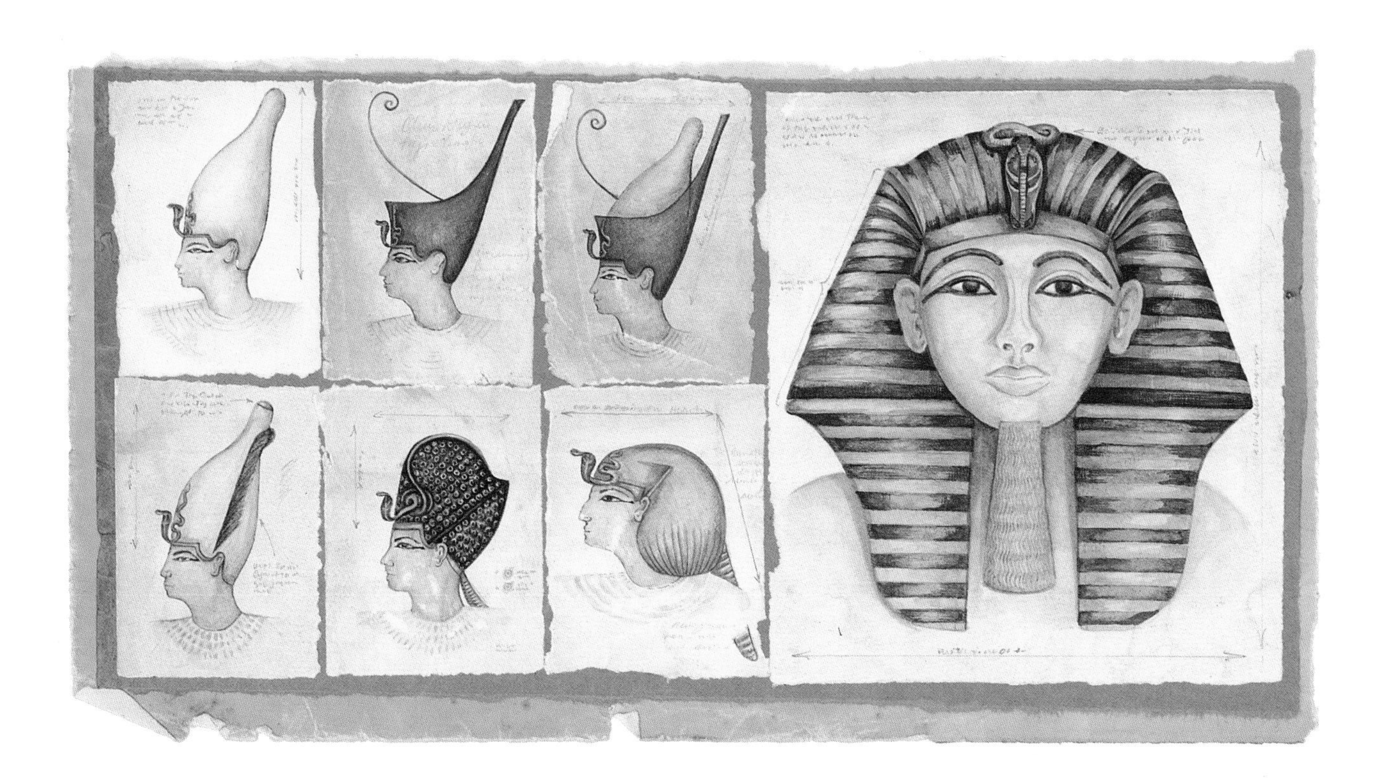

*The most commonly worn royal crowns and headdresses.* MAIN PICTURE: *The blue and yellow striped* Nemes *headcloth, which was worn at most periods but is particularly associated with the kings of the Middle Kingdom. It had two lappets, or flaps, over the shoulders and was gathered into a pigtail at the back.* BOTTOM, LEFT TO RIGHT: *The White Crown, or* Hedjet, *symbolizing Upper Egypt; the Red Crown, or* Deshret *of Lower Egypt; the Red and White Crown, or* Pschent, *representing the "Two Lands" of Upper and Lower Egypt.* ABOVE, LEFT TO RIGHT: *The plumed* Atef *Crown, worn on certain ritual occasions (sometimes with the addition of ram's horns; see illustration, p.106) and associated with the god Osiris; the Blue Crown, or* Khepresh, *also known as the "War Crown", worn especially by 18th-Dynasty pharaohs and associated with the sun god; the* Khat, *a plain variant of the* Nemes.

tles on their behalf; it was they who granted success, and to them that victories were credited. Egyptian casualties and defeats had no place in the official propaganda, which was a necessary tool in the maintenance of *ma'at* and the king's absolute power.

Many rulers actually took an active role in fighting at the head of their troops, and some were even killed in battle, such as Seqenenre Tao, whose mummy displays the multiple fatal skull wounds that he sustained against the Hyksos rulers of the Delta ca. 1543BCE (see p.31). Other kings only narrowly avoided defeat, for example Ramesses II (ca. 1279–1213BCE), who in the fifth year of his long reign fought an inconclusive battle against the armies of the Hittite empire at Kadesh in Syria. In monumental temple inscriptions and reliefs at Abydos, Thebes and Abu Simbel, Ramesses II unsurprisingly trumpeted this confrontation as a glorious victory, in which he single-handedly forced a passage through the Hittite army that had completely surrounded the Egyptian forces. In reality it was probably only the intervention of a relief force that prevented a near-disaster for the young pharaoh.

A number of different terms were used to denote the king, some of them referring to him personally, others indirectly. *Nyswt*, translated as "king", was used to describe Pharaoh in his administrative, legal and other official capacities, while the word *hm* indicated his physical presence. The title "Pharaoh" derives from the Egyptian phrase *per aa*, which actually means "great house" and was originally a reference to the royal palace. It was only from the late New Kingdom onward that it was employed to designate the sovereign himself.

# THE PHARAOH AFTER DEATH

*Part of a granite statue of ca. 1250BCE depicting Rameses II in "Osiride" form, that is, with the characteristics of the god Osiris: he has his hands crossed over his chest like a mummy and holds the crook and flail, part of the royal regalia. Kings were customarily buried in this attitude.*

Not all Egyptologists agree on the extent and nature of the divinity of the living pharaoh (see pp.112–13). They are unanimous, however, in believing that the king achieved divine status in death. Much of the archaeological record of royalty focuses on the material remains of the rulers' attempts to secure their eternal and divine existence. The earliest pharaohs began the tradition of building a mortuary complex to ensure their immortality, and within a few generations this complex had grown to monumental proportions (see pp.168–191; 210–11). Tombs and mortuary temples (see pp.136–7), a testimony to royal status and power, were an eternal resting place for the dead ruler, who was believed to have become a god, as well as a necessary component for his apotheosis.

Our sources for explicit information about the nature of the dead pharaoh's divinity are texts and scenes decorating the walls of royal tombs and mortuary temples. Pyramid Texts, the earliest of the funerary inscriptions, were carved on the walls of the burial chamber and antechamber of the pyramid of King Unas of the Fifth Dynasty (see p.188). Later pharaohs modified this material and it eventually (in the New Kingdom and later) took the form of detailed illustrated books of the afterlife (see pp.136–7).

All these sources have in common the concept that Pharaoh had transcended his earthly existence: "Raise yourself, O king! You have not died" is a command given in the Pyramid Texts to the newly transformed king. Some funerary spells in the Pyramid Texts specify the many divine identifications and associations of the king – for example, "I am Horus" or "I am Sobek." Pharaoh may also be referred to as the son of a god, as in an address to the sun god Re – "the king is your son" – which also appears in the later *Litany of Re*. Several spells associate parts of the king's body with different deities, and some specify his various functions. He may be the Lord of the Thrones who crosses the sky as the successor of Re (*Litany of Re*), or the head of the heavens (Pyramid Texts), the one who gives orders. Other passages dictate his behaviour, directing him to bestow honours and make offerings.

*Remains of the "Ramesseum", the mortuary temple of Ramesses II in Western Thebes. It was known to the Greeks mistakenly as "the tomb of Ozymandias", a name that derives from Ramesses'* praenomen *(see p.113) of Userma'atre. The pharaoh was actually entombed in the Valley of the Kings.*

In both earlier and later funerary texts, the gods most often mentioned are Re and Osiris. The *Litany of Re* notes that the king had come into being as Re, and the Pyramid Texts state that the king would be assigned his throne so that he could sit as Osiris, the legendary first king on earth, at the head of all the powers. In some passages the king's identification with divinity is directly stated; for example, "The king is Osiris in a whirlwind" and "the Osiris-king Men-ma'at-Re [Sety I] is Re [and vice versa]" (both from the *Litany of Re*). In referring to the deceased pharaoh as "the

Osiris-king So-and-So", the texts indicate his metamorphosis into Osiris, who was reborn as king of the afterlife and was the divine force associated with the fertility of the earth and eternal cycles of growth.

The deceased pharaoh was also identified with Re, the supreme deity who was born every morning and died every evening. The king's identification with the supreme earthly and solar deities of the Egyptian pantheon suggests that the king in death embodied the duality that characterized the ancient Egyptian cosmos. The deified ruler represented both continuous regeneration (as Osiris) and the daily cycle of rebirth (as Re). In their understanding of the cosmos, the ancient Egyptians were accustomed to each of their deities possessing a multiplicity of associations and roles. It was a natural extension of this concept for them to view the deified pharaoh in a similar way.

## THE POWER OF THE DEAD PHARAOH

The funerary complexes built for the pharaohs were vast, labour-intensive projects that took decades to complete and required long-term endowments to ensure the continuation of the mortuary cult of the dead king (see pp.140–41), and thereby to guarantee his perpetual divinity. Investment on such a scale implies great faith and determination amongst the people and suggests that they believed strongly in the power of Pharaoh on earth and in his potential as an important god in the afterlife.

Yet this belief might not have been as literal, nor as all-encompassing, as might be presumed. While the mortuary complexes did survive to a great extent, allowing a lengthy span of years during which the cult of the deceased king could continue, the royal tombs did not fare so well. All were robbed, and the confessions of tomb robbers recorded in official papyri make the modern reader question the depth of belief in the power of the deceased, divine pharaoh. Criminals convicted in the late Ramesside Period (ca. 1120BCE) testified to the theft of objects from tombs, the looting of precious metals from coffins and mummies, and the destruction of royal corpses. Other texts record carousing on royal burial equipment and blasphemous activity by individuals (see p.196). Such behaviour suggests that at least part of the population had little fear of repercussions in this world or from the gods in the next.

*The interior of the temple of Ramesses II at Abu Simbel in Lower Nubia, dedicated to his deified self and the great gods Re-Harakhte, Amun-Re and Ptah. The pillars are in the form of massive statues of the king as the god Osiris.*

# HUMAN OR DIVINE?

*A fragment of a relief from the mortuary complex of King Unas (ca. 2371–2350BCE) of the 5th Dynasty, apparently showing the king being suckled by a goddess. Egyptian concepts of divine kingship may derive from Old Kingdom royal cults that associated the monarch directly with the gods.*

The divine status of the dead king is apparent from the popular cults that arose in the later New Kingdom, when votive stelae depicting deceased rulers such as Amenhotep I (see p.153) were produced by individuals who believed that prayers and offerings to the late monarch would bring them benefits. The practice of the élite class of locating their own burials near the pyramid of the pharaoh whom they had served, or whose mortuary cult they took part in, also indicates great respect for the deceased pharaoh. But not everyone honoured Pharaoh's memory; some people robbed royal tombs and temples (see pp.111 and 196), drew satirical cartoons of the king and even mocked him in writing, in one instance calling him "the old general".

How the people viewed the reigning pharaoh varied in different periods and among different social classes. Two distinct but inextricable elements were constant in the royal ideology: the ruler (the king) and his office (kingship). In the office of king the pharaoh was an immortal and divine presence, a fixed element in the cosmos. Each mortal king became identified with the office he held, and to that extent became immortal as well. Although the "official" picture of Pharaoh was vital to both religious and socio-political concepts, the people still saw a distinction between kings and gods. Rulers had some superhuman attributes, but these were gifts from the gods. Kings were not naturally omniscient but often sought knowledge from others. Subjects sometimes played upon the king's frailties in popular texts, sculpture, graffiti and letters to emphasize his human aspect and even, as we have seen, to satirize him.

*The temple of Abu Simbel, erected in the 24th year of the reign of Ramesses II. The monument was probably intended primarily to impress the Nubian population living on Egypt's southern frontiers with the tremendous power of the divine pharaoh and of the gods of the land.*

## THE ASSERTION OF DIVINITY

The ancient Egyptians expressed their piety through a variety of cults. While votive stelae depicting the deified dead king were sometimes erected by private individuals (see main text), other such cults originated on royal initiative. The concept of the divine ruler may have originated in the official cults of the king that began in the Old Kingdom. However, the worship of the living king as a god appears to have been more explicit in the reigns of those monarchs who struggled to re-establish national unity after the chaos of the first two Intermediate Periods. The king would then emphasize royal divinity by reasserting his divine birth and his link with the gods and the cosmos.

In the New Kingdom, the cult of the living king became more common. Amenhotep III (ca. 1390–1353BCE) established cults and erected statues of the living king in temples primarily situated in frontier areas. Inscriptions at Luxor emphasize his divine birth, a concept perhaps taken to extremes by his son Akhenaten. Ramesses II (ca. 1279–1213BCE) also encouraged his subjects to honour him as a god by putting put up numerous statues of his deified self in temples throughout much of the country and at his city of residence, Piramesse in the Delta.

*No monarch produced more monuments expressing his divinity than Ramesses II. In this relief from the Ramesseum, his mortuary temple at Western Thebes (see also illustration, p.110), he is represented as the god Amun-Re, attended by a goddess.*

The king's divine association was corroborated by his titulary. This set of five names, titles and epithets was used from the Middle Kingdom on (the practice of enclosing royal names inside a cartouche, or oval ring, began with the founder of the Fourth Dynasty, Sneferu, ca. 2625–2585BCE). The king received four names upon his accession to the throne. Of these, the "Horus name", the "Two Ladies name" (or "*nebty* name") and the "Golden Horus name" emphasize the king's divinity, while the throne name or *praenomen*, "King of Upper and Lower Egypt" (*nswt bity*, or "He of the Sedge and the Bee", emblems of Upper and Lower Egypt) refers to the official role of the ruler. The only name received by the pharaoh at birth, termed the *nomen*, was the last in the sequence of five royal titles and is often referred to as the *Sa-Re* ("Son of Re") name, after the epithet that precedes it. This designation is a reference to the pharaoh's divine birth and his association with the sun god. The Horus-name was often inscribed within a *serekh*, a rectangular panel with the design of the palace façade, and with the Horus falcon above it.

● CHAPTER 9

# THE CELESTIAL REALM

According to Herodotus, the Egyptians were "religious to excess, beyond any other nation in the world". Egyptian religion was not a belief system in the same sense way as Christianity or Islam, with a single deity and one fundamental set of explanations for the origin and functioning of the cosmos. Among the most striking aspects of Egyptian religion were its great number of gods and goddesses – each of whom might have several "aspects" – and its readiness to accept the validity of different and even contradictory cosmological accounts.

▲



ABOVE: *An 18th-dynasty image of the god Osiris, the head of the divine family that played a focal role in Egyptian belief and mythology. From the tomb of Sennedjem, Western Thebes. New Kingdom, 20th Dynasty, ca. 1140BCE.*

## THE EGYPTIAN COSMOS

Looking at the sky without telescopes, the Egyptians saw only an undifferentiated background of blue by day, or black by night – the same qualities visible in the river Nile. Understandably, therefore, the Egyptians concluded that the sky, like the Nile, was composed of water. The waters of the sky were thought to surround the earth and extend infinitely outward in all directions. The world existed as a single void inside this endless sea, with only the atmosphere to keep the heavenly ocean from falling onto the earth – much like a balloon kept inflated by the air inside it.

All life existed inside this cosmic bubble: the universal waters themselves were devoid of life. By day, the sun sailed across the surface of the sky-ocean, animating those who lived on the earth below; after sunset, while the stars sailed through the sky, it descended into a region called "the Duat". Because the Egyptians recognized that the sun was composed, in some manner, of fire (the source of light and heat), they realized that it had to remain within the cosmic void, but in a place not visible to those on earth. The Duat was generally thought to lie under the earth, a counterpart to the sky and atmosphere of the known world. In Egyptian cosmology, therefore, the world consisted, as the ancient texts themselves tell us, of "sky, earth and Duat".

This picture of the cosmos is reflected also in images from temples, tombs, papyri and sarcophagi. However, perhaps the clearest and most comprehensive illustration is found on the ceilings of two Ramesside monuments: the cenotaph of Sety I (ca. 1290–1279BCE) at Abydos, and the tomb of Ramesses IV (ca. 1156–1150BCE) in the Valley of the Kings, Western Thebes. The ceilings are remarkable not so much for their images (which occur elsewhere) as for the texts that accompany them: these are the subject of analysis and commentary in two papyri of the second century CE – some fifteen hundred years after their Ramesside originals. The scene depicts the surface of the sky (the goddess Nut, "watery one") held above the earth (the god Geb, "land") by the atmosphere (the god Shu, "dry" or "empty"), while along Nut's body the sun is depicted

*The modern Western zodiac derives from the ancient Egyptian view of the night sky. The best-preserved depiction of the Egyptian zodiac is on the ceiling of the small chapel (*naos*) of the temple of Hathor at Dendera (left). Ptolemaic Period, 323–30BCE. (See also pp.120–21.)*

at various points in its daily cycle. The text above her describes both the universe outside the cosmic void and the structure of the cosmos itself: "The upper side of this sky exists in uniform darkness, the limits of which ... are unknown, these having been set in the waters, in lifelessness. There is no light ... no brightness there. And as for every place that is neither sky nor earth, that is the Duat in its entirety." Texts elsewhere in the scene describe the Duat as lying within the body of Nut, the sky. This reflects the Egyptian concept of the sky "giving birth" to the sun each morning. In Egyptian thought, these images were complementary, not contradictory. Fundamentally, the concept of the world as a cosmic void within a universal ocean remained consistent and essentially unchanged throughout the three millennia of recorded ancient Egyptian history.

The Egyptian image of the cosmos was usually depicted by using the "mythological" counterparts of its elements – Nut stretched above the recumbent body of Geb, with Shu in between (see illustration, p.126). However, the concept of the world also appears as a standard element bordering most reliefs and paintings. Traditionally, the ceilings of Egyptian tombs and temples would be decorated with yellow stars on a blue ground; the floors were paved with basalt, evoking the fertile black soil of Egypt; and columns supporting the ceiling were carved and painted in imitation of lotus or papyrus stalks.

**ROUND WORLDS**

**The rather "box-like" image of the cosmic void conveyed by Egyptian texts (see main text) is somewhat misleading. Arched versions of the "sky" hieroglyph in very early reliefs indicate that the Egyptians recognized the void as round. The same vision is reflected in later metaphors such as the circle often held by gods – whose extended version, the cartouche surrounding royal names ( ⊂⊃ ), indicates the pharaoh's kingship over the universe.**

**Depictions of a round world occur late in Egyptian history. One of the earliest and most complete is on a sarcophagus of ca. 350BCE, now in the Metropolitan Museum of Art, New York. The world is framed by the body of the sky and a compounded hieroglyph (arms with feet) spelling the name of Geb. Between them, two concentric discs depict the known world, with Egypt (represented by the signs of its nomes) inside, surrounded by the peoples of other lands. A third circle, with two winged sun-discs, is meant to be seen at 90° to the others and depicts the sun's journey above and below the earth. The famous Dendera zodiac (above) clearly implies that the sky covers a round earth.**

**REALMS OF GODS AND BIRDS**
**The Egyptians saw the sky as the gods' primary domain, although they could be associated with all regions of the cosmos. The Pyramid Texts tell of the time "when the sky was split from the earth and the gods went to the sky". Birds were also thought to come from the sky, particularly its northern regions – probably reflecting both the annual migration of birds from the north and the rich fauna of the Nile Delta in ancient times. The dual association of gods and birds with the sky is often reflected in Egyptian images of the sun, stars and planets as birds.**

# THE HEAVENLY DOMAIN

In keeping with their view of the sky as the surface where the waters of the universal ocean met the atmosphere of the world, the ancient Egyptians envisaged the motion of celestial bodies as a journey by boat. During the day, the sun sailed across the sky, and at night the stars did the same. The text accompanying one scene of the solar journey describes it as follows: "When this god [the sun] sails to the limits of the sky-basin, she [Nut, the sky] causes him to enter again into night, into the middle of the night, and as he sails inside the dark these stars are behind him. When the incarnation of this god enters ... inside the Duat, it stays open after he sails inside it, so that these sailing stars may enter after him and come forth after him."

Ancient texts describe a number of celestial regions, especially in the night sky. The Egyptians were keenly aware of the nocturnal heavens, and recorded nearly every visible aspect of them. Apart from individual stars and planets (see box, opposite), several features attracted particular attention, and were interpreted as celestial counterparts of the kinds of environments found along the ancient Nile.

The earliest substantial source of Egyptian cosmological texts, the Pyramid Texts of the Old Kingdom (dating from the Fifth to Sixth

*A scene from the Book of the Dead of Ani. At left, the deceased enjoys the pleasures of the "Field of Reeds" (see main text); at right, he greets the sun god. New Kingdom, 18th or 19th Dynasty, ca. 1300BCE.*

dynasties, ca. 2350–2170BCE), are a rich source for this ancient "geography" of the sky. Among the phenomena they describe is the Milky Way, which the Egyptians called "the beaten path of stars". Like other parts of the sky, it could be navigated by boat, and was apparently viewed as a series of islands in the midst of celestial waters.

The texts pay more attention, however, to the "Field of Offerings" and the "Field of Reeds". (The latter term is the ancestor of the Classical Elysian Fields, a term – now commemorated in the Champs Elysées, Paris – derived from a Greek rendering of the Egyptian word for "reeds".) Both these areas were associated with the northern rim of the sky, the domain of the circumpolar stars (which the Egyptians called "imperishable" because they never set). Like the Milky Way, they could be navigated by boat, and the texts speak of a "winding waterway" through them. As the North Celestial Pole lies approximately 30° above the horizon in Egypt, these "fields" were apparently thought to lie along the edge of the celestial ocean, much like the marshes that lined the banks of the ancient Nile. The sky above them seems to have been viewed as relatively empty, except for the Milky Way, which the Pyramid Texts locate in "the height of the sky".

## THE STARS AND PLANETS

Like most agricultural societies, the Egyptians observed and tracked the stars as harbingers of the change in seasons. In this respect, the most important celestial body was the brightest star, Sirius, which the Egyptians called "the sharp one" (*spdt* or Sopdet, vocalized by the Greeks as Sothis): its annual reappearance in the morning sky, after an absence of some seventy days, coincided with the beginning of the yearly inundation of the Nile, the chief determinant of life in ancient Egypt.

The Egyptians identified five of the nine planets: Jupiter, Saturn and Mars (all associated with various aspects of the god Horus), Mercury and Venus (called both "the travelling star" and "the morning star"). They also recognized many of the same constellations that we do, although they saw in their patterns images different from those familiar to us. Among the most important was Ursa Major, perceived as the leg and haunch of a bull; and Orion, identified as the god Osiris and seen as a man holding a staff. Based on their observations of stars and stellar motion, the Egyptians divided the night and day into twelve hours each. This division produced our 24-hour day, although in ancient Egypt, the hours varied in length, like the durations of the day and night, over the year. Toward the end of pharaonic history, Egypt also produced the first zodiac (see illustration, p.119).

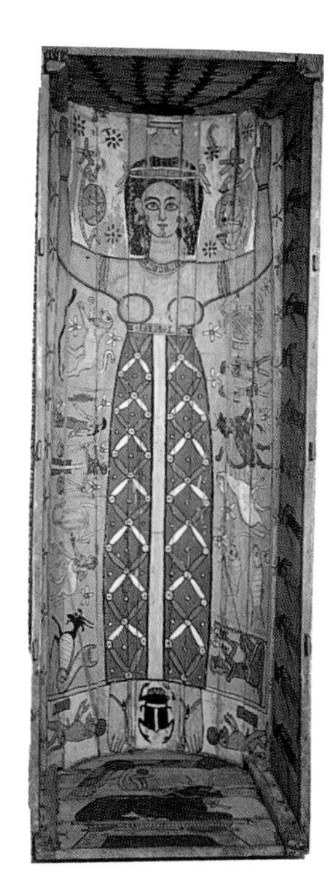

*From the inside of this coffin lid, the sky goddess Nut, surrounded by figures representing the signs of the zodiac and the hours of the day, looked down on the deceased, a woman called Soter. Roman Period, 2nd century CE.*

# THE SOLAR CYCLE

*Re-Harakhte ("Re, Horus of the Horizon"), an aspect of the sun god assimilated with the sky god Horus (see box, opposite), is worshipped by baboons at sunrise. From the Book of the Dead of Hunefer. New Kingdom, 19th Dynasty (ca. 1292–1190BCE).*

For the ancient Egyptians, the day began at sunrise, when Nut, the sky, "gave birth" to the sun in the east. The sun, envisioned as a male deity, sailed the celestial waters in his "day-boat", before descending in the west into the Duat, the region beneath the earth, and the womb of his mother Nut (see pp.118–19). At night, he sailed from west to east through the Duat in his "night-boat", to be reborn again in the morning.

While the sun's daytime journey could be observed as a serene progression through the sky, his trip through the Duat could only be imagined. The Egyptians saw this – like the night itself – as a time of uncertainty and danger. Concepts of the nightly voyage appear in the very earliest religious texts, but they are best seen in a series of "netherworld books" composed at the beginning of the New Kingdom. The most detailed of these is the composition known as the Amduat ("He who is in the Duat"), depicting the sun's progression through the night.

The dangers associated with the Duat were personified in the form of a gigantic serpent, called Apep (Apophis in the Greek rendering), who inhabited the entire length of the netherworld and sought to impede the sun's journey at the gates marking the entrance to each of the night's twelve hours. As the sun passed within each region, his light awoke the inhabitants of the Duat, who were thought to include both demons and the souls of the damned. A typical passage in the Amduat describes the sun "calling out to their souls ... and a sound is heard in this cavern like the sound of people wailing, as their souls call out to the sun".

**ASPECTS OF A SINGLE GOD**

**Egyptian thought recognized the validity of many different explanations of natural phenomena, even where we might perceive these as contradictory. As a result there is an often bewildering profusion of names and images associated with Egyptian deities (see box, opposite). These were understood not as competing theologies but as alternative explanations of reality, each concentrating on separate aspects of a single force or element of nature. For example, the god Horus could be seen, at one and the same time, as the sun (Horus as Re, king of the universe; see illustration, above, and box), as the current pharaoh (Horus as king of the living, the "Son of Re"), and as a form of the previous pharaoh (Horus, the son of Isis and Osiris). Each of these were understood as an aspect, a manifestation, of the single phenomenon of kingship, embodied in the god Horus.**

**This approach is reflected not only in the multiplicity of Egyptian gods and goddesses, but also in the readiness with which the Egyptians adopted the gods of other cultures (see p.52).**

In the middle of the night, at the deepest part of the Duat, the sun came upon the mummified body of the god Osiris, the power of life and rebirth. At this point in the journey, the two gods became one: "the sun at rest in Osiris, Osiris at rest in the sun". Through this union, the sun received the power of new life, and Osiris was reborn in the sun. Given new life "in the arms of his father Osiris", the sun could then proceed through the remainder of the night toward rebirth at dawn.

When the sun left the Duat, he did not sail immediately above the visible horizon, but into a space lying between the Duat and the sky. The Egyptians called this region the Akhet, which means "the place of becoming effective". In practical terms, this was an explanation for the fact that the sky starts to grow light some time before the sun actually appears. Here the sun received a form capable of life before his actual birth: "Then he is on course toward the world, to be apparent and born. Then he produces himself above. Then he parts the thighs of his mother Nut. Then he goes away to the sky." In the Egyptian view, this daily solar cycle was not merely a natural phenomenon, but a daily affirmation of the triumph of life over death.

## MANIFESTATIONS OF THE SUN

The sun was in many respects the pre-eminent Egyptian god. His prominence is shown in the plethora of gods with solar associations – each representing one or more aspects of the sun itself (see sidebar, opposite):

**Re**: the sun *per se*; depicted as a man, a falcon, a ram, or a man with the head of one or the other of these animals. Re means simply "sun", and in the New Kingdom and later, it was often preceded by the definite article (*pa-Re*, "the sun"). As the physical manifestation of the sun god, the sun was also called the Eye of Re, and in this form was depicted as a goddess.

**Khepri**: the sun at dawn. The name means "the evolving one", and was written with the hieroglyph of a scarab beetle (*khepreri*, ). Khepri is often shown as a scarab (sometimes holding the sun disc) or scarab-headed man.

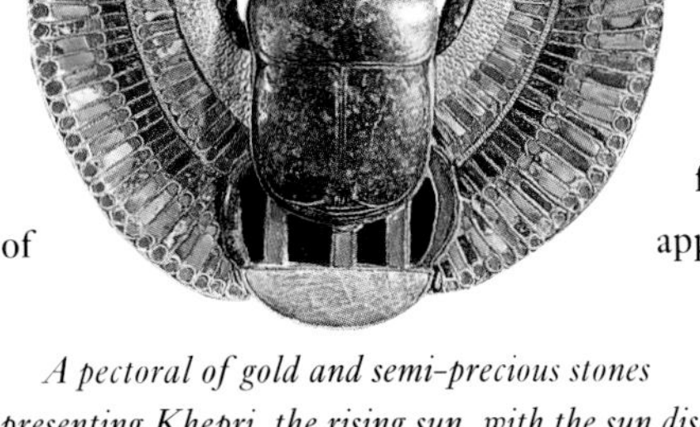

*A pectoral of gold and semi-precious stones representing Khepri, the rising sun, with the sun disc. From the tomb of Tutankhamun (ca. 1332–1322BCE) in the Valley of the Kings, Western Thebes.*

**Atum**: the sun as the culmination of creation. Depicted as a man, Atum is often associated with the sun at sunset, in the combined Re-Atum.

**Horus**: "the far one", the sun as ruler of creation; depicted as a man, falcon or falcon-headed man. He often appears under the names Harakhte or Re-Harakhte ("Horus of the Horizon [*Akhet*]") and Hor-Em-Akhet or Harmachis ("Horus in the *Akhet*"). As with Re, the sun could be called the Eye of Horus.

**Aten**: the visible disc of the sun, depicted as such. It was not so much a god as the medium through which the sun's light comes into the world. It was the focus of the reforms of Akhenaten (see pp.132–3).

**Amun-Re**: the sun as the manifestation of Amun, the first and greatest of all the gods. This aspect is usually depicted as a man crowned with two tall plumes.

# "BEFORE TWO THINGS"

Egyptian speculation about the state of the universe before creation centred on the nature of the universal ocean that was thought to surround the created world. Like all natural phenomena, these cosmic waters were viewed as a god, which the Egyptians called Nu ("watery one", a masculine form of the word "Nut"; see p.118–19) or Nun ("inert one"). Prior to creation, the universe consisted only of Nu's waters: Egyptian texts describe this as the time when the creator "was alone with Nu ... before the sky evolved, before the earth evolved, before people evolved, before the gods were born, before death evolved". As the creation itself was viewed, in part, as the development of multiplicity out of an original oneness (see pp.126–7), the eternity preceding it was known as the time "before two things evolved in this world".

This pre-creation universe was the subject of speculation quite early in Egyptian history. Viewing it as the opposite of the known, created world, theologians codified several of its essential features, in a series of abstract concepts: wateriness (*nwj*), or inertia (*nnw*), the most basic qualities, enshrined in the names of the waters (Nu, Nun); infinity (*hhw*); darkness (*kkw*); uncertainty (*tnmw*, literally "lostness") or hiddenness (*jmnw*). These four qualities first appear as a group in the funerary Coffin Texts

### THE PRIMEVAL HILL

The first texts that deal with Egyptian ideas about the universe and its creation appear nearly a thousand years after the beginnings of recorded Egyptian history. For earlier concepts, we are dependent on pictorial and architectural images, and on what the later texts tell us these may have meant. One of the earliest notions seems to have been that of the primeval hill, the first "place" to emerge from the infinite waters, over which the sun first rose. It is tempting to see in this image a reflection of the environment experienced by Egypt's first settlers: watching the highest points of fertile land emerge as the annual Nile flood receded, these early farmers could easily have pictured the world gradually appearing in the same way at the creation.

Whatever its origins, the image of the primeval hill remained potent throughout Egyptian history. Some temples contained, in their sanctuaries, a mound of earth or sand evoking it. The tombs of Egypt's first dynasties were marked by a similar mound, promising a new creation and rebirth to those buried below it. The image of the primeval mound combines with powerful solar symbolism in the pyramids that housed royal burials from the Old Kingdom onward and also in the obelisks that graced Egypt's temples (see pp.170–71).

Like all features of the Egyptian world, the primeval mound was viewed as a divine force – in this case, a god called Ta-tenen, whose name means "Rising Land". It was also associated with the god Nefertum, who was depicted as a lotus, said in some accounts to have been the first life-form to appear after the primeval waters had receded. It was from this flower that the sun could blossom into the world.

ca. 2000BCE. In reliefs of temples of the Ptolemaic Period (323–30BCE), they are usually depicted as four pairs of gods and goddesses, whose names are masculine and feminine counterparts of each other: Nun and Naunet, Huh and Hauhet, Kuk and Kauket, Amun and Amaunet. Collectively, the eight deities are known as the Ogdoad (Greek for "group of eight"). The Ogdoad was venerated at Hermopolis, part of which was called "Eight-town" (Ashmun, modern el-Ashmunein) in their honour. Although the group as such first appears just before the Ptolemaic Period, it is probably much older: the name "Eight-town" goes back to the Fifth Dynasty (ca. 2500–2350BCE), and two of its divine pairs (Nun and Naunet, Amun and Amaunet) appear in the Pyramid Texts from ca. 2350BCE.

Together with the universal waters, the gods of the Ogdoad were thought to have existed before the creation. The theologians of Hermopolis viewed the qualities that they represented as a negative image of the created world. The pre-creation universe was watery, inert, infinite, dark and uncertain or hidden, in contrast to the created world, which was dry, active, limited, light and tangible. These contrasts formed a dynamic tension between the negative potentiality of the universe before creation and the positive reality of the created world. To the theologians of Hermopolis, this tension contributed to the inevitability of creation itself. As a result, the gods of the Ogdoad were venerated as creator-deities: "the fathers and mothers who were before the original gods, who evolved first, the ancestors of the sun".

*The creation of the world, from the Book of the Dead of Khensumose, a priest of Amun. On the first day of creation, the sun rises in three stages and finally appears above the horizon of the primeval mound (depicted as a circle), which is surrounded by waters dispensed by two goddesses associated with the North (right) and the South (left). On the mound itself are eight creator divinities – the Ogdoad – represented as figures hoeing the soil, symbolizing the first acts of creation. Third Intermediate Period, 21st Dynasty (ca. 1075–945BCE).*

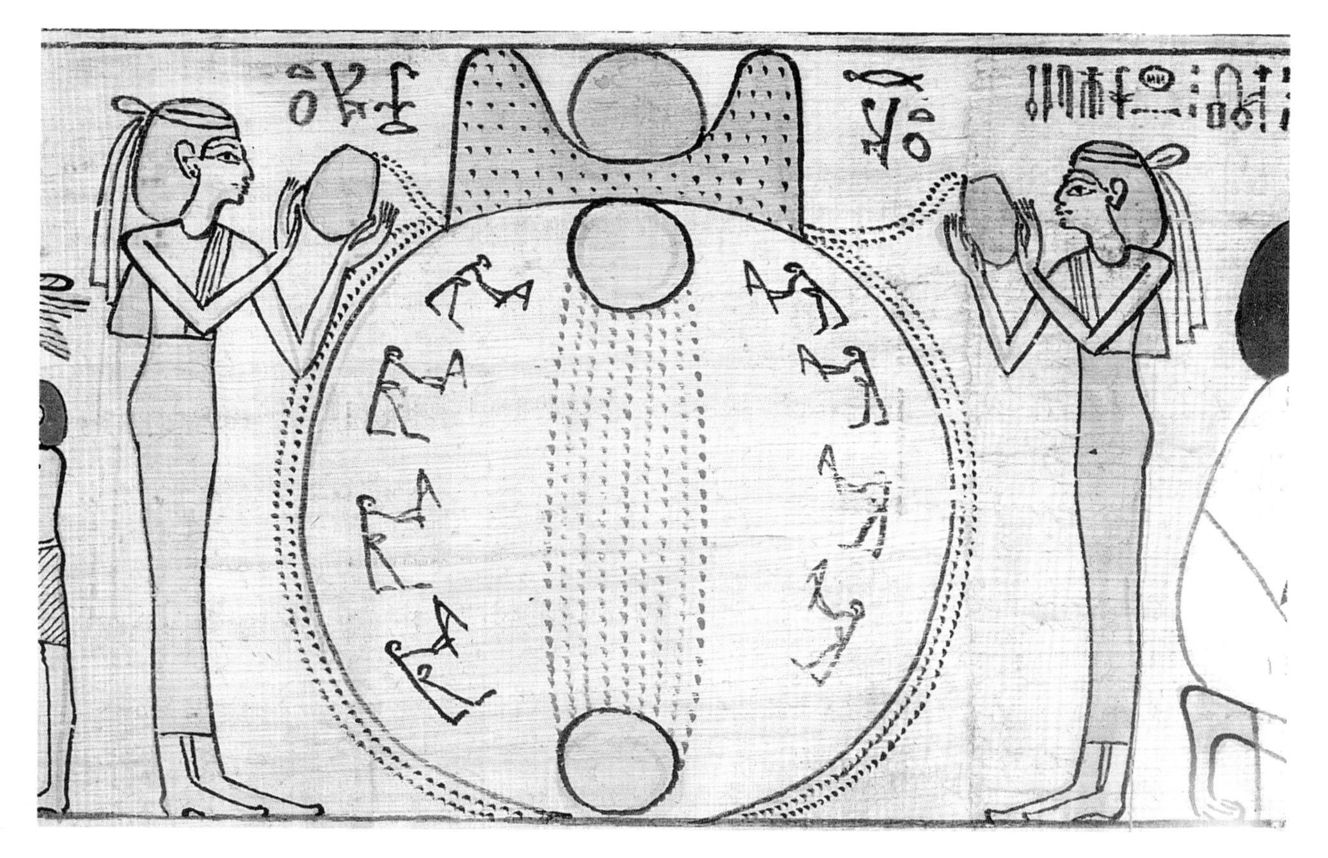

# THE ONE AND THE MANY

No single text or document preserves for us the full range of Egyptian thinking about how the world came into being. This is partly an accident of history: although ancient Egyptian theology remained remarkably uniform throughout the three thousand years of its existence, it did develop and become more complex as the centuries passed. The traditions of Egyptian theology itself also promoted diversity rather than uniformity. Temples throughout the country concentrated on different aspects of the creation story. These were not generally understood as competing theories (see sidebar, p.118), but the separate traditions have nonetheless resulted in a fragmented picture of ancient Egyptian thought.

One of the earliest, richest and most influential of these traditions arose in the city of Heliopolis, whose temple was devoted to the god Atum. Here, creation was viewed as an evolutionary process that has much in common with the "Big Bang" theory of modern physics. However, it was recorded in typical Egyptian metaphors of birth rather than in abstract scientific or philosophical terminology. The theologians of Heliopolis concentrated their attention on the problem of explaining how the diversity of creation could have developed from a single source. Their solution was embodied in the god Atum, whose name means something like "The All". Before creation Atum existed, together with the primeval

*The god Shu separates the sky, Nut, from the earth, Geb, who lies beneath her (see box, opposite). A scene from the Book of the Dead of Nesitanebtashru, daughter of Pinudjem I, high priest of Amun and king of Upper Egypt. Third Intermediate Period, ca. 1065–1045BCE.*

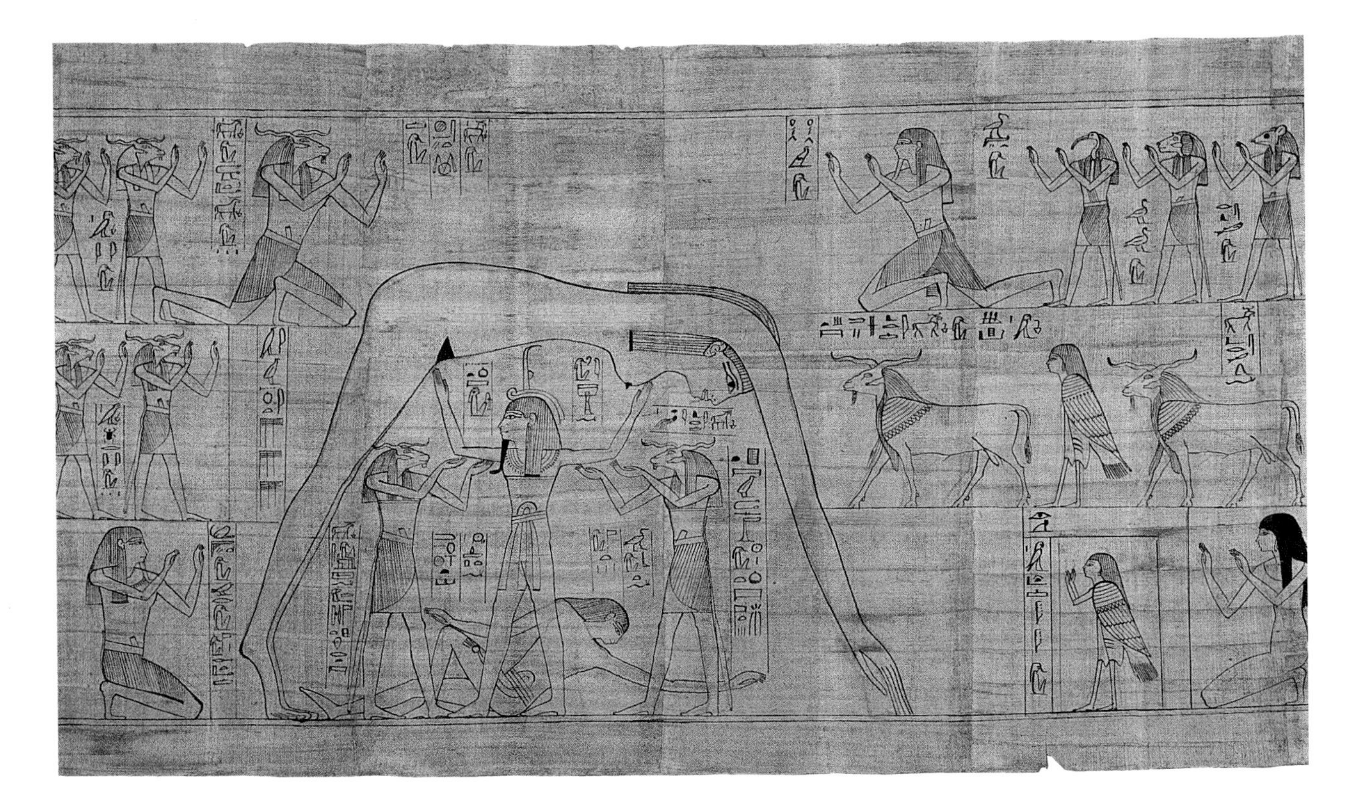

waters, in a state of unrealized potentiality – now recognized as being akin to the notion of a primordial singularity in modern physics. Egyptian texts describe this with the image of Atum "floating ... inert ... alone with Nu".

Creation occurs when Atum "evolves" from his initial state of oneness into the multiplicity of the created world. The first stage in this process is the evolution of a dry void within the universal waters. The void creates a space with the earth and sky as its limits. These, in turn, make possible the process of life in all its diversity, culminating in – and started by – the first sunrise into the new world. Although the process takes place in stages, it was probably envisaged as happening all at once. The texts reflect this thinking by describing the atmosphere as that which "Atum created on the day that he evolved". The process of Atum's evolution is described in concrete metaphors, beginning with him fathering his first "children", Shu and Tefnut (see box, below). But the final product of creation in all its diversity is in one sense nothing more than the ultimate evolution of Atum himself – a relationship reflected in his frequent epithets "Self-evolver" and "Lord to the limit".

## THE ENNEAD

In the "birth" metaphor used to convey the Heliopolitan theory of creation, Atum's evolution into the major elements and forces of the created world is described in generational terms as a group of nine gods, called the "Ennead", with Atum at its head. The first generation consists of Shu (meaning "void"), the atmosphere, and Tefnut, his female counterpart. This divine pair produces Geb, the earth, and Nut, the sky, who lie in close embrace until separated by Shu (see illustration, opposite). Geb and Nut, in turn, give birth to the two divine couples Osiris and Isis, and Seth and Nephthys: the four gods who embody the forces of life, birth and sexuality.

Although the Ennead consisted ideally – and probably originally – of these nine gods, the Egyptians at times seem to have understood the concept more loosely. The texts sometimes speak of an "Ennead" of as few as five or as many as twenty or more gods, or of several Enneads. Quite often, the traditional Ennead includes a tenth god, usually Horus (the son of Osiris and Isis) or Re (the sun). These numerical "aberrations" show that the Egyptians often considered the Ennead as a collective designation for the major cosmic gods rather than a specific group of nine deities. The term "Ennead" is sometimes written with the hieroglyph for "god" repeated in three groups of three. As one group of three is a common way of writing the simple plural word "gods", this may indicate that the Ennead was considered the "plural of a plural" – that is, it was seen to encompass all the gods.

**THE EVOLUTION OF ATUM**

**The story of the births of Atum's "children" is merely a metaphor to explain how the world's major elements and forces derived from a single source: like babies, they receive their life and substance from their parents. However, as Atum existed alone at the beginning, his first "evolution" must be self-generated. The texts explain this through the metaphors of self-impregnation or masturbation; eventually, Atum's hand came to represent his female counterpart.**

**These concrete images are used throughout Egyptian history to convey the Heliopolitan creation theory. However, a papyrus dating from the beginning of the Ptolemaic Period (323–30BCE) – but which may have antecedents as early as the Twentieth Dynasty (ca. 1190–1075BCE) – describes Atum's account of his evolution in a more abstract fashion: "When I evolved, evolution evolved. All evolution evolved after I evolved ... from Nu, from inertness. I surveyed in my heart by myself, and the evolutions of evolutions became many, in the evolutions of children and in the evolutions of their children."**

# THE WORD OF GOD

**THE DIVINE INTERMEDIARY**
**Where most texts are content simply to ascribe the powers of "perception" and "annunciation" to the creator, the theology of Memphis explores more fully the critical link between idea, word and reality – a link that it sees in the god Ptah. When the creator utters his command, Ptah transforms it into the reality of the created world, just as he continues to do in the more prosaic sphere of human creative activity.**

**This concept of a divine intermediary between creator and creation is the unique contribution of the Memphite Theology. It preceded the Greek notion of the demiurge by several hundred years; it had its ultimate expression in Christian theology a thousand years later: "In the beginning was the Word, and the Word was with God, and the Word was God" (John 1.1–2).**

Heliopolitan theology was concerned primarily with the material side of creation. Occasionally, however, Egyptian theologians dealt with the more fundamental question of means: how the creator's concept of the world was translated from idea into reality. Their solution usually lay in the notion of creative utterance (see box, opposite) – the same concept underlying the story of creation in the Bible ("God said: Let there be light"; Genesis 1.3). Some of the earliest Heliopolitan texts ascribe this divine power to Atum: they relate how the creator "took Annunciation in his mouth" and "built himself as he wished, according to his heart".

The link between concept and physical reality was seen differently by theologians in the city of Memphis: their chief god Ptah embodied it in the normal human activity of artistic creation. Crafts such as building and sculpture involve an initial concept in the artisan's mind; eventually, through the artist's skill, this concept takes shape as a finished building, or a statue carved from stone. To the Memphite theologians the link between the artist's concept and the ultimate transformation of his raw material was the force embodied by the god Ptah.

Ptah's theologians united the two concepts of craftsmanship and the creative word into a single theory of creation. The result was one of the more remarkable witnesses of human thought to survive from ancient Egypt. Written originally on papyrus or leather, probably no earlier than the reign of Ramesses II (ca. 1279–1213BCE), it has survived only because it was transferred to stone during the reign of the Nubian pharaoh Shabaka (ca. 716–702BCE; see illustration).

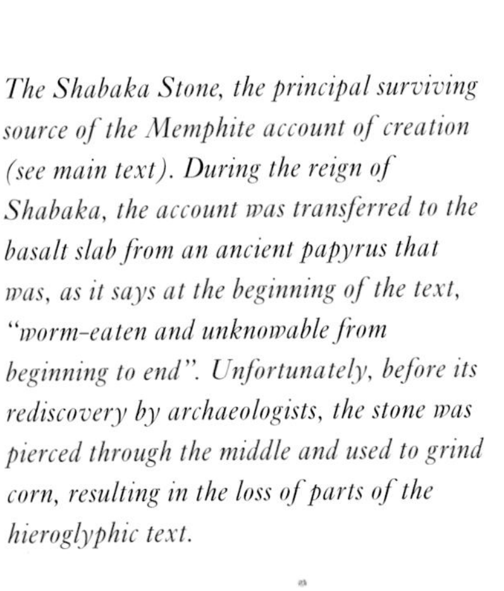

*The Shabaka Stone, the principal surviving source of the Memphite account of creation (see main text). During the reign of Shabaka, the account was transferred to the basalt slab from an ancient papyrus that was, as it says at the beginning of the text, "worm-eaten and unknowable from beginning to end". Unfortunately, before its rediscovery by archaeologists, the stone was pierced through the middle and used to grind corn, resulting in the loss of parts of the hieroglyphic text.*

The "Memphite Theology" makes a carefully reasoned connection between the processes of "perception" and "annunciation" on the human plane and the creator's use of these processes in creating the world. It ascribes the power behind Atum's evolution to the mind and word of an unnamed creator: "Through the heart and through the tongue evolution into Atum's image occurred." The word used to describe Atum's "image" is one that normally refers to reliefs, paintings, sculptures and hieroglyphs (called "divine speech" by the Egyptians). All these are "images" of an idea, whether pictorial or verbal: in the same way, the world itself is an "image" of the creator's concept. Atum, however, is only the raw material through which this image took shape: "Atum's Ennead evolved through his seed and his fingers, but the Ennead is teeth and lips in this mouth that pronounced the identity of everything ... So were all the gods born, Atum and his Ennead as well, for it is through what the heart plans and the tongue commands that every divine speech has evolved."

These passages reproduce, at a sophisticated level, the standard theology of creative utterance. The document goes on to link this concept with the action of Ptah, "who gave life to all the gods ... through this heart and this tongue ... So was made all construction and all craft: the hands' doing, the feet's going, and every limb's movement, according as he governs that which the heart thinks, which emerges through the tongue, and which facilitates everything ... So has Ptah come to rest after his making everything and every divine speech as well."

*Ptah, chief god of Memphis, patron of skilled artisans and the medium through which the creator's concept of the world became physical reality. From the tomb of Tutankhamun (ca. 1332–1322BCE) in the Valley of the Kings, Western Thebes.*

## IDEA AND REALITY

Like all ancient cultures, Egypt believed in the creative force of the spoken (and written) word. This power had two essential components: the formation of an idea in the mind (called "perception" – the Egyptians viewed this process as occurring in the heart rather than the brain), and the creative expression of that idea (called "annunciation"). Like all forces of nature, these were understood as divine: the gods representing them are often depicted accompanying the sun god in his barque. Perception and annunciation were primarily the property of the gods and the king, both of whom could "speak and it is done".

The link between annunciation and reality was usually seen as a third force, "effectiveness" or "magic". Possession of this power made the difference between a normal utterance and one that had true creative force. Because, like perception and annunciation, it was used by the creator in making the world, it is "older" than the other gods, "the one whom the Sole Lord made before two things evolved in this world".

# AMUN THE UNKNOWABLE

The creation theologies of Heliopolis and Memphis were each based on the pre-eminent Egyptian understanding of the gods as the forces and elements of the created world. Atum's evolution explained where these components came from, and the notion of creative utterance explained how the creator's will was transformed into reality. However, Egyptian theologians realized that the creator himself had to be transcendent, above the created world rather than immanent in it. He could not be directly perceived in nature like other gods. This "unknowability" was his fundamental quality, reflected in his name: Amun, meaning "Hidden".

The theology of Amun is most closely associated with the city of Thebes and its two great temples, Karnak and Luxor. Although Amun first appears in the Pyramid Texts from ca. 2350BCE, the full implication of his transcendence seems to be a later development, beginning perhaps as late as the New Kingdom (from ca. 1539BCE). Once Amun had been established as the greatest of all gods, his theology quickly assimilated those of the other religious centres, whose gods were seen as manifestations of Amun himself. As a result, Theban theology is better represented than any other major school of thought in surviving Egyptian texts.

A papyrus now in Leiden, written during the reign of Ramesses II (ca. 1279–1213BCE) and composed in a series of "chapters", is the most sophisticated expression of Theban theology. Chapter ninety deals with Amun as the ultimate source of all the gods: "The Ennead is combined in your body: your image is every god, joined in your person." Chapter two

*A view (looking west) of the great temple complex of Amun at Karnak, Eastern Thebes. (See also pp.208–9.)*

hundred identifies Amun, who exists apart from nature, as unknowable: "He is hidden from the gods, and his aspect is unknown. He is farther than the sky, he is deeper than the Duat. No god knows his true appearance ... no one testifies to him accurately. He is too secret to uncover his awesomeness, he is too great to investigate, too powerful to know." As he exists outside nature, Amun is the only god by whom nature could have been created. The text recognizes this by identifying all the creator gods as manifestations of Amun, the supreme cause, whose perception and creative utterance, through the agency of Ptah (see pp.124–5), precipitated Atum's evolution into the world.

The consequence of this view is that all the gods are no more than aspects of Amun. According to chapter three hundred: "All the gods are three: Amun, the sun and Ptah, without their seconds. His identity is hidden as Amun, his face is the sun, his body is Ptah." Although the text speaks of three gods, the three are merely aspects of a single god. Here Egyptian theology has reached a kind of monotheism: not like that of, say, Islam, which recognizes only a single indivisible God, but one more akin to that of the Christian trinity. This passage alone places Egyptian theology at the beginning of the great religious traditions of Western thought.

*Thutmose III (ca. 1479–1425BCE) before the enthroned figure of Amun-Re, the form of Amun assimilated with the sun god (see p.119). From a shrine close to the mortuary temple of Hatshepsut, Western Thebes.*

**"WONDERFUL GOD OF MANY EVOLUTIONS"**

**Despite being "unknowable", Amun is most persistently shown as a man crowned with two tall plumes. One of the earliest images of Amun, perhaps derived from the god Min, depicts a man with an erect penis, reflecting both creative power and "self-engendered" origin; the latter is embodied in the name Amun-Kamutef ("Bull of his mother"). Besides Min, Amun is most often assimilated to the sun god Re. The combined Amun-Re expresses the transcendence of Amun and his immanence in the sun. Amun-Re is often called "Lord of the thrones of the Two Lands" and "King of the gods", reflecting his supremacy. In animal association, Amun was linked with the ram (reflecting his creative force and his association with Re) and the goose (his role as sole creator).**

**Amun's image emerged several times a year in processions around Thebes, during which he could be approached for personal petitions. Stelae, ostraca, papyri and graffiti with prayers to Amun attest to his approachability; his state temple at Karnak was provided with a chapel specifically for private prayer.**

# THE HERESY OF AKHENATEN

*A painted limestone statuette of King Akhenaten and his chief queen, Nefertiti, in the more naturalistic style that appeared at the end of the king's reign. The couple's pose illustrates the intimacy that marked royal portraiture during this period. New Kingdom, Amarna Period, ca. 1340BCE.*

Egyptian religion remained remarkably uniform throughout its more than three thousand years of history – with one exception. For about two decades at the end of the Eighteenth Dynasty (the "Amarna Period"), the pharaoh Akhenaten (ca. 1353–1336BCE) promoted a view of the world that challenged the very foundations of Egyptian belief.

Akhenaten began his reign fairly traditionally, as Amenhotep IV. His earliest inscriptions, however, already suggest innovation, showing particular devotion to the sun god, who appears with a new epithet: "The living one, Re-Harakhte, who becomes active in the Akhet, in his identity of the light that is in the sun disc." By the king's fifth year, a radically new iconography had been introduced. Re-Harakhte's name and epithets were now inscribed in two cartouches, like those of the pharaoh, and his time-honoured image of a falcon or falcon-headed man was replaced by that of the rayed sun disc (the Aten). At the same time, the king changed his own name from Amenhotep ("Amun is content") to Akhenaten ("Effective for the Aten"). He also began construction of a new capital, called Akhetaten ("Place where the Aten becomes effective"), at the site now known as el-Amarna, midway between the traditional capital of Memphis, in the north, and the dynastic seat of Thebes, in the south.

Akhenaten's changes affected most aspects of Egyptian culture (see p.227) but none more profoundly than religion. In place of the hundreds of natural elements and forces that the Egyptians had always honoured as their gods, Akhenaten recognized only one: the supreme force of Light, which came into the world and gave it life every day through the sun disc. Unlike traditional deities, this god could not be depicted: the symbol of the sun disc with rays, dominating Amarna art, is nothing more than a large-scale version of the hieroglyph for "light" (𓇶).

Toward the end of Akhenaten's reign, two new pharaohs appear in the historical record, one or both of whom ruled alongside him. One of them, Smenkhkare, reigned until ca. 1332BCE and was married to Akhenaten's eldest daughter; the other (ca. 1336BCE onward) bears the name Nefernefruaten, which had been used by Akhenaten's chief queen Nefertiti (see p.90), and may have been Nefertiti herself. A graffito from Thebes, dated to Nefernefruaten's third year, contains a prayer to Amun, suggesting that Akhenaten may have bowed to political pressure by appointing a co-regent to rule in traditional fashion over the rest of Egypt while he restricted himself, and his new religion, to Akhetaten. Eventually, all three pharaohs were succeeded by Tutankhamun (ca. 1332–1322BCE), who abandoned Akhetaten and restored the older religion. Later generations repudiated Akhenaten's vision and sought to erase all memory of his reign by destroying his monuments throughout Egypt. Akhenaten himself was henceforth mentioned only as "the heretic of Akhetaten".

Much is still unclear about this remarkable period in Egyptian history, not least the reasons for its very existence. Some scholars have suggested that Akhenaten's reforms were the first true intellectual revolution in recorded history; others have seen them as a more prosaic attempt to curb the political influence of Amun and his temples. Both views have some validity, as religion and politics were essentially identical in the Egyptian mind. However, it is clear that Akhenaten's ideas were not accepted by most Egyptians. This may be partly owing to the powerful influence of tradition. But it also reflects a basic flaw in Akhenaten's theology itself: by promoting the sole divinity of a single force of nature, it denied the rich complexity of the traditional Egyptian world-view. More importantly, the impersonal nature of Akhenaten's god left his subjects no deity to whom they could directly relate. This, perhaps more than any other reason, is why Akhenaten's intellectual revolution was doomed to failure.

**TOLERANCE AND CENSORSHIP**

**Akhenaten tolerated the worship of the traditional gods until some time after his ninth year on the throne, when he began to establish the supremacy of his new religion. Reference to the older deities was avoided in reliefs and inscriptions and the name of Amun, the former supreme god, was erased, along with the word "gods", from monuments. The cartouches of the new god became more abstract: "The living one, sun of the *Akhet*, who becomes active in the *Akhet*, in his identity of the light that comes in the sun disc."**

*Akhenaten worships Light in this relief from an altar. Each beam of sunlight terminates in a hand proffering an* ankh (☥), *the hieroglyph for "life". The much smaller figure of Queen Nefertiti can be seen at bottom left. Amarna Period, ca. 1350BCE.*

# THE HUMAN SPHERE

**_MA'AT_: THE DIVINE ORDER**
**The Egyptian vision of an ideal world established at creation is summarized in the concept known as _ma'at_ ("order", "right"), which was personified as a female goddess. _Ma'at_ involved the control of differences rather than their elimination. On the cosmic level, creation itself – like the rising of the sun each day – was seen as the establishment of order within the chaos represented by the darkness of the pre-creation universe, and of night. For the king, _ma'at_ meant the responsibility of keeping order within Egypt and of holding Egypt's enemies at bay. And for the Egyptians themselves, _ma'at_ was both social justice and moral righteousness: the need for the powerful not to exploit the weak, and for all people to live in harmony with their environment – that is, with the gods – and with one another. (See also pp.148–9.)**

Whereas the biblical account of Creation culminates in the appearance of humankind, Egyptian creation ends with a cosmic event: the first sunrise, beginning the ongoing cycle of life. Where Egyptian myth refers to the creation of humanity at all, it is usually content with a passing play on words, deriving the origin of people (*rmt*) from the creator's tears (*rmyt*). For the Egyptians, the origin of living things was of less interest than that of the conditions needed for life to exist at all: the appearance of a dry void within the universal waters (Shu and Tefnut, Geb and Nut); the principles of sex and birth (Osiris and Isis, Seth and Nephthys); and the first sunrise to set these forces in motion. From the surviving documents, it seems that life itself, in all its diversity, was viewed as an automatic consequence of these cosmic events.

However, at least one text offers a more detailed look at the place of humanity. In Spell 1130 of the Coffin Texts, the creator recounts "four deeds ... I have made the four winds, so that every person might breathe in his environment: that is one of the deeds. I have made the great inundation, so that the poor might have control like the rich: that is one of the deeds. I have made every person like his fellow; I did not decree that they do disorder, but it is their hearts that break what I said: that is one of the deeds. I have made their hearts not forget the West." These "deeds" reflect the Egyptian vision of the ideal state in which the world was created – abundant resources, and humans at peace with one another, with their destiny ("the West", land of the dead) and with the gods.

Human beings were seen as the primary source of unrest in the world: "it is their hearts that break" the ideal order, *ma'at* (see sidebar, left), of

*The goddesses Ma'at (wearing on her head the feather that traditionally represents her) and Renpet before Osiris (left). The scene reflects the dependence of world order (*ma'at*) and time (*renpet*, literally "year") on the principle of daily rebirth (Osiris). From the cult temple of King Sety I (ca. 1290–1279 BCE) at Abydos.*

## TIME AND ETERNITY

The Egyptian language had two words for "eternity". One of these (*dt*) denoted the changeless pattern of existence; the other (*nhh*), its continual renewal. These words reflect the Egyptian view of time as both linear and cyclical. Much like a play, which has a single written form but can be performed a countless number of times in different ways, life was set in eternal motion at the creation but was continually created anew at every sunrise and in each generation of living beings.

Few Egyptian texts deal with the end of time. Those that do, however, envisage it as a cataclysmic destruction followed by a return to the state of the universe as it was before creation. In Spell 175 of the Book of the Dead, the creator god Amun describes how, after "millions of millions" of years, "I will destroy all that I have made: this world will return to Nu, to the limitless waters, like its original state." Following this apocalypse, only two cosmic forces will survive: "I and Osiris will be the remainder ... then I will come to sit with him in one place" (Spell 1130 of the Coffin Texts). According to Egyptian cosmology, Amun was the original source of all the world's elements and forces (*dt* eternity), while Osiris embodied the principle of daily rebirth (*nhh* eternity). The vision of their joint survival after the end of the world thus carries with it the promise of a new creation and the beginning of a new eternity.

*The limitless waters of Nu (the figure on the left) and the earthly Nile and Mediterranean (the figure with the blue rectangles, right). From the Book of the Dead of Ani, ca. 1300BCE.*

the creator. Recognition of this flaw in humanity gave rise to the most prolific and enduring of all Egyptian textual genres: the "wisdom" literature, ranging from the instructions of kings for their successors to prosaic advice on proper social behaviour. For Egyptian religion, the need to act "in accordance with *ma'at*" derived from the practical realization that any other behaviour resulted in disorder. It is probably for this reason that ancient Egypt, for most of its history, never developed a codified system of either civil laws or religious commandments.

The recognition that *ma'at*, with its primary application to humans, was instituted by the creator at the beginning of the world implies that the world was created for the benefit of humankind. This belief appears in one of the earliest "wisdom" texts, *The Instruction for King Merikare*: "Provide for people, the herd of God. For their sake he made the sky and the earth; for them he drove away the darkness of the waters; he made the air of life so that they might breathe. They are his likeness, who came from his flesh. For their sake he rises in the sky; for them he made the vegetation, small animals, birds and fish that feed them." And like a true father, the creator's care for his children did not end with creation: "When they weep, he hears ... for God knows every name."

● CHAPTER 10

# THE CULT OF THE DEAD

Few aspects of ancient Egyptian culture come more immediately to mind than the elaborate customs associated with burying the dead. Egypt's greatest surviving monuments, the pyramids of Giza, are tombs, and the concept of the living dead has spawned the popular twentieth-century idea of the "mummy's curse". However, to think of the Egyptians as a people obsessed with death is a misconception. Their funerary rituals were primarily concerned not with the pangs of death itself, but with the blessed continuation of one's earthly existence in a paradisial afterlife.

▲



ABOVE: *In this vignette from the Book of the Dead of the scribe Ani, the deceased and his wife stand before a table of offerings intended for Osiris, the god of the dead. Early 19th Dynasty, ca. 1290BCE.*

## ATTITUDES TO DEATH

While all societies must confront death, few – if any – have confronted it so directly or so elaborately as did the ancient Egyptians. Painstakingly embalmed mummies accompanied by costly gifts were sheltered in ingeniously devised tombs. Such exotic funerary practices dominate the present-day popular image of ancient Egypt as a morbid, death-obsessed society. However, despite such notions, Egyptian funerary religion was primarily life-affirming, with buildings, rituals and prayers designed to maintain an individual's life and status beyond the transition of death, which was regarded as an unpleasant necessity. In sharp contrast to the early Christianity that would replace it, Egyptian theology entailed neither rejection of earthly life nor willing martyrdom in the name of an ideal paradise. Rather, the Egyptian desired to continue his or her earthly life as far as possible after death – with personality, social ranking, family and even possessions intact – albeit with newly acquired divine status. Egyptian funerary religion failed only in the Roman Period, when daily life became so miserable for most of the native population that a desire to affirm it and to continue it after death became obsolete, and the old beliefs were replaced by a particularly fanatical and violent adherence to a faith that promised a paradisial hereafter.

Frequently termed "the enemy" of the living, death was considered a termination for enemies of the gods as well as for those who failed to provide proper cultic ritual. Even for the most virtuous and best-prepared, the transition of death was fraught with many dangers, and the spirit's survival depended on the deceased's knowledge of arcane theology and his or her command of potent magic spells. When the spirit left the body, it was thought to wander the pathways and corridors of the underworld in search of the Hall of Judgment of Osiris, lord of the West – the place of the setting sun (see pp.118–19) and thus the land of the dead. At each stage of the soul's journey, it risked destruction by hostile serpents and demons and by ferocious doorkeepers who yielded passage only to those who knew their names. Once it had arrived at the Hall of Judgment, the

*This scene from the painted coffin of the priest Djedhoriufankh, who lived ca. 925BCE, shows the god of mummification, Anubis, embalming the body of the deceased. Anubis was depicted as a jackal or a jackal-headed man, thus converting this animal from an ancient despoiler of graves into a guardian of the blessed dead.*

soul was obliged to name not only doorkeepers but doorbolts and floorboards as well. The perceived complexity of the underworld and its dangers (see pp.118–19) necessitated the production of funerary literature (Pyramid Texts, Coffin Texts, Books of the Dead ) to accompany the deceased and ensure his or her success (see pp.136–7).

Central to the Egyptian conception of the underworld was the notion of a divine tribunal presided over by the "great god", explicitly identified as Osiris from the later Old Kingdom onward. The soul was ushered into the presence of Osiris and his retinue of forty-two judges, and the heart of the deceased was weighed in a scale against the feather of Ma'at, goddess of truth, harmony and justice (see p.139). Only if the heart and feather were in perfect balance could the deceased attain the state of blessedness as an effective spirit, or *akh*. Some texts instruct such spirits to transform themselves into plant or animal forms. Moreover, all spirits could visit the living, whether as the ghostly *akh* or in the form of a winged *ba*, a being that was depicted with the head of the deceased and the body of a bird (see illustration, p.143).

The deceased person became a distinct aspect of the god of the underworld, and was formally addressed as the "Osiris [name of the deceased]". Through this merger, he or she attained divine status and powers, while retaining an individual human personality. In exceptional cases the revered dead might acquire the status of "sainthood", with formal cults and even temples. Within individual families, important forebears were often the focus of ancestor cults, and a deceased relative might regularly receive "letters" from surviving family members (see pp.142–3).

*A* shawabti, *a mummy-like servant figurine, from the tomb of Ptahmose, royal vizier during the reign of Amenhotep III (ca. 1390–1353BCE). In Predynastic times real servants were sometimes buried with their masters, but human sacrifices soon gave way to stone servant figures and, from the First Intermediate Period, wooden models. By the New Kingdom, people were buried with the all-purpose* shawabtis *("persea-wood figures"), later called* ushebtis *("answerers"). In the afterlife, a* shawabti *was thought to come to life when called and perform any task demanded of it.*

# THE THEOLOGY OF DEATH: ISIS AND OSIRIS

**THE BRINGER OF LIFE**
**Although commentators have often compared Osiris to the "dying and rising" gods common to the ancient Near East, Osiris is not resurrected from the underworld. Rather, he is a deity who creates life from death in the depths of the earth, being thus the force that is inherent in sown grain and rising flood waters. These aspects of Osiris are incorporated in funerary cult through the use of the "grain Osiris", a box in the shape of the god, filled with earth and seeds that germinate in the tomb.**

Few Egyptian myths are as well known, or have had such an impact on Western speculation, as the cycle of tales regarding the salvation deities Isis and Osiris. References to the actions of these gods abound in surviving Egyptian hymns, prayers and funerary literature, yet perhaps because their story was so familiar to a native audience, it is the Greek adaptation of Plutarch (*De Iside et Osiride*) that preserves the longest exposition of their myth, composed some 2,500 years after the formation of the cult.

The origins of the Osirian cult well precede the first mentions of the god's name. Ritual imagery later associated with Osiris has been recovered from the First Dynasty, while the god's epithets and connection with the holy site of Abydos derive from a fusion with the early funerary jackal deity Khentimentiu, "Foremost of the Westerners". First securely attested in the Fifth Dynasty (ca. 2350BCE), Osiris is a central figure in the mythological tradition associated with the prominent cult centre of Heliopolis ("On" of the Bible).

As members of the "Ennead", or first nine gods, Isis and Osiris were two of the five siblings (along with Seth, Nephthys and Horus the Elder) born on successive days to Nut, the goddess of the sky, and Geb, the god of the earth. As Geb's eldest son, Osiris attained kingship of the earth, and married his sister Isis, whom he had loved even in the womb. His brother Seth, in a loveless marriage to Nephthys, coveted the throne and schemed to obtain it by stealth. In the classical rendition, the unsuspecting Osiris was betrayed at a grand feast for the gods, where Seth offered a novel object – a coffin – as a "party favour" to whomever it should fit. Although various gods sought to claim the prize, the coffin had been carefully made to fit Osiris alone. Once the god was securely inside, Seth and his confederates promptly sealed the coffin and cast it into the Nile. Osiris drowned, and death was introduced to the world. With much labour, Isis then sought and retrieved the body of her slain husband, but Seth again seized the corpse and cut it into many pieces. These he scattered across Egypt, so that each province could later claim a relic and shrine of the deceased god.

*Horus and Isis flank the squatting figure of Osiris, who is depicted with the facial features of King Osorkon II. A gold pendant of the Third Intermediate Period, reign of Osorkon II (ca. 874–835BCE).*

In company with her sister Nephthys, Isis sailed through the marshes or flew as a kite in search of the scattered parts, and at length they reunited the dismembered body of Osiris with the aid of Anubis, the god of mummification. While still a corpse, Osiris was reinvigorated through the magical abilities of Isis, so that she conceived a son and heir to the throne, Horus the Child. The orphan Horus is assaulted repeatedly by Seth and his emissaries, but through the protection of Isis he is cured of all injury. The image of the wounded Horus became a standard feature of healing spells, which typically invoke the curative powers of the milk of Isis. To be effective, these spells require recitation over "milk of a woman who has borne a male child", an ingredient adopted by Western folk medicine and used as late as the fourteenth century CE.

In an extensive series of contests and trials well documented in literature and art, Horus and Seth compete for the vacant throne of Osiris, until Horus is ultimately victorious and Osiris avenged. In one of these many contests, Horus lances Seth, who has taken the form of a ferocious hippopotamus or crocodile. This particular contest forms the subject of an elaborate dramatic re-enactment ("The Play of Horus") at the sacred lake of the Ptolemaic temple of Edfu in what is arguably the earliest example of ritual theatre.

Following the conquest by Alexander the Great in the fourth century BCE, Isis was formally adopted as the patron of Ptolemaic Egypt, and temples and iconography dedicated to the Osirian cycle spread throughout the Mediterranean region. Although it was initially suppressed by Rome, this Hellenized mystery cult of Isis was welcomed by Caligula and rapidly became a primary religious force across the Roman empire, with significant influence on contemporary and later cults (see p.57).

*A Late Period bronze and gold statuette of Isis suckling the infant Horus, ca. 600BCE. Such images may have been the prototype of Christian depictions of the Virgin and Child.*

## THE EGYPTIAN CULT OF THE DEAD AND CHRISTIANITY

Christianity, which was traditionally brought to Egypt in the late first century CE by the evangelist St Mark, spread initially among the Greek-speaking élite of Alexandria and other cities, but began to be adopted by non-élite Egyptians from ca. 200CE. Egyptian Christians, or Copts, used a Greek-based alphabet (see pp.232–3) and were vitriolic in their denunciations of Egypt's ancient pagan beliefs. However, a number of elements appear to have been carried through from the old cult of the dead and from Osirian myth into the iconography and mythology of the new faith.

As depicted in schematic "Guides to the Underworld" commonly illustrated in New Kingdom royal tombs, some of those who failed the judgment of Osiris would be thrown into hellish pits and tortured. When early Christian monks came to inhabit the abandoned tombs, these depictions were surely influential, and Coptic visions of the domain of Satan (which was styled *Amente*, "the West", after the realm of Osiris) retain the ancient doorkeepers with their animal heads and threatening knives.

Some late representations of one of the contests between the god Harpocrates (Horus the Child) and Seth show Horus as a mounted Roman warrior spearing Seth in the form of a crocodile. These are considered to have been the inspiration for subsequent icons of St George and the Dragon. Equally, the widely disseminated cult image of the goddess Isis suckling Horus the Child (see illustration, above) probably serves as the artistic precedent for later Christian representations of the Virgin nursing the Child.

Other striking survivals from earlier Egyptian religion include the Coptic representation of the Christian cross as something very close to the ancient *ankh* (☥), the hieroglyph for "life", and the depiction of a worshipper with hands uplifted in adoration, a common motif on pre-Christian monuments of the pharaonic and later periods.

# FUNERARY LITERATURE

**"MAGIC" AND MORTUARY CULT**
**All mortuary texts were designed to ensure a positive outcome and so may be classed as "magical" compositions. In Egyptian thought, "magic" (*heka*) was a legitimate divine force produced by the creator for the benefit of humankind. The use of funerary magic is therefore appropriate, although a "Dives and Lazarus" episode in the second demotic tale of Setna Khamuas from the Roman Period teaches that the wealthy and ritually prepared may yet be damned, while the righteous poor are vindicated.**

In the complex world of the afterlife, the deceased was equipped with a series of hymns and actions that acted as magical spells for his or her protection and rejuvenation. First codified solely for the benefit of the king, the earliest collections of such spells were inscribed inside Old Kingdom royal pyramids from the reign of Unas (ca. 2350BCE). Comprising some eight hundred incantations in total, these "Pyramid Texts" appear in varying numbers in the pyramids of nine kings and queens of the Sixth to Eighth dynasties. It is in these texts that Osiris first appears clearly as lord of the dead, patron and partner of the dead king (see pp.110–11).

With the breakdown of central authority at the end of the Old Kingdom and the establishment of powerful local rulers (nomarchs) during the First Intermediate Period (ca. 2130–1938BCE), the promise of immortality, originally extended only to the king, became available to ever wider social classes. During this period and through the Middle Kingdom, expanded collections of funerary incantations (numbering in total more than eleven hundred spells) were copied onto the interior surfaces of coffins, and sometimes onto tomb walls and ceilings. These "Coffin Texts" included new "guide books" to the underworld that described and illustrated the paths of the wandering spirit. In the New Kingdom, the Coffin Texts were replaced by collections of spells copied onto long papyrus scrolls and illustrated by painted vignettes. These papyri were quite costly, but were available to anyone who could afford them. Popularized in Western literature as the Egyptian "Book of the Dead", they were known to the Egyptians as the "Book of Going Forth by Day". Not standardized until the end of pharaonic rule, the scrolls vary greatly in the number and sequence of their spells, depending upon local custom and the example available to the copyist. The spells number just under two hundred, but this does not represent a decrease from earlier collections, as the individual texts are frequently combinations of several early sources.

*Part of the Book of the Dead of Pinudjem I, high priest of Amun at Thebes, who is shown in the vignette on the left adoring the god Osiris. Pinudjem is depicted in the costume of a pharaoh, because at this period (ca. 1065–1045BCE) the Theban high priests had declared themselves kings of Upper Egypt (see pp.38–9).*

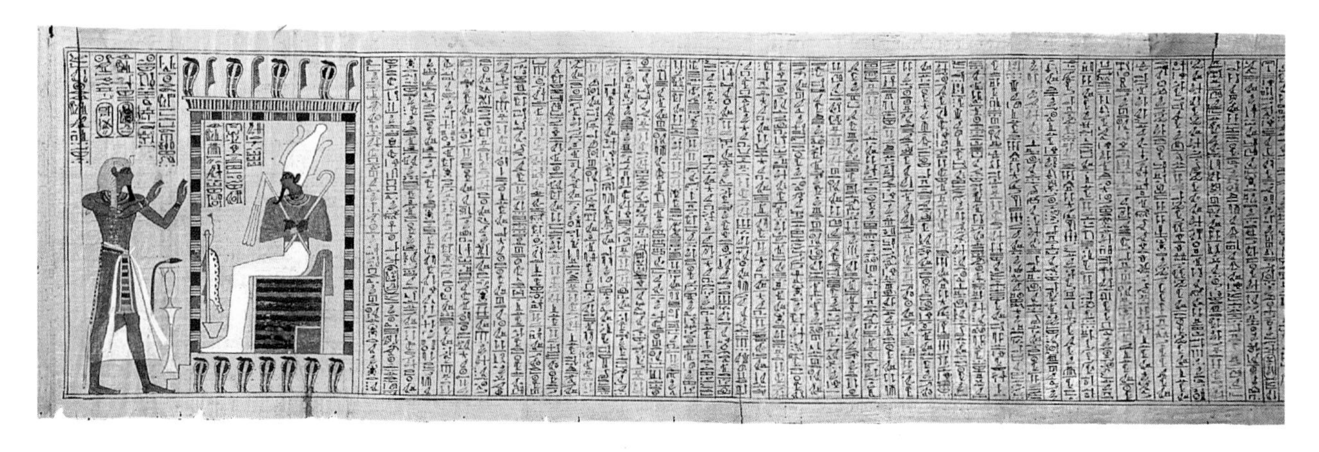

The Book of the Dead continued in use into the Roman era in company with several newly devised compositions of the Late Period. Most prominent among these late texts are the "Book of Breathings", perhaps originating in the Saite age, and the Ptolemaic "Book of Traversing Eternity". The former derives from the Book of the Dead, and includes a "Negative Confession" (see box, below) and recitations by Isis. In the early Roman Period, an additional section was appended with spells for the preservation of the owner's name. The "Book of Traversing Eternity" consists of a prolonged address to the deceased, who is assured of his powers to visit Osiris as well as the earthly cult centres and festivals of Egypt. Dating to the first or second century CE, the "Ritual of Embalming" is one of the last Egyptian mortuary compositions, although its spells are traditional.

## THE WEIGHING OF THE HEART

The most important of the "core" spells that recur consistently in different versions of the Book of the Dead is Chapter 125, which details the "weighing of the heart" of the deceased against the feather of Ma'at, the goddess of truth. Before Osiris and a panel of judges, each representing an Egyptian nome (province), the deceased denies a series of offences (the "Negative Confession"). Next, the heart – centre of thought, memory and personality – is weighed in a balance by the god Anubis, while the divine scribe Thoth records the verdict. If heart and feather are of equal weight, the deceased is declared "true/justified of voice" and accorded a portion in the domain of Osiris. He or she might also join the sun god in his celestial circuit, or dwell among the circumpolar stars. Miscreants faced annihilation by the "Swallowing Monster", a hybrid of crocodile, lion and hippopotamus, that crouches by the scale.

*The judgment scene from the Book of the Dead of the royal scribe Hunefer (ca. 1285 BCE). From left: Anubis brings Hunefer into the judgment hall; his heart is weighed and Thoth notes the favourable result; Horus conducts Hunefer into the presence of Osiris.*

**THE ORGAN OF THE SOUL**
**The heart, considered the seat of reason, emotion, memory and personality, was the only major organ intentionally left in the body during mummification. A "heart scarab" placed on the mummy was inscribed with a spell that sought to secure the heart's silence regarding past transgressions during the ritual of the "weighing of the heart" (see p.137). Of those who failed the weighing rite, the most reprobate might face hellish torture pits. Usually, however, the heart was cast to a "Swallowing Monster", which devoured it, thereby erasing the deceased's personality. For the Egyptians, this irrevocable loss of self was the ultimate horror.**

*The mummy mask of a princess, made of gilded* cartonnage *(layers of linen stiffened with plaster). Early New Kingdom, ca. 1500BCE.*

# MUMMIFICATION

In Egyptian belief, the preservation of the corpse was fundamental to the continuation of life after death. The process of mummification evolved over the millennia from the natural desiccation of corpses buried in shallow pits hollowed from the desert sands to the intricate wrapping of prepared corpses in yards of linen and the addition of portraits of the deceased.

The earliest artificial technique for the preservation of the corpse dates to the late Predynastic Period, when it began to be interred in brick- or wood-lined tombs, enclosed within reed or wooden coffins. Separated from the drying effect of the sand, the body was wrapped in resin-soaked linen bandages, one of the first methods by which natural decay was retarded. By the Third Dynasty, attempts at greater naturalism entailed padding the wrapped body to retain lifelike proportions. True mummification was devised only in the Fourth Dynasty, with the discovery of a process of desiccation by natron, a naturally occurring compound of sodium carbonate and sodium bicarbonate, often in combination with sodium chloride (salt). With slight modification, this process served as the basis of chemical mummification for the next three millennia.

Egyptian records are generally reticent about the embalmer's techniques, but details of the procedure are preserved in the Greek writings of the historians Herodotus and Diodorus Siculus, and in the first demotic tale of Setna Khamuas (Ptolemaic Period; see p.147). According to these sources, the standard period of embalming lasted seventy days. The first step was the evisceration of the corpse, with the surgical extraction of the lungs, liver, stomach and intestines. These viscera were desiccated and wrapped separately, then placed in a container, the first example appearing in the Fourth Dynasty. Shortly thereafter, the organs were placed in individual vessels now termed "canopic jars". The heart remained within the corpse (see sidebar, left). Evisceration involved the removal of those organs particularly subject to decay according to Egyptian conceptions of natural disease. *Wekhedu*, the detritus of unabsorbed food, was thought to clog the internal vessels to produce disease and ageing, and was believed to cause the corpse to decompose. Among the less expensive treatments to eliminate *wekhedu* were injections to dissolve the infected organs and purges of the bowels.

After evisceration, the corpse was packed internally and externally with dry natron for a period of forty days to complete desiccation. Thereafter, the body was washed, the internal cavity packed with resin and linen, and the whole corpse carefully wrapped in hundreds of yards of fine linen bandages. Facial features were restored to the mummy by painting, by applying a coat of moulded plaster, or, from the First Intermediate Period, by the addition of a separate funerary mask.

Variations on this standard scenario did occur. From the Middle Kingdom to the Ptolemaic era, the brain was pierced through the nasal cavity and drained off. Placement of the embalmed viscera was subject to fashion: they were put back into the body in the Twenty-First Dynasty, again stored in canopic jars in the Twenty-Sixth Dynasty, and placed in packets between the legs thereafter. In the Roman era, the funerary mask was replaced by realistic painted wooden panels known as "Faiyum portraits" (see pp.226–7). Mummification survived even the initial conversion of Egyptian society to Coptic Christianity, although by the fourth century CE its pagan associations led to official denunciation and discontinuation.

## PALEOPATHOLOGY

The often remarkable preservation of Egyptian mummies has spurred the development of a recent branch of medical history: paleopathology, the study of ancient health and disease. Although early studies of mummies entailed destructive autopsies, newer non-invasive techniques are now preferred, including CT scans and X-rays. From such examinations, a clearer picture of Egyptian pathology has emerged, with occasional evidence of such diseases as pneumonia, tuberculosis, smallpox and poliomyelitis, and frequent incidence of parasitic disease such as bilharzia (liver-fluke). An autopsy on the mummy of a fourteen-year-old girl at Manchester University, England, in 1975 revealed that both her legs had been amputated shortly before her death, giving rise to the theory that she had been the victim of a Nile crocodile.

Only eight cases of cancer have been identified, while ten to twenty percent of adult mummies show signs of arteriosclerosis. Studies of the teeth have revealed few cavities, but they generally show intense wear that had often led to abscesses. The probable source of this common abrasion was grit, unknowingly incorporated into bread as a result of using millstones to grind flour.

An examination of the mummies of some of Egypt's kings has thrown fascinating light on their medical conditions. Ramesses II (ruled ca. 1279–1213BCE), who probably lived to nearly ninety, suffered from hardened arteries and arthritis, common problems even today among those advanced in years. X-rays revealed that, doubtless like many of his subjects, the red-headed pharaoh also had dental abscesses and worn molars.

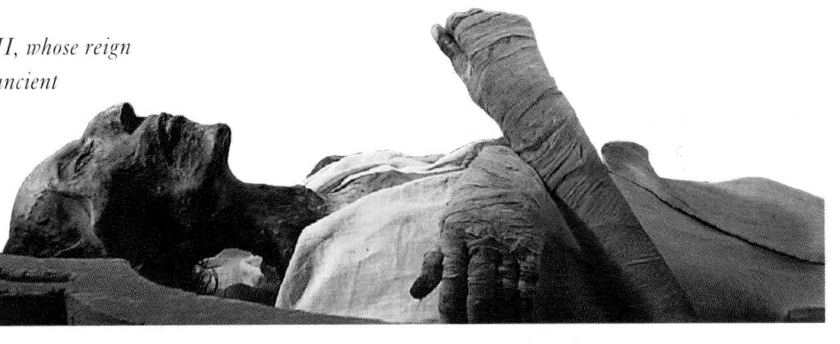

*The well-preserved mummy of King Ramesses II, whose reign (ca. 1279–1213BCE) was one of the longest in ancient Egyptian history. An examination of his body (including X-rays) during conservation work in Paris in 1975 revealed that, unsurprisingly, the king had suffered from a number of ailments associated with advanced old age – Ramesses II was almost certainly in his late eighties when he died.*

# THE FUNERARY CULT

In the belief that the dead, like the living, required continuing care in the afterlife, an Egyptian would have begun to make careful preparations for his or her future existence well before death. These preparations included not only the construction of a tomb and the acquisition of its contents, but the endowment of a mortuary cult designed to perpetuate the owner's name among the living and his divine status among the dead. The funerary cult that developed from the very earliest Predynastic burials, which were accompanied by offerings of food and personal possessions, closely paralleled (on a reduced scale) the great temple rites performed on behalf of the gods. As in a temple, the ritual actions of hymns, prayers and offerings required a priestly staff attached to the cult and a continuing source of income to fund both personnel and offerings.

*The painted wooden funerary stela of the priestess Deniuenkhons, who is depicted adoring the falcon-headed sun god Re-Harakhte-Atum. She stands next to a table heaped with food offerings to the god. Twenty-Second Dynasty, ca. 945–712BCE.*

The staff, known as "*ka*-priests" or "servants of the *ka*-spirit" of the deceased, varied in number according to the wealth of the donor. After the New Kingdom, these priests were often termed "water-pourers", in reference to their most common function of pouring water for the refreshment of the dead. Ideally, the office of *ka*-priest was performed by the eldest son and heir of the deceased, echoing the services undertaken by Horus on behalf of his slain father Osiris. Other family members might also participate, so that the cult provided occupation and some financial security for descendants. The funerary cults of the nobility were far more extensive, with many priests bound to the cult by signed contracts of service. The financial support for these cults was derived from the endowments of agricultural "mortuary estates", tracts of farmland that provided income for the cult as well as produce for the altar. Food remained on funerary altars only until the spirit was felt to have obtained its benefit. Thereafter, it served as part-payment for priestly service.

*In this wall painting from the 20th-Dynasty tomb of Inerkha in Western Thebes, a priest wearing a mask of Anubis performs the "Opening of the Mouth", the last funeral rite before final offerings were made and the body was entombed. The priest touches the mummy's mouth, eyes and ears with an adze to ensure that the deceased will be able to use them in the afterlife. At more ostentatious funerals, the ceremony came after a great procession of the corpse and grave goods to the tomb, accompanied by paid mourners, purification rites and ritual dancers.*

Over time, funerary cults invariably lapsed, due either to family extinction and the lack of *ka*-priests or to the redistribution of land and the loss of financial support. The Egyptians devised a supplementary system to ensure, by the magic of image and word, that the deceased would be provided with the full range of necessary items. Actual menus are often inscribed beside the altar within the tomb chapel, in association with the standard funerary prayer. The elaborate decoration and inscriptions of the open chapel were intended to entice visitors, who might leave offerings, pour out water, or recite the funerary prayer, thereby acting themselves as *ka*-priests and extending the life of the cult. Simply uttering the name of the dead person in prayer was sufficient to guarantee his or her continued existence.

## "AN OFFERING WHICH THE KING GIVES"

If any prayer from Ancient Egypt could be considered as a culturally unifying "scripture", it is surely the standard funerary prayer, which appears over some two thousand years from the early Old Kingdom through to Hellenistic times. Opening with the phrase *Hetep di nysut*, "An offering which the king gives ...", the prayer reflects the notion that only the king might make divine offerings; in practice, such offerings were made in his name. In typical versions of the prayer, the offerings are said to be made to, or in company with, the funerary deities Osiris and Anubis, joined on occasion by local gods. This corps of divinities in turn transmits the offerings to the cult and spirit of the deceased, in a practice known as "reversion of offerings". Royal offerings first exposed on temple altars were subsequently removed and used in the mortuary cults of individuals affiliated with that temple. The prayers, styled "invocation offerings" ("the going forth of the voice"), request for the dead person's *ka*-spirit a thousand each of loaves of bread, jugs of beer, oxen, fowl, alabaster vessels and bolts of cloth, as well as "everything good and pure on which a god lives".

# COMMUNICATION WITH THE DEAD

By virtue of their justification in the underworld tribunal of Osiris, the beatified dead were believed to serve as intermediaries between their surviving relatives and the court of the gods. Now equipped with divine powers themselves, these spirits (*akh*s) retained their personal interests, allegiances and family bonds, and could be swayed by petitions and prayers. Within individual families, particularly prominent ancestors sometimes became the focus of a formal "ancestor cult", as documented by preserved busts from New Kingdom household shrines (see p.152). In most cases, however, interaction between the living and the dead would have been more casual, with spoken prayers that have left no trace. Nevertheless, a wide variety of entreaties does survive in a small number of documents known as "Letters to the Dead", attested from the Old Kingdom through to the Late Period.

Whether inscribed on pottery bowls, linen or papyrus, these documents take the form of standard letters, with notations of addressee and sender and, depending upon the tone of the letter, a salutation: "A communication by Merirtyfy to Nebetiotef: How are you? Is the West taking care of you as you desire?" (Such questions about the condition of the deceased are probably not simply rhetorical, since spells 148 and 190 of the Book of the Dead contain elaborate rituals enabling the deceased spirit to "make known to you what fate befalls it".) Deposited at the tomb, the letters were typically written on bowls probably filled with offerings. The pious (and hopefully persuasive) effect of the offering is coupled with petitions that are variously anxious, cajoling or even indignant. In all cases, the deceased is urged to take action on behalf of the writer, often against malignant spirits who have afflicted the author and his or her family (see pp.144–5). Such requests frequently refer to the underworld court and the role of the deceased within it: "you must instigate litigation with him since you have witnesses at hand in the same city [of the dead]". The principle is stated succinctly on a bowl in the Louvre in Paris: "As you were one who was excellent upon earth, so you are one who is in good standing in the necropolis". Despite this legalistic aspect, the letters are never formulaic, but vary in content and length, and may be by turns obsequious, chatty or chiding.

The concerns to which Letters to the Dead usually give voice include matters of family property and inheritance, personal guilt, the fertility of spouses and

**LETTERS TO THE DEAD**

**Letters addressed to the dead were deposited in the tomb or coffin. An example now in the Louvre, Paris, greets the "noble chest of the Osiris [name of deceased], who lies at rest beneath you, hearken to me and transmit my message". The living did not, however, anticipate any correspondence in return. Rather, the dead was expected to become manifest in a dream, performing the desired task. Such dream communications represent an early form of what is termed "incubation", in which the petitioner seeks a cure through the dreams that come to him or her while sleeping at a shrine. In Hellenistic Egypt, such incubation was common at shrines of the deceased Imhotep, the 3rd-Dynasty architect later deified as a god of healing (see p.178).**

*An artwork of a "letter" to a dead relative, written in hieratic script on a simple pottery bowl formerly in Berlin. The addressee is depicted in the centre of the bowl, which would have been filled with food offerings and left in the tomb. The bowl, one of only 20 "letters to the dead" so far discovered, was destroyed during the Second World War.*

*Egyptians believed that the dead could fly between their tombs and the world of the living in the form of a* ba, *a creature with the body of a bird and a human head bearing the features of the deceased. In this vignette from the Book of the Dead of the scribe Ani (ca. 1290BCE), the* ba *leaves Ani's body.*

daughters and affliction by ghosts – including the addressee. To gain the sympathy of the recipient, writers often alluded to past instances of kindness or diligence, as when a son reminds his mother on the Kaw Bowl that she once desired seven quails to eat and he dutifully obtained them for her. Would she now allow him, her ideal son, to be injured in her presence? If so, he chides, there would be none to pour water for her and her funerary cult would expire. Elsewhere, authors protest their innocence of wrongdoing, including their careful avoidance of "garbling" any spell during the funerary ritual. While the requests made to the dead are most often defensive in nature, asking the ghost to litigate or fight on the author's behalf, the spirit's more creative powers are invoked in matters of fertility (see p.85). On a jar-stand in Chicago, the writer requests that the deceased "let a healthy son be born to me, for you are an able spirit".

Less positive aspects of the deceased are found in the accusations written by troubled survivors. In a Nineteenth-Dynasty papyrus at Leiden in Holland, a widowed husband details at length his devotion to his dead wife. But she is of an "evil disposition", "disregarding" how well she was treated, and refusing to allow her husband's "mind to be at ease". In a repeated refrain, the husband demands: "What have I done against you?"

# GHOSTS AND EXORCISM

**"BREAKING THE RED POTS"**
**Before the appearance of execration figurines (see illustration, opposite) there existed the funerary rite of "Breaking the Red Pots", designed to repel or destroy enemies of the tomb owner. Described in the Pyramid and Coffin Texts, the destruction of pots concluded the offering meal, perhaps using the vessels that had just contained the meal. At a later period in the execration ritual, the use of such pots continued alongside that of figurines that were more realistic representations of the victim. A possible borrowing of the Egyptian practice is to be seen in the Bible (Jeremiah 19.1–11), in which pottery is ceremonially broken by priests to curse political enemies.**

If the blessed dead became the recipients of cult and correspondence, less favoured spirits were widely feared for their potential destructive wrath. Medical spells frequently cite the unholy dead as the enemy afflicting the patient, the source of illness and disease. As is evident from the "Letters to the Dead", even favoured spirits might inflict injury when angered, and petitioners were quick to beg for leniency and favour. Threatening to all – both pharaoh and commoner alike – the danger that these malignant spirits posed required some form of response, and a series of rituals was devised for their suppression.

Individuals who were thought to harass the living after death could be overcome by attacking their tomb, images or name. The defacement of the image or name constituted a direct attack upon the spirit's existence, intended to kill the ghost in the underworld. This motivation lay behind the well-known state-sponsored defacement of figures of the heretical pharaoh Akhenaten (see pp.128–9).

The most explicit ritual for the destruction of hostile ghosts appears in the collections of inscribed pots and figurines collectively known as the "Execration Texts". Now numbering in excess of one thousand examples,

*On this gateway in the temple of Hathor at Dendera, images of the king and the gods have been carefully chiselled out, probably by early Christians, in the belief that this would destroy their power and "exorcise" the pagan temple.*

*Destruction of a person's tomb deprived the deceased of cultic ritual and support, thereby restricting or eliminating his or her power. This previously unpublished photograph from the tomb of King Ay (ca. 1322–1319BCE), the successor of Tutankhamun, in the Valley of the Kings, Western Thebes, shows images of Ay hunting in the marshes that have been deliberately defaced, as has the king's name. Like all the immediate successors of Akhenaten (ca. 1353–1336BCE), Ay may have been reviled because of his association with the discredited Amarna regime (see pp.128–9).*

these cursing formulae date from the Old Kingdom through to the Late Period. Although the texts and rituals described by these texts vary widely, the standard pattern of the exorcism is clear: a "rebellion formula" listing the names of Egypt's potential enemies is inscribed on a series of red pots or figurines, which are subsequently broken, incinerated and buried. Although this is obviously a state ritual, there seems to have been some local input on textual choices. Most sections of the formula list the names of living rulers of lands neighbouring Egypt, based on information that was surely provided by the royal chancellery. To these sections, however, is appended a list of Egyptians, all qualified as "dead", who also pose a threat. A number of these individuals seem to have been implicated in harem conspiracies (see p.89) and other crimes against the state. In certain instances, the same individuals are also known to have suffered the defacement of their tomb images. Some exorcised ghosts, however, may have been personal enemies of the sponsors of the state rite. Supplementing these "private" curses are the many finds of individual figurines naming specific deceased Egyptians and their families.

The hostile actions of the execration ritual were intended to destroy the malign spirit by breaking, burning and burying its named image. To ensure the ghost's defeat, the practitioner enlisted the assistance of other deceased spirits. Execration figurines (see sidebar, opposite) are typically found buried within older, abandoned cemeteries, whose formerly blessed inhabitants were perceived to have become angry or vengeful because they lacked funerary offerings. Just as the bowls of the "Letters to the Dead" (see pp.142–3) held offerings for the blessed, so those of the "Execration Texts" offered the ghosts of the damned as sacrifices to the vengeance of neglected spirits. The practice of using the disgruntled dead to "ward off" threatening spirits continued throughout the Roman period and became a feature of wider Hellenistic magic.

*Execration figurines were given the name of the victim, then ritually mistreated by smashing and burning. Many examples have been found to have been pierced by small metal knives or nails, a process analogous to the use of pins in later Western sorcery. This late 12th-Dynasty terracotta figurine bears the names of several Asiatics.*

# THE MUMMY'S CURSE

*One of the "magic bricks" from the tomb of Tutankhamun, photographed* in situ *shortly after the tomb's opening. Four such "bricks", each inscribed with a portion of chapter 151 of the Book of the Dead, were placed in niches at the cardinal directions of a tomb to form a defensive perimeter and support. Spells from the Book of the Dead are also found on other tomb objects. Examples include chapter 6, for the activation of* shawabti *servant figurines (see p.133), and chapter 30, with its spell for the "heart scarab" that replaced the heart in the thoracic cavity.*

The sudden death of the earl of Carnarvon (1866–1923) – the aristocrat who had funded Howard Carter's search for Tutankhamun's tomb (see p.196) – within six months of the tomb's opening, produced a spate of journalistic hysteria. Newspapers reported a supposed "curse of King Tut", promising "death on swift wings" for any desecrator of his tomb. Although Egyptologists have repeatedly shown that no such text was inscribed on Tutankhamun's tomb or its furnishings, the "mummy's curse" took root in the public imagination, inspiring numerous tales and a series of horror films. The Tutankhamun curse may be discounted as a mistranslation of a protective spell (Book of the Dead, chapter 151) inscribed on a magic brick from the tomb. However, genuine tomb curses have been found at other sites. Such curses are almost invariably from private tombs; royal tombs were protected by more physical means.

Tomb curses are most frequent in the Old Kingdom, with the dead owner invoking judgment in the underworld against potential violators. In the tomb of Ni-ka-ankh at Tehne, dating from this era, the deceased declares: "As for any man who will make disturbance, I shall be judged with him". A surviving block from a lost Old Kingdom tomb threatens more immediate punishment: "A crocodile be against him in the water; a snake be against him on land, he who would do anything against this [tomb]. Never did I do a thing against him. It is the god who will judge".

The notions of retribution visited upon offenders by the empowered dead are combined in the tomb curse of Ankhmahor (Old Kingdom) from Saqqara: "As for anything that you might do against this tomb of mine of the West, the like shall be done against your property. I am an excellent lector priest, exceedingly knowledgeable in secret spells and all magic. As for any person who will enter into this tomb of mine in their impurity, having eaten the abominations that excellent *akh*-spirits abominate, or who do not purify themselves as they should purify themselves for an excellent *akh* who does what his lord praises, I shall seize him like a goose [that is, wring his neck], placing fear in him at seeing ghosts upon earth, that they might be fearful of an excellent *akh* ... But as for anyone who will enter into this tomb of mine being pure and peaceful regarding it, I shall be his protective backer in the West in the court of the great god." Ankhmahor's threats to become manifest on earth and throttle his transgressor closely prefigure the vengeful mummy of Hollywood films.

Although less common, the protective curses of later periods are more colourful, condemning the violator to the "wrath of Thoth" or the "flame of Sekhmet", with destruction visited upon his tomb and descendants. From Ramesside times derives the "donkey curse", invoked for several centuries to safeguard wills and property endowments by threatening the violator with rape by a donkey, the animal of Seth. The most extensive curse dates from the Twenty-First Dynasty, in a retrospective decree for the funerary estate of the deified Amenhotep, son of Hapu (see p.33). The deceased Eighteenth-Dynasty administrator threatens anyone who would damage his tomb or funerary cult with a list of punishments: they would lose their earthly positions and honours, be incinerated in a furnace in execration rites, capsize and drown at sea, have no successors, receive no tomb or funerary offerings of their own, and their bodies would decay "because they will starve without sustenance and their bones will perish".

The most novel curse, from the twelfth year of Alexander IV (312BCE), would be viewed with sympathy by librarians and book collectors: "As to anyone of any country – whether of Nubia, Kush or Syria – who will displace this book or remove it from me, they will not be buried, they will receive no libations, they will not smell incense, no son or daughter will arise for them to pour out water for them, their names will not be remembered anywhere on earth, and they will not see the rays of the sun."

## PRINCE SETNA AND THE MUMMIES

If tomb curses only rarely suggest that the deceased might appear to the living, Egyptian literature provides several instances of the "walking dead". Set in a ruined tomb, a fragmentary Ramesside "ghost story" recounts the summoning of its resident spirit by a priestly magician. Having related his life story and misfortunes, the deceased is placated by reburial and renewed offerings.

The first demotic tale of Setna Khamuas pits its princely hero, an antiquarian son of Ramesses II, against a family of animated mummies. In a contest for the sacred Book of Thoth, Setna loses at *senet* (a common Egyptian board game) against the dead husband but then steals the ancient volume from the tomb. Cursed for his theft, Setna falls victim to the mysterious seductress Tabubu, a manifestation of the deceased wife. Now penitent, Setna returns the stolen volume and is sent to seek out and transport the bodies of the dead husband's wife and son from Coptos, where they had been buried. Under the supervision of the dead husband, risen from his tomb in the form of a wizened, age-old man, Setna finally does so and re-inters the whole family together in the tomb in Memphis.

*This scene from the tomb of Sennedjem in Western Thebes shows Sennedjem and his wife Iyneferti in a pavilion before a table of offerings. Like the mummy in the story of Setna Khamuas, they are playing* senet, *a popular board game symbolizing the passage of the dead through the underworld. The object was to move pieces around a board of thirty squares, or* perw *("houses"), avoiding various hazards and accumulating blessings. The winner was the first to pass into the blessed afterlife (see p.165).*

● CHAPTER 11

# THE LIFE OF RITUAL

The ancient Egyptians lived in a world permeated by ritual. Their religious beliefs were underpinned by a host of repeated actions and utterances that aimed to bring about a specific goal or objective, such as to bring health or to ward off adversity. To the Egyptians of the élite, for whom we have the most evidence, worship of the divine was exercised through rituals composed of formalized actions documented in texts and pictorial representations.

▲



ABOVE: *Titiw, a singer in the temple of Amun, with a sacred rattle (*sistrum*). The sound of the* sistrum, *which was carried by several classes of priests during rituals, was believed to please the gods. The ivy is a symbol of rebirth. From the Book of the Dead of Anhai, 20th Dynasty, ca. 1150BCE.*

## SECURING THE COSMIC ORDER

The prominence of ritual in ancient Egypt is intimately related to the Egyptians' world-view, which perceived the universe in dualistic terms. A permanent tension existed between cosmic opposites, such as good and bad, light and dark, barrenness and fertility, and, above all, between cosmos, or harmonious order (*ma'at*), and chaos (*isfet*). The universe itself participated in cycles of repeated patterns: the daily rising and setting of the sun, the round of the seasons, the annual flooding and recession of the Nile. When *ma'at* broke down, *isfet* ensued: when the Nile flood failed to materialize, the country suffered from famine. It followed in the Egyptian mind that mortals should seek to ensure, through their rituals, the continuation of cosmic order and the benevolence of the gods and goddesses who controlled the universe.

Egyptian rituals were primarily concerned with maintaining the image of the deity and offering it food and sustenance. The idea that an act of worship could propitiate a god is reflected in the fact that the Egyptian word for "offering" (*hetep*) is the same as "to be at peace".

Because rituals are repetitive, each one echoes all those that have gone before. Ritual gave structure to the past, which the Egyptians viewed with profound reverence. The state of the world was considered to have been perfect at its creation; change was not necessarily viewed as progress, but more likely as an undesirable deviation. As unchanging re-enactments of ancient events, actions and utterances, rituals contributed to the preservation of the ideal condition of the universe.

The king was the primary focus of Egyptian ritual life and, in theory, it was he who enacted all the sacred rites that were performed in the temples. Although in practice it was his religious deputies – the priests – who carried out the daily devotions, the pharaoh is depicted as the officiant in reliefs and paintings on temple walls. Even funerary offerings for the soul of a deceased private individual would be made in the king's name,

*A divine personification of one of the Egyptian nomes (provinces) in the mortuary temple of Ramesses II at Western Thebes. She is shown presenting the bounty of her part of Egypt, including bread, figs and pomegranates, and the hieroglyphs for "life" and "dominion". The offering of food was central to many religious rituals.*

*A 19th-Dynasty silver and gilt statuette depicting the presentation of the goddess Ma'at. The liturgy of the daily offering service included material offerings – such as bread, cloth, incense, milk – as a part of* ma'at, *which was called "the food of the gods". Presenting a figure of Ma'at, therefore, symbolized everything, material and immaterial, that was due to the gods.*

regardless of who actually presented the offerings (see p.141). From the Middle Kingdom onward, the king's role is clarified by his assumption of the title "Lord of [Ritual] Action".

The most important of all rituals in ancient Egypt was the "presentation of Ma'at", a goddess who personified the concept of *ma'at* (truth and universal order). Scenes depicting this rite from the reign of Thutmose III (ca. 1479–1425BCE) into the Roman Period show the donor – almost always the king – offering up a feather (𓆄, the hieroglyph for *ma'at*) or a small figure of the goddess (see illustration, right). This gesture expresses the pharaoh's promise to the gods to uphold the divinely ordained cosmic order (see also p.130).

# THE DAILY OFFERING

The daily temple ritual was the most common of all the rites that were enacted in ancient Egyptian religion. It took place three times a day in every temple in the land: at dawn, at noon and in the evening. The ritual was considered essential for the sustenance of the god, whose image, in the form of a cult statue, was kept in the temple sanctuary. Although the statue was not believed to *be* the deity, its spirit was thought to reside within it. The actions performed by the priests in the temple sanctuary were intended not only to honour the god or goddess, but also to purify the cult image in order to encourage the continued residence within it of the divine spirit.

The various activities that made up the daily offering ritual are shown in fullest detail at three sites: in the reliefs in the temple of Sety I (ca. 1290–1279BCE) at Abydos; the great hypostyle (columned) hall of the

## OFFERING SCENES

Most representations of ritual offering scenes follow a distinct pattern. The king, who acts as the primary officiant, faces the god to whom the offering is made. He may stand or kneel before the deity who in turn may be standing or seated. Hieroglyphic "captions" divide the single scene into three actions. The first caption refers to the power and success that the god has given to the king; for example: "Words said by Amun-Re, King of the Gods, 'It is to you that I have given every victory. It is to you that I have given all life and stability'." The next caption refers to the actions of the king and is phrased in the active present form ("Giving incense to his father ..."): the king is depicted in the act of handing the offering to the god. The final caption, appended to the king's dedication of the offering, is a wish: "May he [the pharaoh] be given life."

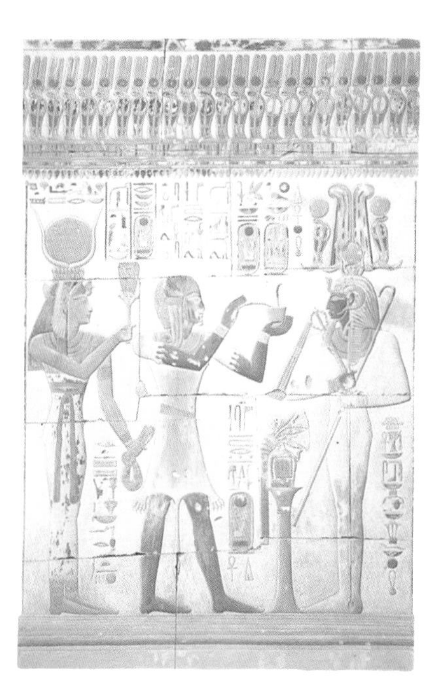

*The three-part offering formula captions this relief of Sety I (ca. 1290–1279BCE) from his temple at Abydos. Wearing the panther skin and sidelock of a priest, he offers incense to his deified self. The goddess Isis, shaking a* sistrum *and a beaded necklace, witnesses the ritual.*

Before the Nineteenth Dynasty, only the king is shown making offerings to the deity, but from the time of Ramesses II (ca. 1279–1213BCE) some stelae show the king before the deity in the upper register, while the (non-royal) man or woman who commissioned the stela appears in the lower register, usually in a gesture of adoration before an offering inscription. Later scenes might omit the image of the king entirely, but the traditional form showing Pharaoh before the god or goddess continued to be used in temple reliefs.

temple of Amun at Karnak; and the temple of Horus at Edfu. While it is possible to distinguish individual ritual actions in these representations, their sequence is less obvious because of the way in which the scenes are arranged on the walls. The king is always shown as the officiant, but the accompanying texts indicate that the high priest, who enacted the ritual in the name of his royal master, was instructed to declare to the god: "It is the king who sends me."

*King Nectanebo I (381–362BCE) offers a statuette of the goddess Ma'at to Thoth, who is shown in the form of a baboon, his head topped with a solar disc. Another god, Onuris, stands to the right. From a temple at Abydos. Late Period, 30th Dynasty.*

After a number of purification rituals that involved washing himself with water and natron, the officiant (in some periods accompanied by other priests of the requisite level of ritual purity and a choir of priestesses) broke the seals on the door of the sanctuary and entered the holy place. Prayers were said and an offering of incense was made to the *uraei*, the protective but also potentially dangerous images of rearing cobras that adorned the shrine.

The priest then lit a torch, which served to awaken the deity from his or her slumber, and also symbolized the rising of the sun, hence the renewal of the cosmos. After the recitation of further prayers, the air was purified with incense. The priest took the cult statue from the shrine and placed it on a pile of clean sand representing the primordial mound from which all life sprang (see pp.120–21). He then removed the clothes from the image and cleansed it of unguents left after the previous ritual. After purification with more incense, the god was adorned with clean clothing, and anointed with perfumes and cosmetics. Bolts of white, green and red cloth, necklaces, perfumes, crowns and sceptres were presented to the statue.

At this point, the priests uttered an invocation to encourage the spirit of the god to enter the statue and partake of food that was placed before it: "Come to your body! Come to the majesty [that is, the king], your servant who does not forget his part in your feasts! Bring your power, your magic and your honour to this bread which is warm, to this beer which is warm, to this roast which is warm!"

After the symbolic meal, the statue would be purified anew with incense, unguents and liquids, and enveloped in fine white linen. Then, clean sand was scattered on the floor of the sanctuary (it is not known who performed this action) as the priest bowed and backed out of the room, brushing his footprints away as he went with a reed broom. Finally, the sanctuary doors were closed and the seals replaced, allowing the deity to rest until the following day.

# ANCESTOR WORSHIP

Either collectively or individually, the kings of the past were honoured in ritual. The worship of royal ancestors involved presenting food offerings to the spirit of the deceased monarch and reciting prayers, which were thought to enable the *ka*, or spirit, of the king to partake of the food. Royal mortuary temples (see pp.210–11) were centres of such worship, and other, non-mortuary temples such as those at Luxor and Karnak also played a role. Most New Kingdom temples incorporated chapels designated for the worship of the king's father, as well as for the veneration of the gods Amun and Re.

Our clearest information about the royal ancestor cult comes from texts that were incorporated in some versions of the daily offering service (see pp.150–51). Another specific reference is contained in a decree from the reign of Thutmose III (ca. 1479–1425BCE), which mentions that twelve heaps of offerings were to be prepared for the daily service in the temple of Ptah at Karnak, and that six of these were to be placed before a statue of the king. The reliefs and inscriptions at the temple of Sety I at Abydos indicate that after the image of a deity or king was judged to have drawn sufficient sustenance from the food offerings placed before it, the offerings were removed to an altar in front of a list of all the kings of Egypt – with the exception of "disgraced" monarchs such as Hatshepsut and the "heretic" Akhenaten and his immediate successors – with the object of nourishing and honouring every monarch of the past. Similar lists, such as that from the temple of Thutmose III at Karnak (now in the Louvre), and from the temple of Ramesses II at Abydos (now in the British Museum; see illustration, p.20), should be regarded as relics of royal ancestor worship rather than simply records of who ruled Egypt. The food offerings that were presented to these lists of royal ancestors were, after a suitable period during which the kings were thought to have "consumed" them, given to the priests as a part of their compensation for temple duty.

Mortuary temples of the New Kingdom were designed primarily for the royal cult, which was celebrated after the death of the king. However, features such as a chapel for the worship of the living king and a royal "palace" indicate that these temples were also used during the king's lifetime. Although it has been suggested that the living king, represented by one or more statues, was revered as a deceased pharaoh in anticipation of his actual death, it is more likely that he was worshipped through his association with Amun, the supreme state god. There is evidence for this identifica-

*The 2nd-century CE mummy case of a man called Artemidorus, whose portrait in encaustic (coloured beeswax) takes the place of the three-dimensional mummy masks of earlier times. Such lifelike representations were popular in the Roman era, particularly in the Faiyum, hence the use of the term "Faiyum portrait" for such images. They were commissioned during the owner's lifetime, which accounts for the youthful appearance of many sitters who were actually quite advanced in years when they died. It was perhaps believed that the deceased would be more effective ancestral spirits if they were reborn as their youthful selves. Although Artemidorus is shown in Roman dress, the mummy case bears traditional Egyptian motifs.*

tion in New Kingdom Thebes, where chapels dedicated to dead kings refer to them as forms of Amun. Sety I, Ramesses II and Ramesses III were called "the Lord, Amun-United-with-Eternity within the Temple".

In contrast, the cult of King Amenhotep I (ca. 1514–1493BCE), who was venerated as the patron deity of workers on the west bank of Thebes, was instituted some two centuries after his death. Statues of Amenhotep and his mother, Ahmose Nofretari, were paraded through the small communities of Western Thebes – hauled on sledges or conveyed in sacred boats – surrounded by men and women beating drums and chanting. During these processions, people would address questions to the king's statue, which would give answer "yes" or "no", indicated by a perceived slight movement of the statue. The popularity of the worship of this king and his mother in the Ramesside Period is indicated by calendars, which list at least seven days devoted to festivals of the king each year, and also by numerous stelae and tomb paintings of the statues of the divine pair.

## "ABLE SPIRITS OF RE"

The worship of non-royal ancestors was enacted mostly in private homes rather than in the temples. Stelae, offering basins and stone bust-like statues discovered at the workers' village of Deir el-Medina in Western Thebes show that certain non-royal ancestors were called *akh iker n Re*, or "able spirits of Re". "Able" describes their effectiveness as intermediaries between the living and the dead. These effective spirits were considered to have a special relationship with the sun god: they were entitled to take up an honoured position in his barque, in which he travelled through the land of the dead at night, and were sometimes represented as rays of sunlight.

Domestic ancestor worship involved leaving food offerings in front of the *akh iker n Re* monument. Most examples of such monuments have been recovered from, or near, a niche within the house. The provision of sustenance propitiated the "able spirit" and encouraged it to intercede with the gods on behalf of the living. Texts associated with the practice show that the living asked the deceased for assistance in a variety of matters, from help in conceiving a child to resolving legal disputes. One text implores, "Become an *akh* for me before my eyes so that I can see you in a dream fighting on my behalf. I will then deposit offerings for you ...."

*A bust representing the "able spirit" of Mutemonet, the mother of Amenmose, a scribe in the reign of Ramesses II (ca. 1279–1213). Food offerings were left before such statues in the hope of persuading the spirit to intercede on behalf of the living.*

# RITUAL AND HISTORY

Egyptians believed that depicting an action, for example in a relief or a wall painting, was as effective as actually performing it. Thus, when the king was shown officiating at a sacred ritual, he was deemed to have conducted the ceremony in person, even if in reality his priests had acted on his behalf. The same applied to images of an apparently historical nature: for example, a scene of the king defeating foreign enemies was believed to have the effect of keeping his real enemies at bay.

Scenes of royal triumph may commemorate real historical events, and distinguishing them from symbolic representations of conquest is not always simple. However, some images are obviously ritualistic because the actions depicted are unrealistic, as in scenes in which the king holds an impossibly large group of captives by the hair, while the god Amun looks on approvingly. Similarly, standardized depictions of the king smiting his foes (see box) and of the captive "Nine Bows" (the collective name for Egypt's traditional enemies) often appear on temple pylons.

Other types of representations can prove problematic when attemping to separate fact and ritual (see box). References to the *sed*, the jubilee festival commemorating the thirtieth year of a king's rule, present difficulties for our understanding of Egyptian chronology. A number of kings claimed the celebration of a *sed*, including the Eighteenth-Dynasty pharaohs Hatshepsut (ca. 1479–?1458BCE), Thutmose III (ca. 1479–1425BCE) and Amenhotep II (ca. 1426–1400BCE). However, according to accepted chronologies, Thutmose III was the only one of these kings actually to rule for three decades or more, so the historicity of the *sed*s of Hatshepsut and

*On this chest from his tomb, Tutankhamun (ca. 1332–1322BCE) leads his orderly troops against Syrian enemies, who are shown in disarray. Another panel of the same box shows the king defeating Nubians (see illustration, p.41). Such scenes are not to be taken as depictions of historical events, because there is no evidence that the young pharaoh, who was probably no more than 18 when he died, actually undertook such campaigns. However, by having himself portrayed as Egypt's victorious champion, he stimulated divine forces to protect the country from its enemies.*

## PHARAOH TRIUMPHANT

Scenes of the pharaoh defeating Egypt's enemies are frequently depicted in Egyptian art. For example, the "Smiting Scene", which shows the king raising a mace to strike a cringing foe, appears in a remarkably unvarying form on monuments and artefacts throughout the entire span of pharaonic history. The scene is too much a standard part of royal iconography to be interpreted as anything other than a symbolic statement of the king's supremacy. However, other scenes of royal domination are presented in such a realistic manner that it may be tempting to take them for historical documents.

A scene first noted at Abusir in the temple of the Fifth-Dynasty king Sahure (ca. 2485–2472BCE) shows the king in the form of a sphinx trampling Libyan enemies. To the right stand a woman and two men in Libyan dress, perhaps the foreigners' families or members of the Libyan campaign retinue, each clearly identified by his or her personal name. Other information includes exact numbers of captured livestock. The relief was once considered to be certain evidence of a historical campaign against the Libyans. However, a very similar group of figures, bearing exactly the same names, appears in scenes depicting the pharaoh Pepy II (ca. 2288–2194BCE) in his temple at South Saqqara. Two more instances, featuring different kings but the same group of Libyans, appear in the temples of Osorkon I (ca. 920–889BCE) at Bubastis and Taharqa (690–664BCE) at Kawa in Nubia. These later scenes cannot be records of historical events, but are ritualistic commitments to protect the land against its traditional aggressors, whether the king actually campaigned against Libya or not. Such documents have been termed "ritual affirmations of conquest".

The king's traditional enemies, the "Nine Bows" (see main text), are also portrayed on his walking sticks, footstools and the soles of his sandals, enabling the king to perform a ritual crushing of his enemies by simple everyday acts.

*"Smiting Scenes" varied little over the centuries, as seen in a 1st-Dynasty ivory label of King Den (top) and a 20th-Dynasty relief of Ramesses III (above).*

Amenhotep II is questionable. In the same way that depictions of victories might bring about real ones, perhaps a king celebrated the *sed*, or ordered its representation, in the hope that this would ensure his own longevity.

Another mingling of ritual and history can be seen in the battle reliefs of Ramesses III (ca. 1187–1156BCE) on the walls of his temple of Medinet Habu in Thebes. He is shown defeating the Hittites at Tunip and Arzawa, although those towns were defeated nearly a century earlier, as indicated by reliefs dated to year ten of Ramesses II. Scholars have concluded that the Medinet Habu scenes are copies of the reliefs of his revered predecessor. The fact that the Ramesses III images bear no date and are adjacent to scenes of religious processions suggests that they were intended to be recognized at once as scenes of purely ritual significance.

*The creator god Ptah holds Senwosret I (ca. 1919–1875BCE) in a protective embrace in this scene from the pharaoh's barque chapel at the temple of Karnak.*

# RITUAL GESTURES

Representations of rituals in Egyptian art often incorporate gestures and poses that convey vital symbolic information. Some gestures are readily recognizable to a present-day observer. For example, a figure kneeling before another person or a deity is inferior in rank or status; bowing expresses honour and respect; arms outstretched in front, palms outward, signify adoration; and being embraced by a deity symbolizes protection.

Other gestures are less easily interpreted across the millennia. For example, holding a hand to the cheek normally denotes a singer; throwing dirt over one's head indicates sorrow; an outstretched arm with palm upward is a greeting; while one hand clenched and held to the breast with the other hand raised, palm outward, is a sign of veneration. Some gestures may not be immediately recognizable as ritual acts at all. For example, a scene of a man or woman sniffing a lotus flower alludes to the process of rejuvenation in Egyptian funerary rites: the lotus, whose petals open in the warm morning light, close in the evening and reopen the following morning, is associated with the cycle of birth, death and rebirth.

A number of gestures relate to common everyday rituals, as in Old Kingdom boating scenes that show one of the boatmen pointing an extended index finger and thumb at a crocodile. This action was intended to ward off the creatures, which were a common hazard on the Nile. An apparently similar gesture is encountered in temple reliefs, where the king points his index and little finger at the face of a deity. This act, which today might be assumed to be disrespectful, or even vulgar, in fact simply indicates that the king is applying fragrant ointment to the divine figure.

*The act of smelling the lotus flower, depicted in many private tombs, is a reference to rejuvenation after death. Two such scenes are shown here in the tomb of Sennefer, the mayor of Thebes under Amenhotep II (ruled ca. 1426–1400BCE).*

The use of hieroglyphs as a form of shorthand could enrich what at first sight were straightforward devotional acts and elucidate those whose significance is not immediately apparent. For example, the hieroglyph for "cloth" (𓋳𓋳) placed between a figure and a god or goddess, indicates the presentation of bolts of fine cloth to the deity. In the scene referred to as "the Baptism of Pharaoh", deities are depicted in the act of pouring a ritual libation over the king. The "waters" of the libation sometimes take the form of a stream of hieroglyphs, which include most frequently the *ankh* (☥, "life") and the *was* sceptre (𓌀, "dominion"): in addition to its more apparent meaning, the scene therefore also represents the gods granting these gifts to Pharaoh. A further layer of significance is added by the slender vases from which the "water" flows, which take the form of the hieroglyph for "honour" (𓏘).

The king and the god Osiris are often shown holding a sceptre that resembles a shepherd's crook (𓋾), which is in fact the hieroglyph for "to rule" and serves as a visible affirmation of their authority. In certain scenes where a goddess greets the deceased in the afterlife, she holds a wavy line (𓈖), which resembles water, above two oblique lines ( \\ ) in each hand. These signs are the hieroglyphs for "to greet".

## OFFERINGS TO THE SUN DISC

Religious practices were greatly modified in the Amarna period, when the pharaoh Akhenaten promoted the worship of the Aten, or sun disc (see pp.128–9). Although the Aten had no human manifestation, in ritual scenes he interacts more directly with devotees than do the deities in traditional offering scenes. Images show not only the worshippers with hands raised to the Aten, but also the Aten's rays reaching out to bless the offering and the donor, each ray terminating in a hand.

The most common Egyptian religious ritual – the daily offering to the god – continued during this time, but with modifications. For example, the venue of the ritual shifted from the dark inner sanctuary of a traditional temple to an open courtyard flooded by the sunlight that emanated from the Aten. In the pre-Amarna period, there was one altar in each temple, but Akhenaten's temples had several altars: the omnipresence of the Aten's rays was believed to necessitate more areas from which the disc could receive sustenance.

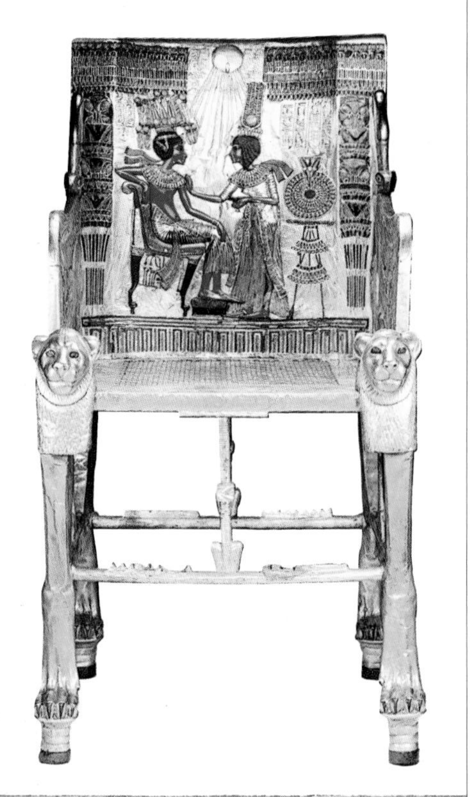

*The cult of the Aten was abandoned under Tutankhamun (ca. 1332–1322BCE). However, this throne dates from before the restoration and shows the arm-like rays of the Aten reaching out to touch the pharaoh and his queen, Ankhesenamun, in a gesture typical of depictions of the Aten. At the end of each ray is a tiny hand holding an* ankh *(☥), the hieroglyph for "life".*

# SACRED BOATS

*Khufu's reconstructed barque in the Giza Museum. A clue to the function of four of the five barques buried next to Khufu's pyramid (see main text) may lie in their orientation. The two known dismantled barges are oriented east–west, and were probably intended as boats in which the deceased king, as the sun god, would cross the sky on his daily voyage (see pp.118–19). Two pits flanking the mortuary temple are oriented north–south and may have been intended for use by Khufu-Horus in the afterlife to patrol his domains along the Nile. The fifth pit may be connected with the cult of Hathor, a deity worshipped at Giza in this period.*

Reflecting the importance of boats in daily life (see pp.14–15), the gods of the Egyptians were said to be transported in celestial barques. From the Early Dynastic Period onward, an image of the deity would regularly be carried from the temple sanctuary in a model boat by priests in procession, either on a visit to another temple or on a circumambulation of the deity's own sanctuary. More elaborate sacred boats appeared in the reign of the pharaoh Hatshepsut (ca. 1479–?1458BCE) and are portrayed in reliefs at Thebes carrying the king as well as the Theban triad of deities: Amun (see pp.126–7), his wife Mut and their offspring Khonsu.

Although the clearest evidence for sacred vessels dates from the New Kingdom, imagery of boats from the Predynastic Period onward suggests that ritual ships may have been housed in special chapels in the temples from the earliest times. When the deity was due to depart on its sacred voyage, the cult statue was placed in a portable shrine, made in the form of an ancient shrine of Upper Egypt, its roof ornamented with a frieze of protective *uraei* (rearing cobras) and its sides covered with royal names and a hieroglyphic frieze. The shrine was placed in the boat's "cabin". The bow and stern were decorated with the head of the god whose statue

*A vessel bearing the body of the deceased; a wall painting from the tomb of Menna, Western Thebes; 18th Dynasty, ca.1395BCE.*

## "USERHAT" AND THE OPET FESTIVAL

One of the greatest boat processions took place as part of the feast of Opet at Karnak. First attested in the reign of Hatshepsut (ca. 1479–?1458BCE), the procession is known to have continued at least into the Twenty-Fifth Dynasty (reign of Piye, ca. 747–716BCE). Held annually in late summer, it celebrated the journey of the god Amun from Karnak to the nearby temple of Luxor. An extraordinary series of reliefs at Karnak, some twenty-eight yards (26m) long and dating to the reigns of the New Kingdom kings Tutankhamun and Sety I, record the movement of barques from Karnak to Luxor and back, and the animated ceremonies outside the temples.

As a procession of priests slowly carried the ceremonial boats to the water's edge, singers, dancers, acrobats, musicians, military men and local residents would have swarmed around the procession. At the bank of the Nile, the processional barques were placed on river-going boats, the greatest being the *Userhat* ("Mighty of Prow") which carried the barque of Amun. The *Userhat*'s hull was painted gold with scenes of the king before the god. The ram head of Amun, draped with necklaces and costly pectorals, decorated the bow and stern posts, and obelisks adorned its deck. Once the statue of the god was loaded onto the *Userhat*, the boat was drawn southward, against the current, by gangs of sailors heaving on tow ropes. A modern survival of Opet is seen in the Islamic festival of Abu Haggag, which is celebrated annually in Luxor. The celebration, in honour of the local Muslim saint, culminates in the procession of small boats around the Luxor temple precinct.

*In this sketch on a limestone* ostrakon, *a procession of priests is shown carrying the sacred boat of Amun (which can be identified by the ram's head on the bow and stern). The statue of the god is inside the shrine-canopy in the middle of the boat. New Kingdom, 19th or 20th Dynasty.*

it carried: a ram for Amun; a man's head for Khonsu or the king; a woman's head for Mut. The protective god Tutu, depicted as a sphinx with erect tail and wearing the *atef* crown (see p.109), stood in the prow, and small statues of the king and other deities were placed around the shrine in the boat's cabin. The whole was borne on poles by slow-moving lines of priests. The procession would periodically stop to allow the local population to pray to the god, or to call upon the deity to pronounce oracles. While at rest, the barque was placed on a stone pedestal. Along some routes stood special barque shrines where the priests could rest and purify the statue with incense.

A number of kings were buried with full-size boats that served a sacred purpose in the afterlife. For example, five boat pits have been discovered alongside the pyramid of Khufu (see sidebar, opposite and pp.180-83). Three were empty, but two contained dismantled barques, one of which has been reassembled. The second remains in its pit, where it was photographed *in situ* in 1987 by a tiny camera inserted through a small hole.

# DIVINE BIRTHS AND MARRIAGES

Egyptian deities were held to possess human qualities to the extent that they might participate in such mortal activities as marriage and the fathering or bearing of children. For example, reliefs in which the god Amun is depicted as the father of the pharaoh occur in the temple of Hatshepsut (ca. 1479–?1458BCE) at Deir el-Bahari in Western Thebes, and there are similar scenes commissioned by Amenhotep III (ca. 1390–1353BCE) in the temple of Luxor. In the Deir el-Bahari scenes, the god is said to have assumed the form of Hatshepsut's father as he impregnated his wife. The queen was alerted to the divine presence by the exquisite odour of incense, under the spell of which she "saw him [Amun] in his form as a god". Although divine conception and birth are depicted in the temple reliefs, there is no evidence that specific rituals were associated with these events in the New Kingdom. Rather, these scenes served the political purpose of emphasizing the legitimacy and authority of the ruler through his

## BIRTH RITUALS

The rituals that were performed at the birth of a child are known from only a few sources. According to one Middle Kingdom text (Papyrus Westcar), Queen Ruddedet, who was credited with bearing the kings of the Fifth Dynasty, "cleansed herself in a cleansing of fourteen days". A period of post-natal seclusion and purification is still practised by women in parts of the Middle East and in numerous other cultures. The text also suggests that, before the birth, the clothes of the expectant mother may have been loosened or put on inside-out in imitation of certain funerary rituals, possibly an allusion to the cycle of birth, death and rebirth.

Several *ostraca* (inscribed potsherds or limestone flakes) from Deir el-Medina in Western Thebes show a woman suckling a child under a canopy that has been interpreted as a special pavilion, where a mother gave birth and spent her period of seclusion. The canopy has slender columns in the form of lotus blossoms and its sides are hung with convulvulus ivy – both plants were associated with rebirth. The woman wears her hair in a style unique to these scenes, pulled up in a central ponytail, from which it falls onto her shoulders. She wears only a thin girdle, a beaded collar and anklets.

*A 20th-Dynasty* ostrakon *from Deir el-Medina, showing a woman suckling her child, probably in a special birth pavilion. In the lower part of the sketch, a girl offers her a mirror and cosmetics.*

Other texts refer to the presence of a goddess, such as Isis, at the birth. In *The Story of the Doomed Prince*, the birth of the prince is accompanied by "Seven Hathors", deities who were thought to foretell the fate of the newborn infant. They predict that the boy will "die through a crocodile, a snake or a dog". In Papyrus Westcar, seven deities transformed into dancing girls predict the futures of the babies who will become the kings of the Fifth Dynasty.

*A colossal figure of the god Horus in the form of a falcon, from the temple of Edfu, the destination of the cult statue of Hathor during the "Beautiful Embrace" festival.*

or her claim of divine parentage. In the case of Hatshepsut, such legitimization of her right to rule was especially necessary because she was assuming a position traditionally held only by men (see p.89).

Certain features of the Eighteenth-Dynasty reliefs are shared by scenes in the Ptolemaic temples of Philae, Dendera, Armant and Kom Ombo. However, in these later representations, the mother of the king is also a deity – usually Hathor or Isis, the goddesses most closely associated with motherhood – and her offspring is depicted in the form not of the king but of another god, such as Ihy, Khonsu or Harpocrates ("Horus the Child"). The "Ritual of the Divine Birth", the actual events of which are not known, was celebrated in a building known by the Coptic word *mammesi* ("birth house"), a small structure within the temple compound. The *mammesi* at Philae demonstrates the continuity between the New Kingdom and later birth scenes, for here the newborn god Harpocrates, the offspring of Amun-Re and Isis, is specifically equated with the king.

### THE MARRIAGE OF HATHOR AND HORUS

Late Period rituals of divine marriage, in the sense of the coupling of two deities, most commonly involved the goddess Hathor and the god Horus. One of the greatest events in the Upper Egyptian religious calendar was an annual fourteen-day festival, known as the "Festival of the Beautiful Embrace" (or "Meeting"), which celebrated the "marriage" of the two deities. Hathor was taken in a boat named *Mistress of Love* from her temple at Dendera by stages to the great temple of Horus more than one hundred miles (160km) to the south at Edfu. The procession stopped at the temple of Mut in Karnak, at the sanctuary of Anukis at Per-mer between Esna and Hierakonpolis, and at Hierakonpolis itself, a centre for the worship of Horus since the Early Dynastic Period. As Hathor's boat approached Edfu, cult statues of Horus and Khonsu, who was also worshipped at Edfu, were brought out to meet it. The wall reliefs and accompanying texts at both Dendera and Edfu deal only with the processions, and not with the associated rituals enacted in the temples, so the festival's full significance remains unclear.

# PIETY AND PRIESTHOOD

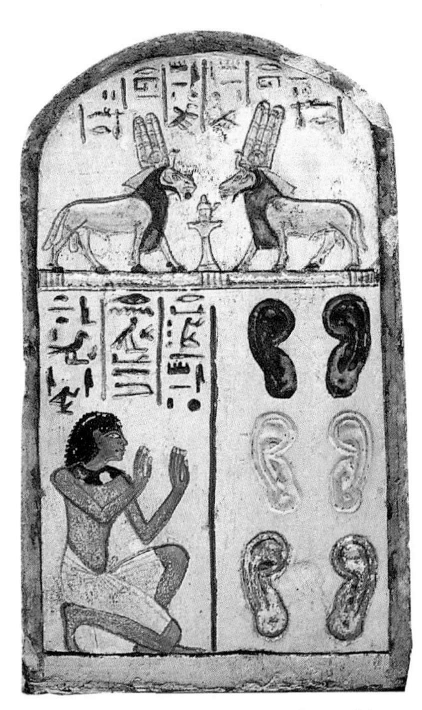

*On this "Hearing Ear" stela of the 19th or 20th Dynasty, Bai, a workman from the village of Deir el-Medina in Western Thebes, adores a representation of six ears of the god Amun, who is depicted twice as a ram in the upper register. Such stelae were thought to channel prayers directly into the ears of the deity. The significance of the colouring of the ears is not known. Worship at "Hearing Ear" shrines persisted well into the Roman Period, when one was incorporated in the temple of Horus and Sobek at Kom Ombo, a thousand years after the erection of the ear shrine at Karnak (see main text).*

Religion and ritual occupied an important place in the lives of ancient Egyptians and great numbers of men and women held priestly titles. Unfortunately, we know little about the exact duties and training of many of the classes of priests, or even if there were any special initiation ceremonies. Priests probably assumed their positions in early adulthood and held them for life. Priestly titles could be passed from father to son and certain posts are known to have remained within a single family for many generations. Priests were also appointed by royal decree.

The most common priestly titles (*wab* and "lector") designate part-time positions. These priests served their temple usually for one month each year. After they had discharged their duty, they returned to their ordinary jobs, perhaps as junior bureaucrats or craftsmen. These men are often pictured officiating in funerary ceremonies, carrying offerings to the tomb or reciting prayers on behalf of the deceased, the latter function suggesting that at least some of them were required to be literate.

In many reliefs the costume of these lower ranks of priests – the common Egyptian knee-length kilt – is indistinguishable from the dress of non-priests. However, the lector-priest was distinguished by a broad fabric sash worn across his chest. In contrast, the full-time professional priests, such as the "prophets" of a particular deity, wore distinctive cloaks made from, or in imitation of, panther skins. The cloak of the High Priest of Memphis was covered with stars.

Although priests may have dressed differently from the ordinary people – at least while officiating at certain ceremonies – for the most part they lived much like non-priests. With few exceptions, male and female priests married and had children, and lived in villages or towns alongside non-priestly neighbours rather than in separate clerical communities.

Even men and women who did not have priestly titles actively participated in religious ritual by praying and presenting offerings at domestic shrines and attending processions in honour of the god. Although a temple's cult statue was hidden in the sanctuary and usually accessible only to the highest-ranking priests, other shrines to the deity were conveniently located for maximum public access at the front gate or back walls of temples. It was here that the common people came to pray directly to the god or goddess of the temple for divine intercession or blessings. The public shrine on the north wall of the temple of Amun at Karnak was designed to ensure privacy of the devotee, with curtains to shelter both the divine image and the petitioner. These places of worship bore names such as the "Shrine of the Hearing Ear" or the "Place where the God hears Petitions" and were decorated with ear motifs that emphasized their function.

The ordinary people might also direct pleas to an "intercessory statue", an image of a god, or a deified human being, thought to have special access to the gods. One such statue, of Amenhotep, son of Hapu (see p.33), is inscribed with the following text: "O people of Karnak who wish to see Amun, come to me! ... I will transmit your word to Amun in Karnak. Give me an offering and pour a libation for me, because I am an intermediary nominated by the king to hear requests of the suppliant."

Egyptians also participated in rituals at many other shrines. For example, the shrine of the cobra goddess Meretseger ("She who loves silence"), guardian of the Theban Peak (Qurn), stands on a hill overlooking the mortuary temples of Western Thebes. Here, people left food offerings and libations to the goddess, together with votive stelae and graffiti imploring her for blessings. Excavations at nearby Deir el-Bahari have recovered thousands of offerings that were probably left at the shrine to the goddess Hathor. These range from miniature garments painted with a scene of the petitioner before the goddess, to figurines or parts of the body (ears, eyes, phalluses) modelled in wood or faience. Many of the donors were élite women: an inscription on one donation exhorts "noble ladies as well as poor girls" to leave offerings.

## POPULAR ANIMAL CULTS

Animal cults were a feature of popular religion in the Late Period and Ptolemaic Period (664–30BCE), reaching their peak in the third and second centuries BCE. In these cults, the animals themselves were not worshipped; rather, they were revered because they were associated with particular deities. For example, baboons and ibises were linked with Thoth, cats with Bastet, crocodiles with Sobek, and dogs with Anubis.

Worshippers sought to honour the deity by donating bronze or faience figurines or the mummified remains of an animal associated with the god's cult. The practice of offering mummified animals at a shrine of the god became a successful business for the temples, which maintained large breeding pens for the animals. When they reached a certain age (for cats this was about ten months) the creatures were killed and mummified, and the mummies were sold to pilgrims. The practice was evidently very lucrative: the ibis catacombs of North Saqqara are believed to hold the mummified remains of around four million birds, and the cemetery at Bubastis, cult centre of the cat goddess Bast, contains many thousands of cat mummies.

Other animal cults were related to the gods of the nomes (provinces): for example, the latus fish was the sacred emblem of the Oxyrhynchus nome.

ABOVE, LEFT: *A bronze and wood statue of the god Thoth as an ibis (ca. 600BCE), which would have been bought by a pilgrim.* BELOW: *A decoratively wrapped mummy of a young crocodile (1st century CE); thousands of such mummies were buried in honour of the crocodile god Sobek.*

# RITUAL GAMES

*In this stick-fighting scene, two men use flowering reed stalks to re-enact one of the battles between the gods Horus and Seth. From the tomb of Kheruef in Western Thebes. New Kingdom, reign of Amenhotep III (ca. 1390–1353BCE).*

Depictions of what at first sight might appear to be sports and games may in fact represent important rituals associated with complex mythological concepts. For example, the Eighteenth-Dynasty tomb of Kheruef, the steward of Queen Teye, at Thebes is decorated with scenes of boxing and stick-fighting contests that are in fact ritual acts. The hieroglyphic captions that accompany the action, and the proximity of these scenes to more readily recognizable rituals, such as the raising of the *djed* pillar (a rite associated with the resurrection of the god Osiris), provide valuable clues to the real significance of these "sporting events". In one scene, six pairs of men, their heads shaved in the manner of priests, engage in what their postures and the hieroglyphic caption designate as a boxing match. However, the caption close to one boxer in each pair proclaims, "Horus has prevailed in truth!", suggesting that this man is taking the part of the god Horus (with whom the living king was identified) in a re-enactment of the battle between the forces of good and evil personified respectively by Horus and his brother Seth (see p.22). As is to be expected in a ritual scene, the outcome of the boxing match is never in doubt: Horus must always triumph over Seth, in accordance with the myth.

In the stick-fighting scenes, the combatants wield batons in the form of a single or double flowering reed, and two of the fighters are labelled "men of Pe", the ancient cult centre in the northwestern Delta. The ritual nature of the competition may be understood through references to stick-fighting in the funerary Pyramid Texts. One such passage refers to stalks of the same reeds depicted in Kheruef's tomb: they are equated with flower sceptres used by Horus to defeat Seth in the form of a hippopotamus. Another passage, part of the liturgy intended to restore life to the dead pharaoh, recounts that during the funeral of Osiris, the men of Pe "clashed sticks for you [Osiris]".

Some scenes depicting wrestling matches are likewise to be interpreted as religious rituals. For example, at the temple of Ramesses III at Medinet Habu in Western Thebes, the area of the "Window of Appearances",

*In this relief from the mortuary temple of Ramesses III (ca. 1187–1156BCE) in Western Thebes, wrestlers and stick-fighters enact ritualized battles between Egypt and its traditional Nubian and Syrian enemies, as members of the court (right) look on. The hieroglyphs declare that "[the god] Amun is the one who decreed victory".*

## CHILDREN'S GAMES AS RITUALS

Certain activities that at first sight appear to be children's games may actually be depictions of religious rituals. One of these is the so-called "hut game" depicted in the tombs of Idu at Giza and of Baki at Beni Hasan, and on a relief in the British Museum. All these scenes show five young boys, four of whom are inside an enclosure, perhaps a hut or a tent. Each boy sports the sidelock traditionally worn by minors. Two boys stand to the right, while another seems to pin the fourth boy to the ground. The pinioned boy extends his hand toward a fifth boy, who is standing outside the enclosure.

The inscription accompanying the scene has been interpreted, "I'll rescue myself from here on my own, friend" or "Rescue your one who is among [them], oh my friend", to which the boy outside replies "I'll save you!"

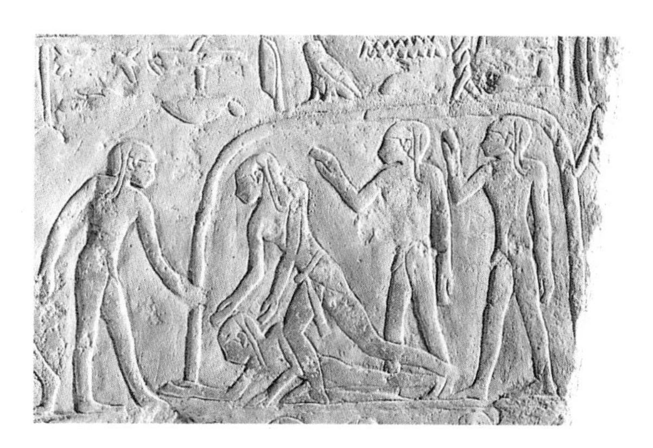

*The "hut game" on a 6th-Dynasty relief from Giza. The "game" was once thought to form part of puberty rites associated with circumcision. Egyptologists continue to agree that some form of ritual is being presented, but are now less certain of its symbolism.*

where the king presented himself to the people on festival occasions, is decorated with images of wrestlers. The competition, between Egyptians and various foreigners, is captioned with such texts as "Woe to you Syrian enemy who boasted! I will hurl you on your side in the presence of Pharaoh!" That the participants are not simply taking part in recreation is emphasized by the fact that the reliefs appear to be an almost exact copy of scenes in the mortuary temple of Ramesses II (the Ramesseum), of which one carved block, showing only Nubian wrestlers, remains. Other blocks from the Ramesseum wrestling scene (which was possibly, but not certainly, from its Window of Appearance) were even used to restore the Medinet Habu reliefs. Such a slavish imitation makes it more likely that the wrestling bouts represent a symbolic victory of Egypt and the king over their enemies, rather than a specific sporting event.

Board games could also have a ritual significance. One of the most common, *senet*, was also known as "Passing through the Underworld". Its rectangular board had thirty squares, some of which were marked with hazards, such as water, or auspicious signs ("life and dominion", "very good", "power"). Completing the game was equated with successfully passing the divine judgment of the soul (see p.137) and being reborn in the afterlife. The symbolism of the game is emphasized by scenes in tombs (see illustration, p.147); some of these show the deceased playing an unseen opponent, who is perhaps to be understood as one of the judges of the underworld.

**THE BAT AND BALL GAME**

**First known from the time of Thutmose III (ca. 1479–1425BCE), representations of the "Bat and Ball" ritual depict the king striking a ball with a club. Texts that accompany some of these scenes indicate that the ball was equated with the eye of the evil underworld serpent Apep, or Apophis. By hitting the ball with the club, which was said to represent "the eye of Re", the king symbolically drove away hostile forces. That this game had a serious purpose did not mean that the king derived no pleasure from it. In one bat and ball scene in the Ptolemaic Period temple of Horus at Edfu, the caption declares that "Pharaoh enjoys himself as a child".**

*Part III*

# ART, ARCHITECTURE AND LANGUAGE

*The great temple of Amun at Luxor, Eastern Thebes. Founded in the reign of Amenhotep III (ca. 1390–1353BCE), Luxor was extended by his successors, including Ramesses II (ca. 1279–1213BCE), who added this court and its colossal statues of himself.*

OPPOSITE: *The exquisite mummy mask of Prince Yuya, the father-in-law of Amenhotep III (ca. 1390–1353BCE). The mask is made of gilded* cartonnage *(layers of linen stiffened with plaster), employed by Egyptian artists from the First Intermediate Period onward.*

● CHAPTER 12

# THE PYRAMIDS

The pyramids are at once ancient Egypt's most striking monuments and its most enduring symbols. So much that lies at the heart of Egyptian culture is encapsulated in these unique structures: the ultimate power of the king; the central role of the cult of the dead; and the importance of the sun god. A pyramid took years to build and involved a huge deployment of labour and materials, although few later pyramids can match the great Fourth-Dynasty structures at Giza in terms of size and sheer mass.

▲



ABOVE: *A reconstructed pyramid on a New Kingdom private tomb at Deir el-Medina, Western Thebes. Kings of the time no longer built pyramids for their own tombs, but small ones were popular among their subjects.*

## MONUMENTS TO AN AGE

Ancient Egypt has left no greater monuments than its pyramids, which stand as towering memorials to the entire pharaonic period. In fact, most pyramids were built in the space of only nine centuries, from the Third Dynasty to the Twelfth (ca. 2675–1759BCE), and the greatest of them all, at Giza, appeared in the seventy-five years (ca. 2585–2510BCE) that span the reigns of Khufu and his sons Khafre and Menkaure.

The first pyramid and its accompanying funerary complex were built by King Djoser (ca. 2650BCE) of the Third Dynasty (see pp.178–9), but their origins have been traced to the tomb of the Second-Dynasty king Khasekhemwy (ca. 2675BCE) at Abydos. Like all the kings of Egypt's first two dynasties, Khasekhemwy was buried in a large underground tomb complex consisting of a flat roof and vertical walls. This particular type of tomb, known as a *mastaba* (see p.197), marks a significant change from the tombs of early rulers. Khasekhemwy's funerary complex includes the earliest boat pits (in which parts of boats were buried), a typical feature of the classic pyramid complex. At Saqqara, King Djoser and his architect Imhotep built the king's funerary monument, the great "Step Pyramid" in several stages (see illustration, p.179), the first of which was a large mastaba. They also set a new trend by using, for the first time, building blocks of limestone rather than mudbrick.

Over the next hundred years, the step pyramid evolved into the celebrated "true" pyramids of the Fourth Dynasty at Dahshur and, most famously, Giza (see pp.180–85). The number of workers needed to build the Great Pyramid of Khufu can scarcely be imagined: the monument contained some 92 million cubic feet (2.6 million $m^3$) of stone. However, over the next two dynasties, architects decreased the sheer mass of their masters' pyramids, using smaller stones, loose rubble or mudbrick for the inner core. It has been estimated that the stone used in Khufu's pyramid alone is almost equal to the entire amount contained in all the pyramids of the Fifth and Sixth dynasties (see pp.188–9).

As the pyramids became less massive, the number of reliefs on the inner walls increased. The decrease in pyramid size toward the end of the Old Kingdom is usually ascribed to economic necessity. Yet a reduced need for the low-paid labour to move stones may have been counterbalanced to some degree by the employment of more skilled artisans – who presumably commanded higher remuneration – to carve the reliefs. The Fifth-Dynasty funerary complex of Sahure (ca. 2485–2472BCE) at Abusir has an estimated 108,000 square feet (10,000m²) of wall reliefs. During this dynasty, the sacred inscriptions known as the "Pyramid Texts" appear for the first time in the pyramid of Unas at Saqqara (see p.188).

Within their limestone casings, pyramids of the Middle Kingdom were built either entirely of mudbrick or of mudbrick with a rubble core. Horrified that even the huge Fourth-Dynasty pyramids had not been safe from thieves, Middle Kingdom rulers created a maze of chambers and corridors to confuse robbers. Nevertheless, despite such ingenious precautions, these pyramids were also later ransacked.

The rulers of the New Kingdom, beginning with the Eighteenth Dynasty pharaoh Thutmose I (ca. 1493–1482BCE), were buried deep in the solid rock of what is now called the Valley of the Kings on the west bank of the Nile at Thebes. The Valley of the Kings was a new departure for royal tombs, but not a complete break with the past. For example, in a return to the practice of First- and Second-Dynasty pharaohs, the architect Ineni built Thutmose I's mortuary temple in a separate location, a practice followed by the king's successors (see map on p.195). Perhaps most significantly, Thutmose I and his successors still lay beneath a pyramid – not one of human design, but the striking natural pyramid known as the Theban Peak or, in Arabic, Qurn.

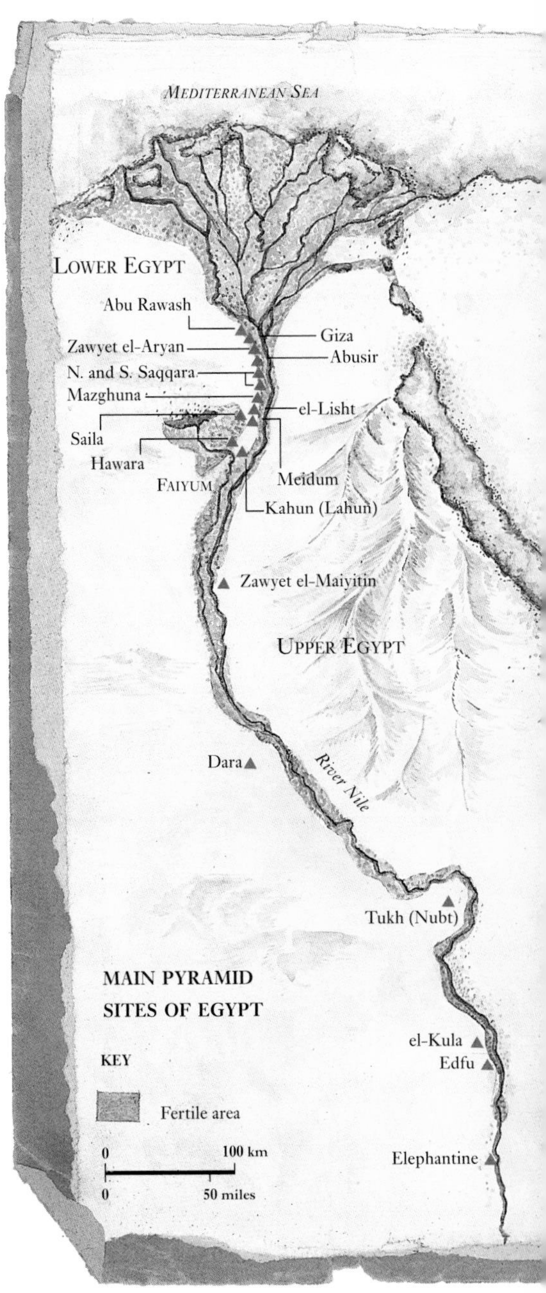

ABOVE: *Most pyramids stand within a twenty-five-mile long (40km) strip with Memphis as its mid-point. In this stretch of land, high above the floodplain on the natural rocky rises at the desert edge, are the Old Kingdom pyramids of Giza, Abusir, Saqqara, Dahshur, Meidum, Abu Rawash and Zawyet el-Arian. Middle Kingdom pyramids are located further south, near the entrance to the Faiyum oasis.*

LEFT: *The pyramids of Khufu, Khafre and Menkaure (foreground) at Giza.*

*An obelisk erected by King Thutmose I (ca. 1493–1482BCE) at the temple of Karnak. The pyramid-shaped tip of the monument, which would probably have been gilded, commemorates the* benben, *the primordial mound upon which the legendary phoenix, an incarnation of the sun god, was said to have perched. Slender, monumental, monolithic obelisks (that is, carved from a single piece of rock) such as this were typical of the New Kingdom; earlier structures, such as those at the sun temples of some 6th-Dynasty kings, were fatter and constructed of large masonry blocks.*

# RAMPS TO THE SKY: THE PYRAMID AND ITS COMPLEX

The pyramid shape is closely associated with the sun and the sun god, Re. According to the Pyramid Texts (see p.188), when the pharaoh died, the sun would strengthen its beams to create a celestial stairway or ramp, by which the deceased king would ascend to the heavens. The true pyramid can be seen to symbolize or represent this solar ramp, and the idea of the staircase is apparent in the design of the early step pyramids.

The pyramid is also connected with the *benu*, or phoenix, the legendary bird that was worshipped at Heliopolis as the incarnation of Re. The *benu* was said to perch upon a pyramidal mound known as the *benben*, represented by the pyramids themselves and also by the pyramidal point of the Egyptian obelisk. Near to their pyramid complexes, six kings of the Fifth Dynasty constructed temples to the sun that incorporated large, squat obelisks in imitation of the original *benben* that stood at Heliopolis (see pp.188–9). The *benben* was supposedly a representation of the primordial mound, which, according to Egyptian creation myths, rose from the

*A funerary stela of ca. 1000BCE depicting a woman before the falcon-headed sun god Re-Harakhte, who radiates beneficial rays toward her. As the manifestation of the rising sun, the deity was associated with the eastern face of the pyramid; he is invoked in an inscription on the pyramidion of Amenemhet III (see box, opposite).*

primeval waters, or Nun, at the beginning of time (see pp.120–21), thereby giving the creator sun god a place upon which to come into existence. It seems that King Khufu, the builder of the Great Pyramid, was regarded as having become the incarnation of Re in his lifetime. He was almost certainly identified with Re after his death, because the Great Pyramid had the Egyptian name *Akhet-Khufu* ("The Horizon of Khufu") – in other words, it was the point from which Khufu would rise each day as the sun. His heirs, the pharaohs Djedefre and Khafre, were the first kings to bear the royal title *Sa-Re* ("Son of Re").

Typically, a pyramid complex contains about fourteen distinct architectural components, each with a specific function and location. Although many of these elements existed earlier, the more complete form of the complex emerged in the Fourth Dynasty and continued throughout the Old Kingdom with little change, except occasionally to accommodate a

## THE PYRAMIDION

The finished pyramid was topped by a distinctive pyramidal capstone known as a pyramidion. Many Old and Middle Kingdom examples have been discovered, the oldest being that from the Red (North) Pyramid of Sneferu, founder of the Fourth Dynasty, at Dahshur. It is thirty inches (78.5mm) high and was found smashed in pieces where it presumably had tumbled. The second oldest pyramidion, found by the author near the recently discovered satellite pyramid of Khufu (see pp.182–3), is one of the few complete examples. The best-known pyramidion is that from the pyramid of Amenemhet III of the Twelfth Dynasty. It is inscribed on four sides with the deities Harakhte (god of the rising sun), Anubis, Osiris, Ptah and Neith. The inscription that relates to Harakhte reads: "May the face of the king be opened so that he may see the Lord of the Horizon [Harakhte] when he crosses the sky; may he cause the king to shine as a god, lord of eternity and indestructible." (I.E.S. Edwards' translation.) In reply, Harakhte says that he has "given the fair horizon to the king".

An inscription from the pyramid of Queen Udjebten at Saqqara suggests that its pyramidion was gilded. That this might have been the case is substantiated by the author's recent discovery at the pyramid of Sahure at Abusir of an inscription suggesting that its pyramidion was encased in fine gold, that is, electrum.

*The grey granite pyramidion of King Amenemhet III (ca. 1818–1772BCE), found in 1900. It is a little over three feet (1.05m) high.*

**MODEL PYRAMID**

**Among the many enigmas surrounding the pyramids is the purpose of the miniature models of these structures that have been found. The British archaeologist Flinders Petrie, the father of modern Egyptology, discovered a miniature step pyramid that is thought to be a model of Djoser's pyramid. He also found a second model at the pyramid temple (known as the "Labyrinth") of Amenemhet III at Hawara (see p.190). The most recent discovery is of a model found by Dieter Arnold near the pyramid of Amenemhet III at Dahshur.**

new cult or for topographical reasons. The first element is the pyramid itself. Often – but by no means always – it serves as the actual burial place for a king or a queen, in which case the burial chamber is usually located beneath the pyramid (those of Sneferu and Khufu are exceptions in being within the body of the pyramid). Another pyramid type, called the subsidiary, or satellite, pyramid, is connected with the cult of the king and is usually located on the south side of the main pyramid. Some subsidiary pyramids are the tombs either of the king's wife or his mother.

Each pyramid has at least two enclosure walls, an inner one marking the boundaries of the pyramid court and an outer one demarcating the area of the pyramid complex as a whole. Enclosure walls dating to the Old Kingdom are not inscribed, but those of the Middle Kingdom bear the king's titles, such as those of King Senwosret I at el-Lisht.

The third component is the mortuary temple, also known as the upper or funerary temple. With one exception – the pyramid of the Fifth-Dynasty pharaoh Userkaf at Saqqara – it is located on the east side of the pyramid. During the Fourth Dynasty, the mortuary temple had a simple plan, but from the Fifth Dynasty onward additional storage magazines were added to the north and south sides. A sloping causeway connects the mortuary temple to the valley, or lower temple, located at the edge of the agricultural floodplain. Of the eight known valley temples, the earliest accompanies the Bent Pyramid at Dahshur, while the most complete is that of Khafre. The causeway between Khufu's mortuary and valley temples was the first to be roofed, in an apparent attempt to protect the reliefs on its walls. From his reign onward, the causeways are decorated with wall reliefs, which are most abundant during the Fifth and Sixth dynasties.

Near the valley temple was the "pyramid city", the home of the personnel who maintained the cult of the king, which was governed by a royal overseer. Remains of pyramid cities have been found at Dahshur, Giza and Kahun (see p.71). "Workshop areas" within the pyramid complex made bread and beer to feed the workers in the pyramid cities, and produced statues, stone and pottery vessels, flint knives and other equipment necessary for the cult of the deceased monarch. The workers involved in the construction of the pyramid itself lived separately from the cult personnel within the pyramid complex. The royal funerary complex also had its own farm estate, or "royal funerary domain", located near the Nile. Half the farm's produce went to people living in the pyramid city; the rest went to the living king and his entourage.

Around the pyramid and the causeway there might be numerous boat pits (see p.168), as in the case of Khasekhemwy, who has twelve near his mudbrick enclosure wall. The number varies: Unas, the last king of the Fifth Dynasty, has two boats, while Khufu and Khafre each have five; yet none are known from the pyramids of Menkaure (see p.185) and others.

A "pyramid harbour" was located in front of the valley temple. During the building of the pyramid, stones, labourers and officials passed through the harbour and along canals leading to the Nile. After construction had been completed, the harbour was used to bring in products needed for the maintenance of the royal cult. The delivery area in front of the valley temple that includes the harbour and canals was known in Egyptian as the "Mouth of the Lake" (see also p.176).

The king may have supervised the building of his pyramid from a neighbouring palace. Certainly, the importance of the pyramid as both a personal and national monument would make it necessary for the king to live nearby. The Predynastic town of Ineb-hedj ("White Walls"), near the royal tombs at North Saqqara, probably took its name from one such royal residence. Ineb-hedj later developed into the capital city of Memphis (see p.74), but there is no concrete evidence that the Old Kingdom pharaohs actually resided there, as is commonly suggested. A large settlement at Giza, covering just over a square mile ($3km^2$), was probably the location of the royal palace and associated administrative buildings during the time of the construction of the three great pyramids of the Fourth Dynasty. According to one source, King Djedkare Isesi of the Fifth Dynasty lived outside the capital city near his pyramid complex at Saqqara. During the Middle Kingdom, King Amenemhet I of the Twelfth Dynasty moved his capital from Thebes in the south of the country to a new city, Itj-Tawy, that was located close to his pyramid complex at el-Lisht (see p.28).

## THE FUNCTION OF THE PYRAMID COMPLEX

There is no clear consensus as to the actual function of the two temples, causeway and other parts of the pyramid complex. According to one theory, the king's body was mummified in the valley temple, where associated rituals took place, then carried in procession along the causeway to the mortuary temple. After further rites, it was borne into the pyramid courtyard and finally into the pyramid itself.

However, in the pyramid complexes of the Fourth to Sixth dynasties, the doorways between the mortuary temple and the pyramid court would have been too narrow for the king's coffin to pass. Nor do any inscriptions in the valley temples remotely suggest that the temples were used for mummification.

A theory that may be more in line with the archaeological evidence is that the pyramid complex served both as a temple to the deified king and as a "ritual palace" that served the needs of the pharaoh in the afterlife. Wall reliefs in the pyramid complexes resemble those found in temples elsewhere and were of a type that probably also appeared in palaces: "smiting" scenes (showing the king dominating his enemies); scenes of the king among the gods; and scenes of the royal *sed* (jubilee) festival.

Assuming that the various elements of the pyramid complex in fact comprised a great palace for the dead pharaoh and were not used for his mummification or funeral, it is likely that the king's body would have been mummified in the royal workshops. The embalming rituals may have taken place outside the valley temple, in a specially erected structure known as a "purification tent". The funeral cortège would then have proceeded directly to the pyramid court – bypassing the two temples and the causeway – before entering the pyramid.

**THE CURSE OF THE BUILDER'S WIFE**

**Inside the tomb of one of the men who worked on the construction of the Great Pyramid and his wife are two curses that illustrate that it was not only the pharaohs who were anxious to prevent robbers from entering their burial places. The wife's curse reads: "O all people who enter this tomb who will make evil against this tomb and destroy it; may the crocodile be against them on water, and snakes be against them on land; may the hippopotamus be against them on water, the scorpion against them on land." The husband's curse is similar, invoking the crocodile, lion and hippopotamus. (See also pp.144–5.)**

# BUILDING THE PYRAMIDS

Of the nearly one hundred pyramids in Egypt, the greatest is that of Khufu (see pp.180–83). It has lost little from its original height of 480 feet (146m) and base width of 755 feet (230m), formed by 2.3 million limestone blocks. Each side of the pyramid rises at an angle of precisely 51° 52'. Such statistics are all the more impressive if we remember that this extraordinary building feat was achieved four and a half thousand years ago.

The massive labour force required to build a pyramid came under the direction of one man, the Overseer of All the King's Works. His position required him to be a man of science, an architect and a figure of commanding authority and outstanding leadership abilities. He was responsible for a monumental undertaking of national importance. When completed, the pyramid complex incorporated several different architectural elements, all of which were essential to ensure the pharaoh's safe journey into the afterlife. Every Egyptian household had to help in the project by providing either food or manpower for the project's work crew.

The Overseer's first decision was critical: where should the monument be located? Tradition required that the site be on the west bank of the Nile, close to the land of the dead (known in Egyptian as "the West").

## THE WORKERS' COMMUNITY AT GIZA

The author recently took part in excavations to the southeast of the Giza Plateau and the Sphinx that have shed light for the first time on the daily lives of those who struggled for decades to build the pyramids, as well as those who served the pyramid temples long after construction was complete. There are four major sites: the tombs of labourers and their overseer; the tombs of the skilled artisans (connected to the workers' cemetery by a ramp); what scholars call the "institution areas", including a bakery and grain silos; and the workers' town, discovered during salvage archaeology before the construction of a sewage system. Scenes on the tomb walls provide us with a view of the workers' tasks, dress, titles and beliefs. Most importantly, the excavations have shown that they built their own tombs in the shape of a pyramid, albeit of mudbrick rather than stone.

*A stonemason's tools: a wooden mallet and two bronze chisels dating from the New Kingdom.*

Perhaps eighteen thousand people lived in the workers' town (which was separated from the pyramids by a wall, the Heit el-Ghorab), while others returned to homes elsewhere at the end of a day's work. No doubt the labour was hard: an analysis of the workers' bones reveals that fractures and crushing injuries were common.

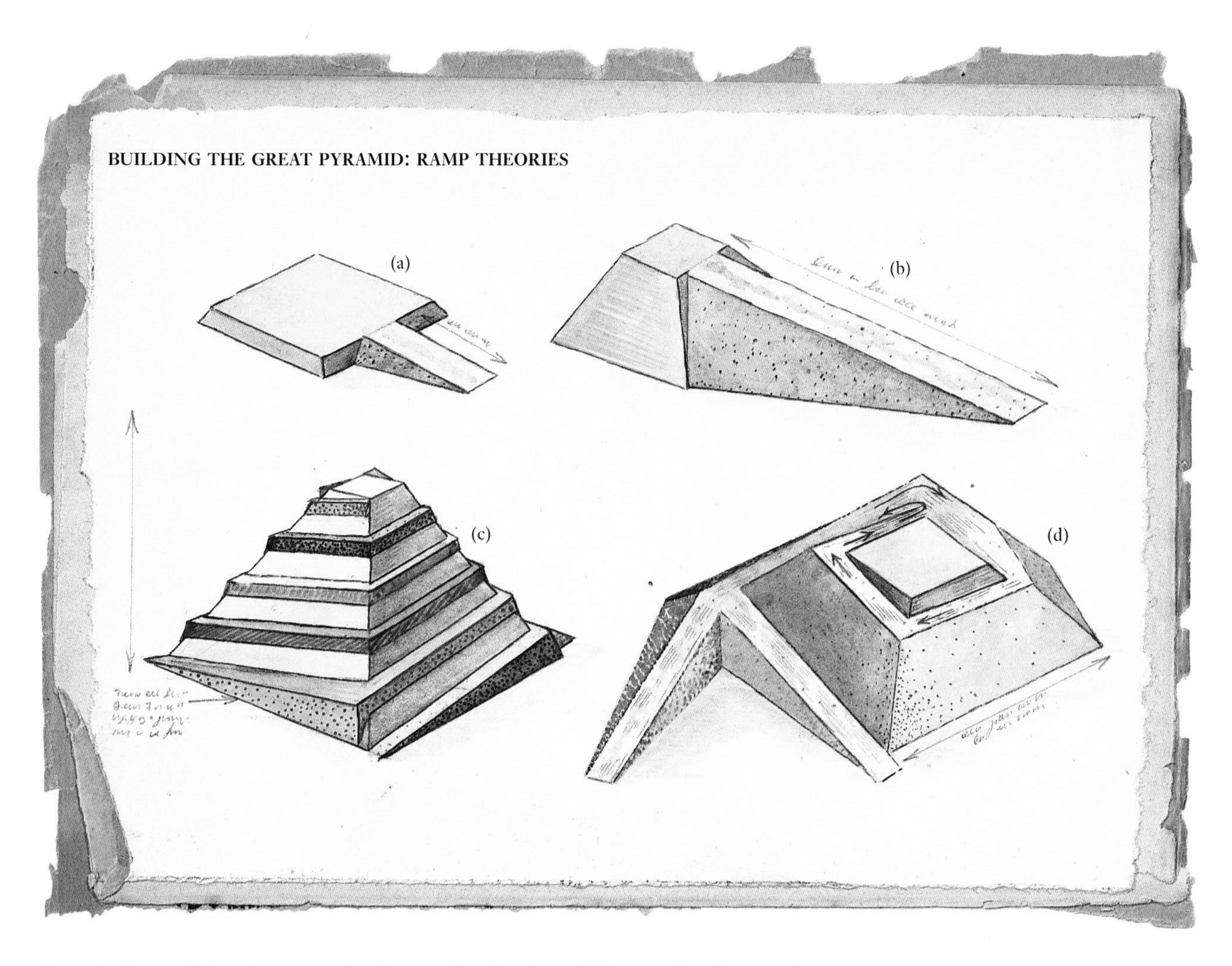


*The principal theories of how the Great Pyramid was constructed (see main text). Diagrams a and b: a single straight ramp; diagram c: four parallel ramps spiralling around the pyramid; diagram d: Mark Lehner's theory, in which the straight and spiralling ramps are combined.*

Practical considerations required that the site be within reach of a good supply of limestone. Next, the Overseer had to establish the quarry (see p.177), the supply ramp, the pyramid harbour (see p.172) and the settlement for the thousands, perhaps sometimes tens of thousands, of workmen. Each element had to be placed in the natural terrain in such a way as to ensure an efficient flow of men and materials.

The question of how such a massive edifice as the Great Pyramid was constructed has long intrigued Egyptologists, and was the subject of speculation even in antiquity. According to the archaeological evidence, it is most likely that a ramp was used – what sort of ramp is not clear (see illustration, above), but there are two basic theories: a straight ramp or a spiral ramp. The first theory (diagram a) proposes a single large ramp at right angles to one face of the pyramid. This would have the advantage of leaving all four corners and three sides of the pyramid clear during construction, allowing builders to check the rise of the sides and the diagonals. Careful surveying during construction was essential, otherwise a "twist" might occur and the edges would not meet at a point.

However, to maintain a functionally low slope – no steeper than 1:6 – the ramp's length would need to be increased every time a rise in height

**THE PYRAMID HARBOUR**

**Non-local materials for the building of a pyramid, such as granite, basalt, alabaster and fine limestone from Tura, would have been brought by river to a harbour and thence to the construction site. In the case of the Great Pyramid, materials would have reached the site along the great wadi (dry river-bed) that lies between the pyramid plateau and the southern rise of the Ma'adi formation. Khufu's pyramid harbour (see p.176), therefore, must lie buried at the mouth of the wadi. This has been confirmed by excavation, which has located the harbours of Khufu and Khafre at Giza.**

**More recently, the author and his colleagues have found a basalt wall just over half a mile (800m) south of Khufu's lower temple which could be the harbour wall. Another huge wall, known as Heit el-Ghorab, "The Wall of the Crow" and constructed of limestone blocks as big as those in the pyramids, extends from the mouth of the wadi and possibly surrounded the harbour. A great gate in the wall may have been the main entrance leading from the quayside to the area where the pyramid workers lived (see box on p.174).**

was desired (diagram b). This would cause all work on the pyramid to cease temporarily, because the design does not allow for the concurrent construction of both pyramid and ramp. The final ramp to the top of the pyramid would have had to be so long that it would have extended over and beyond the quarry, which is clearly impossible.

Another idea postulates a ramp spiralling around the pyramid in some way. The most popular form of this proposal has a ramp starting at each corner to create four parallel ramps spiralling upward and resting on the unfinished outer casing blocks for support (diagram c). These blocks would be smoothed as the ramps were dismantled after the apex of the pyramid had been reached. This theory leaves most of the pyramid's face clear during construction for checking lines and corners, and the ramps remain confined to the pyramid's immediate area. Nonetheless, the hypothesis has its own problems. First, it seems unlikely that the rough faces of the pyramid could have borne the weight of ramps which, the theory proposes, were made of mudbrick or debris. Also, a spiralling ramp increased the distance over which the blocks had to be hauled and entailed great strain for the teams dragging each multi-ton block when they had to turn this ramp's sharp corners.

A compromise between these two principal ideas, suggested by the American Egyptologist Mark Lehner, proposes a ramp that combines the two previous theories (see diagram d). It starts at the mouth of the quarry and rises to about one hundred feet (30m) above the pyramid's base at its southwest corner. Also spiralling upward as the pyramid rose, the ramp encased the entire pyramid. The weight of this ramp would have been borne by the ground around the pyramid and traffic would have moved along a broad roadway with very sharp turns. According to this theory, the initial slope would be a feasible 6.5° from the quarry, rising with each turn to as much as 18° for the final ascent to the apex. Although this last slope may seem too steep, it would only be about 110 feet (40m) long, and by this stage the number and size of the blocks would have been much smaller – at two-thirds of its height, the pyramid was already about ninety percent complete.

The chief advantage of Lehner's proposal is that the ramp would reach the top of the pyramid in the shortest possible distance and, because most of it was supported by the ground, the ramp would not weigh upon the faces of the pyramid. However, if the ramp enveloped the entire pyramid, it would have been difficult for surveyors to check all the lines and angles. Also, it is doubtful whether such a huge ramp could have risen to the full height of the Great Pyramid without collapsing.

Most Egyptologists assume that the pyramid ramp was built of mudbrick. But a mudbrick construction as massive as the pyramid's supply ramp would certainly have left stains or deposits of mud that do not exist in the area south of Khufu's pyramid. However, this area and the quarry

## QUARRYING THE STONE

By carefully studying the ancient clues that the pyramids have left for us, we are coming closer to solving many of their puzzles. The discovery of the main quarry for the Great Pyramid is a case in point. It could not have been on the east side of the pyramid, because there are tombs here from the twelfth year of Khufu's reign, before the pyramid was finished. Other tombs, begun in year five, rule out the area to the west. On the north side of the pyramid, there is no evidence for a quarry.

On the south side, however, are a boat pit dated to the reign of Djedefre and tombs from the time of Menkaure. Both kings ruled after Khufu, indicating that this area was clear during Khufu's reign, and was the probable location for the ramp leading from the quarry to the pyramid. There are usually small subsidiary pyramids on the south side of the main monument, but Khufu's Overseer set his subsidiary pyramids on the east side. This is further evidence that he wanted the south side kept clear for the construction of the ramp. The evidence points to the Overseer siting the quarry low on the Giza Plateau, about half a mile (750m) due south of the proposed position of the pyramid. This area supplied good quality stone for the pyramid's larger blocks. The Great Pyramid's quarry was much larger than that required by the Overseers of the Fifth and Sixth Dynasties and the Middle Kingdom, because their pyramids have cores mostly of rubble or mudbrick.

The stone was prepared by inscribing squares onto the quarry face. To remove the stone, wooden wedges were driven into the top of the rock face and soaked with water, causing them to expand and split the rock. Blocks were then cut out and hauled on sleds up the pyramid ramp by another workgang. The quarry labourers and hauling gangs worked all year round, and probably consisted of conscripts who served in rotation for a few months at a time.

are filled with vast quantities of a particular kind of debris: limestone chips, gypsum and a chalky clay called tafla. It would be reasonable to assume that this must be the material from which the main supply ramp was constructed, and that it was pushed away as the ramp was dismantled. Recently, excavations by the author and his colleagues have revealed two sections of the ramp that prove not only that it linked the quarry with the southwest corner of the pyramid, as Mark Lehner has suggested, but that it was indeed made of rubble and tafla.

The Great Pyramid is nearly a perfect square at its base and it is accurately oriented to true north, while the north and south sides of its 13.5-acre (5.4ha) base are off true parallel by just one inch (2.5cm). How did the ancient builders achieve such accuracy? Around the base of the pyramids of Khufu and Khafre there are a series of holes, each about the size of a dinner plate, set at regular intervals and forming lines that run parallel to the sides of the pyramids. These were almost certainly for a series of stakes that served as a reference line for the builders. The idea of such a line to accurately align the bases of the pyramid is substantiated by evidence of similar lines and stakes used by the architect of a pyramid for Queen Henutsen. In addition, some of the fifteen-ton blocks that form the bases of the pyramids are marked to denote the central axes of the pyramid's faces and diagonals.

# THE EARLIEST PYRAMIDS

**THE ENIGMA OF THE SIX STEP PYRAMIDS**

**Six step pyramids – at Abydos, Elephantine, el-Gonamia, el-Kula, Naqada and Zawyet el-Maiyitin – perplex Egyptologists. They are all built of limestone – most are clad with it too – and are conventional in appearance, consisting of a core and two or three steps. But none are sited in a cemetery, and two, at Zawiet el-Maiyitin and Elephantine, are not even in the Western Desert. Moreover, they do not have burial chambers, passages or attached structures.**

**Some experts believe that they are simply monuments to the king's power in the provinces and attribute all six to Huni, the last king of the Third Dynasty, whose name has been found at Elephantine. From an examination of the structures themselves, however, it seems likely that they were local royal palaces or lodges that the king used for the supervision of tax collection and other provincial business.**

Djoser, who ruled ca. 2650BCE at the beginning of the Third Dynasty, chose Saqqara as the site of the first pyramid: the famous Step Pyramid, which he built from a series of successively smaller mastaba superstructures set one on top of another. It soars two hundred feet (60m) skyward and spans 467 feet (140m) along its east–west axis and 393 feet (118m) along its north–south axis. With such great dimensions, hitherto unheard of in Egypt, the Step Pyramid dominates Djoser's large funerary complex, which is impressive in itself. Within an enclosure wall similar to that of Djoser's father Khasekhemwy near Abydos (see p.168), the complex includes an area for the royal *sed* (jubilee) festival (together with its associated chapel), "Houses of the North and South" (models of temples in Upper and Lower Egypt), mortuary temples and a *serdab* (see p.197).

Djoser's successor, Sekhemkhet, attempted to build his own large step pyramid southwest of Djoser's. It is known as the Buried Pyramid and its base survives to a height of just twenty-three feet (7m). The pyramid was discovered in 1951, together with its 1,820- by 633-foot (545m by 190m) enclosure, by the Egyptian archaeologist Zakaria Goneim. The masonry of the enclosure wall is the same as that of Djoser's pyramid, suggesting

## IMHOTEP, ARCHITECT AND GOD

Imhotep, the architect of Djoser's Step Pyramid (see main text), is one of the few non-royal Egyptians who became a legend (see also p.35). For centuries, he was renowned in Egypt for his achievement in constructing the Step Pyramid and for using stone for the first time as a building material rather than mudbrick.

His name and titles, written on the base of a lost statue of Djoser, indicate how high he rose in the royal administration. He was a vizier; "Overseer of the Seers" (a unique epithet that perhaps links him with the seer-priests of Heliopolis); "First for the King"; "Director of Public Works in Upper and Lower Egypt"; "Keeper of the Seals of Lower Egypt"; "Recorder of the Annals"; and "Supervisor of the Great Palace". Even the names of Imhotep's parents are recorded in an inscription from Wadi Hammamat in the Eastern Desert: his father was Kaneferu and his mother Ankh-kherdu.

*A Late-Period bronze votive statuette of the deified Imhotep, the architect of the great Step Pyramid.*

Long after his death, educated Egyptians considered Imhotep a great wise man and revered him as their patron: scribes would recite his name before embarking on a piece of writing. Late Period Egyptians of all classes revered him as a god, and the Greeks associated him with Asklepios, their god of wisdom and medicine. They built a chapel of Imhotep at Philae and worshipped him at the Asklepion, the healing temple at Memphis.

*Successive generations of Egyptians admired Djoser's Step Pyramid: visitors left graffiti expressing astonishment at the monument more than a millennium after it was built.*

that the architect may also have been the famous Imhotep (see box).

Khaba, who succeeded Sekhemkhet, built the "Layer Pyramid" at Zawyet el-Arian, four miles (7km) north of Saqqara. This step pyramid, which was excavated in 1900 and 1910, takes its name from the fact that it was built outward in layers rather than upward in mastaba-like blocks (Sekhemkhet's pyramid was built in the same way). Now fifty-three feet (16m) high, it was originally about three times this height and had five steps. A little to the northwest is the "Unfinished Pyramid", of which nothing remains but traces of a square burial chamber. Its owner is unidentified, but may have been Nebka, Khaba's successor, although some experts ascribe it to the Fourth-Dynasty king Baka, son of Djedefre.

Sneferu, the founder of the Fourth Dynasty, is considered the greatest Old Kingdom pyramid builder because he constructed four great edifices: one each at Sila and Meidum and two at Dahshur. The Meidum pyramid apparently began as a step pyramid. Sneferu then began the construction of a true pyramid at Dahshur, but it nearly collapsed and the architect had to reduce its 54° angles to just over 43°, creating the unique profile of the Bent Pyramid. In the event, Sneferu built the first true pyramid – the Red or North Pyramid – just over a mile (2km) to the north; it also has angles of 43°. In his last years, Sneferu encased the Meidum pyramid to make a true pyramid, and it is the only example showing the transition from the step pyramid. Apparently, it collapsed in the nineteenth century, after farmers took casing stones for building. It now consists of a mound of debris and an inner core of three steps.

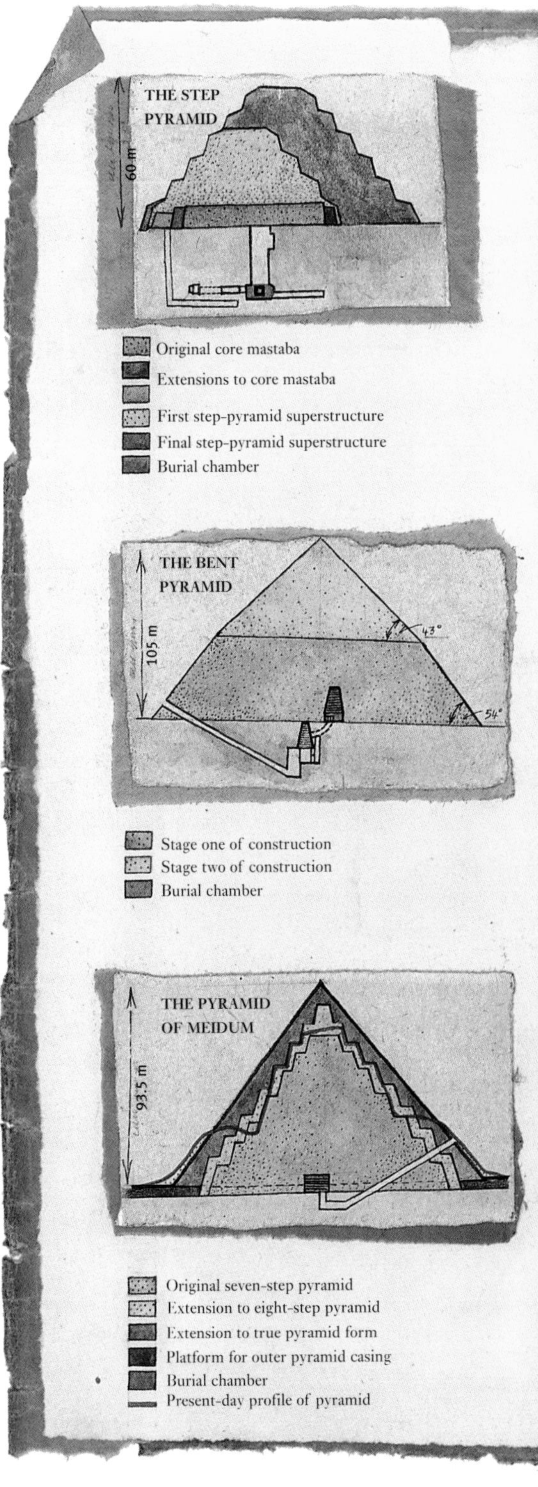


*Cross-sections of the Step Pyramid, the Bent Pyramid and the pyramid at Meidum, showing the stages in which they were constructed.*

# GIZA: THE GREAT PYRAMID

Standing at the northern end of the Giza Plateau, the Great Pyramid of King Khufu, or Cheops, (ca. 2585–2560BCE) is the most massive of all the pyramids and also the most famous, overshadowing the reputation of its Fourth-Dynasty builder. Already in ancient times, those who saw the pyramid were so awestruck by the scale of the monument that they assumed Khufu must have employed inhuman methods to build it, and so began his reputation as a tyrant (see p.25).

Apart from ancient tradition, there are several pieces of concrete evidence to show that the pyramid was actually built by Khufu. His cartouche occurs, together with the names of the workgangs involved in the pyramid's construction, on the inner walls of the second weight-relieving chamber above the king's burial chamber (see pp.178–9). Inscriptions that refer to Khufu and the pyramid also occur in the nearby mastaba tombs on the eastern and western sides of the pyramid. Khufu's monument is the focal point of a large necropolis for the king's family and officials. His mother, chief queens and other family members are buried to the east of the Great Pyramid in three subsidiary pyramids and a series of mastaba tombs. Officials lie in mastabas to the west of the king's tomb.

Khufu's mortuary (upper) temple, on the eastern side of the pyramid, was a rectangular building with a basalt pavement and an inner courtyard;

**THE TOMB OF THE PRIEST**

**Nearly seventy tombs have recently been discovered on the west side of Khufu's pyramid, among the field of mastabas built for his officials. One of these belongs to Kay, a priest who served under the first four pharaohs of the Fourth Dynasty: Sneferu, Khufu, Djedefre and Khafre. The tomb is painted with beautiful scenes of daily life. Among them are processions of boats led by the deceased, an offering scene and a list of offerings. On the left side of the tomb's entrance is a touching and unique scene of a lady embracing Kay with both arms.**

**To the right of the entrance, Kay left a biographical detail that provides us with some idea of how Egyptian tomb workers were categorized at the time: "It is the tombmakers, the draftsmen, the craftsmen and the sculptors who built my tomb. I gave them beer and bread. I made them take an oath that they were satisfied."**

## THE DISCOVERY OF KHUFU'S SATELLITE PYRAMID

In 1991 the author and his colleagues began excavating an area on the eastern side of Khufu's pyramid in order to prepare the site for visitors. Unexpectedly, they discovered the remains of what was later confirmed to be a fourth subsidiary pyramid that had stood at the southeast corner of the main pyramid. The ruins cover an area of 365 square feet ($34m^2$). The surviving superstructure consists of two courses of crude blocks with rubble in-fill on three sides and part of the original outer casing of fine Tura limestone on the east and south sides. A graffito scrawled in red paint on the inside of one set of blocks in the south wall reads *imy rsy sa*, "on the south side" – an instruction informing the construction workers where to position the blocks.

The substructure of the pyramid displays the T-shaped intersection of the passage and inner chamber that is usual in satellite pyramids dating to around the end of Khufu's reign. Interestingly, it appears that the masonry of the pyramid was plundered in pharaonic times, leaving the passage and chamber open to the sky.

The pyramid does not seem to have been a tomb and its purpose is much debated, inspiring several theories: it was a residence for the king's *ka*, or spirit; it housed his viscera; it had some solar function; or it had a role in the king's *sed* (jubilee) festival in the afterlife. In the author's view, the pyramid may have served as the deceased king's "changing room" for the *sed*.

ABOVE, LEFT: *Three of the Great Pyramid's five boat pits and traces of the mortuary temple and causeway can be seen in this photograph from the top of the pyramid toward the town of Nazlet es-Samman, at the foot of the Giza escarpment. Beyond the southernmost boat pit (on the right) are the remains of the small subsidiary pyramids of Khufu's queens, and the cemetery of mastaba tombs belonging to other members of the king's family.*

ABOVE: *A view up one edge of the pyramid, showing the massive limestone blocks that formed the pyramid's core. The smooth blocks of white Tura limestone that once faced the finished pyramid were stripped in early Islamic times for mosque-building projects in nearby Cairo.*

only the pavement remains. The temple was connected to the valley (lower) temple by a causeway that runs just north of due east from the mortuary temple and down the gradual slope of the Giza Plateau for 918 feet (280m), before apparently ending abruptly at the edge of the escarpment overlooking the Nile Valley. Where it went next was not known until the recent construction of a new sewerage system for the town of Nazlet es-Samman at the base of the escarpment. Salvage excavations involving the author and the archaeologist Michael L. Jones uncovered the continuation of the causeway (with some of its paving slabs still in place), together with remains of the valley temple itself. The total length of the causeway, from the east face of the Great Pyramid to the site of the valley temple, is now known to have been at least 2,630 feet (810m). After 2,435 feet (750m), the causeway reaches the bottom of the escarpment and veers slightly further to the northeast, as it descends across the Nile floodplain toward the valley temple. At the valley temple site, the author and his team found a black-green basalt pavement 180 feet (56m) long, about fifteen feet (4.5m) below the present ground level. Along the edge of the pavement is part of a massive mudbrick wall that when first built may have been as much as twenty-six feet (8m) wide.

Back up on the plateau, to the east of the Great Pyramid, are four subsidiary pyramids. The first is the tomb of Khufu's mother, Hetepheres. It has a boat pit on its south side and the remains of a chapel on its east side. The second is similar and contained the body of Queen Mereyites. The third, belonging to Queen Henutsen, has a mortuary chapel on the east side that was converted during the Twenty-Sixth Dynasty into a temple of Isis, with whom Henutsen came to be identified at this period. The author and his colleagues recently discovered the remains of the fourth satellite pyramid (see box), and there are traces on the south and east sides of a further two unfinished subsidiary pyramids.

# Inside the Great Pyramid

The entrance to the Great Pyramid leads to a corridor that slopes downward to a first, unfinished, burial chamber in the limestone bedrock beneath the pyramid. After this chamber was built, the architect, Hem Iwno, changed his original plan – presumably on the king's instruction – and began tunnelling upward into the roof of the descending corridor at a point approximately 65 feet (20m) from the entrance. This new ascending corridor, hewn through the pyramid's limestone core, opens about halfway up onto a level passageway leading to a second burial chamber, also unfinished, known – erroneously – as the "Queen's Chamber". Continuing upward, the corridor opens into a magnificent corbel-roofed passageway, the "Grand Gallery" (see illustration, opposite page). This ends in a low, narrow passage that leads to the third, and final, burial chamber, lined with red granite from Aswan. Khufu's granite sarcophagus lies at the western end of the chamber.

1 Great Pyramid of Khufu
2 Pyramid of Khafre
3 Pyramid of Menkaure
4 Mortuary (Upper) temples
5 Causeways
6 Valley (Lower) temples
7 Great Sphinx and Sphinx temple
8 Mastaba cemeteries
9 Subsidiary pyramids
10 Boat pits of Khufu

N

THE PYRAMIDS OF GIZA : PLAN

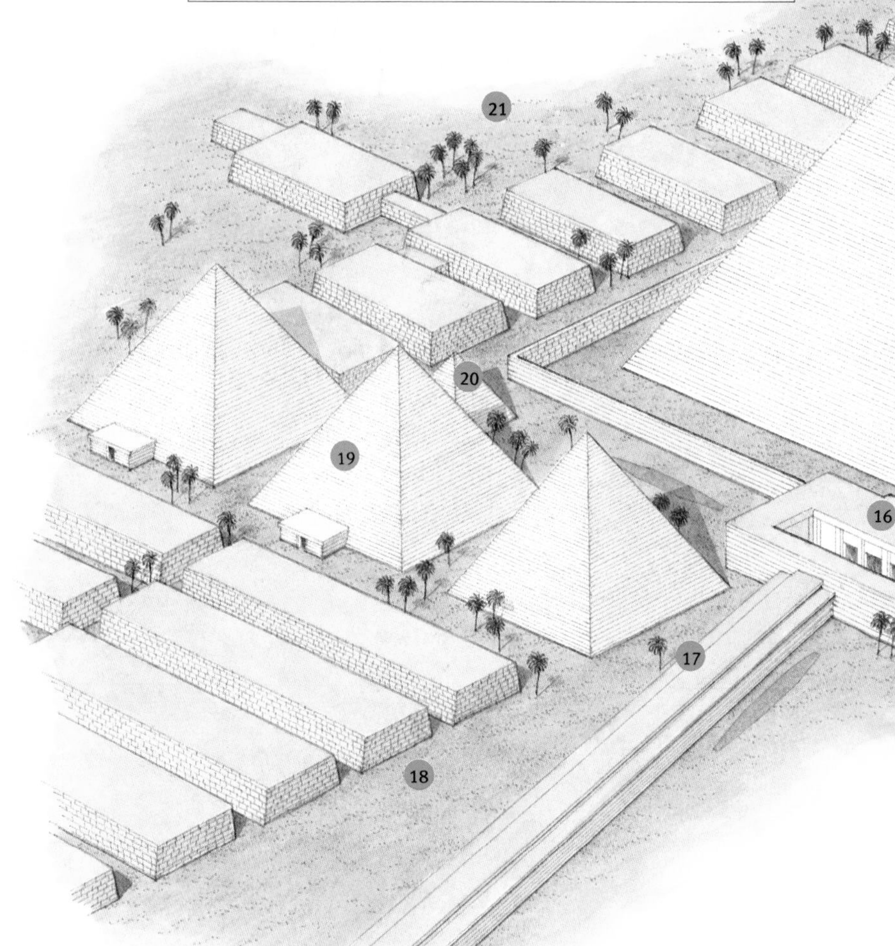


1 Pyramid of Menkaure
2 Pyramid of Khafre
3 Pyramidion of Khufu
4 Weight-relieving chambers
5 "Air channel"
6 Final burial chamber and sarcophagus ("King's Chamber")
7 "Grand Gallery"
8 Second burial chamber ("Queen's Chamber")
9 Ascending corridor
10 Entrance to Khufu's pyramid
11 Enclosure wall
12 Descending corridor to first burial chamber
13 Exit passageway used by tomb workers
14 First burial chamber, in bedrock beneath the pyramid
15 Boat pit
16 Mortuary (Upper) temple
17 Causeway
18 Eastern mastaba cemetery
19 Subsidiary pyramids
20 Newly discovered fourth subsidiary pyramid
21 Southern mastaba cemetery

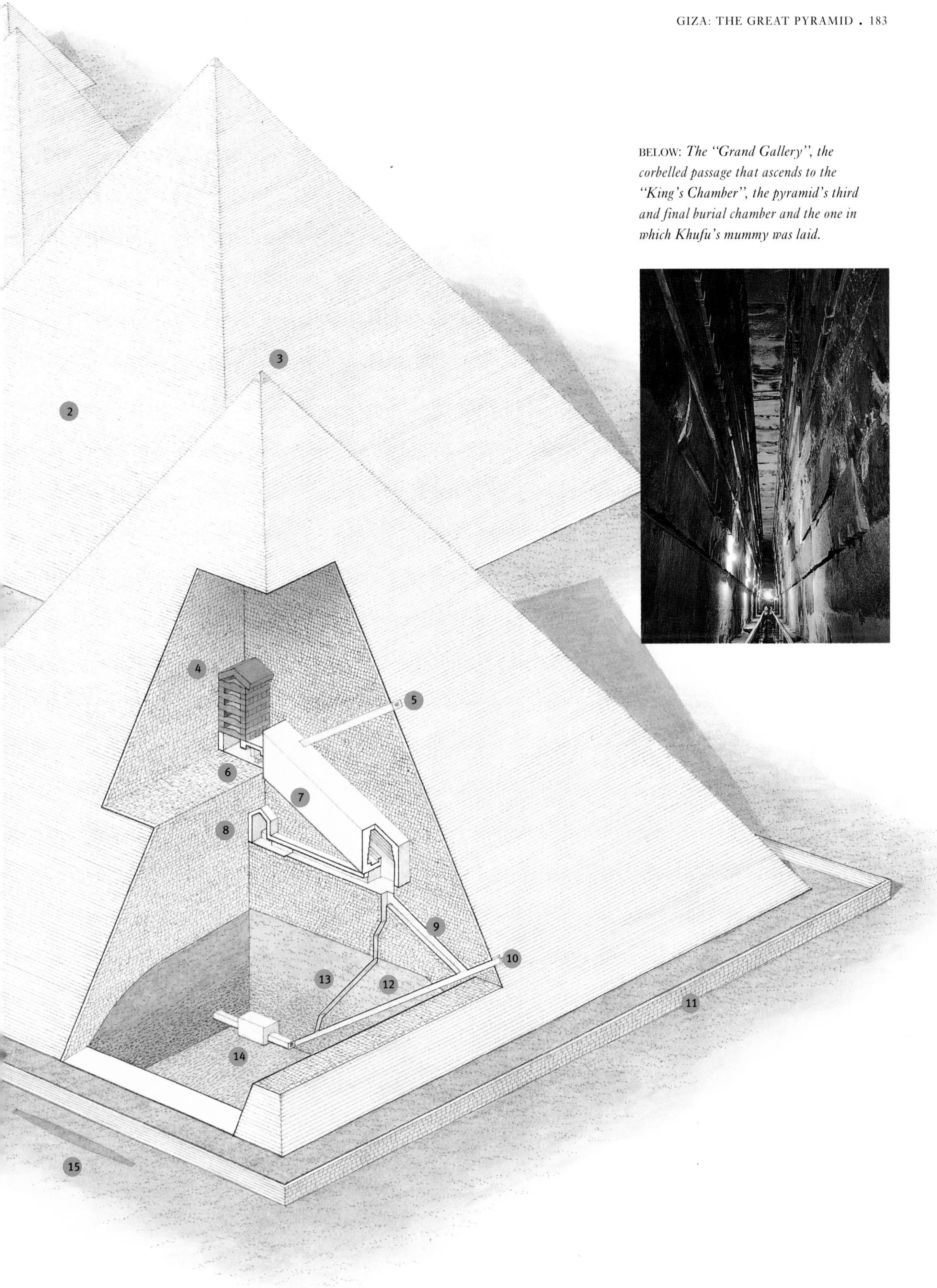

BELOW: *The "Grand Gallery", the corbelled passage that ascends to the "King's Chamber", the pyramid's third and final burial chamber and the one in which Khufu's mummy was laid.*

2
3
4
5
6
7
8
9
10
11
12
13
14
15

# GIZA: THE PYRAMIDS OF KHAFRE AND MENKAURE

The funerary complexes of the Fourth-Dynasty kings Khafre (ca. 2555–2532BCE) and Menkaure (ca. 2532–2510BCE) – Chephren and Mycerinus to the Greeks – are unique among Old Kingdom pyramid sites in their completeness. Khafre's complex, with its pyramid, mortuary temple, causeway and valley temple almost intact, affords us our most comprehensive understanding of an Old Kingdom pyramid complex. Systematic excavations of this huge site were first carried out under Auguste Mariette, the founder of the Cairo Museum, in 1869 and are continuing today.

Khafre's pyramid, called "Khafre is Great", rose from a 705-foot (215m) wide base to a height of 471 feet (144m) at an angle of 53°7'. It has two entrances, each opening onto a descending passage that leads to a chamber. The lower, earlier, passage, begins two hundred feet (68m) north of the pyramid and was hewn entirely from the solid rock of the plateau.

*Like its two companions, Khafre's pyramid was stripped of its outer casing of fine dressed stone in the Middle Ages, although the bottom course of red granite and the upper courses of Tura limestone remain, so that the pyramid has lost little from its original height of 471 feet (144m). It is built on slightly higher ground than the Great Pyramid of Khufu, and thus appears to be somewhat taller. In fact, the Great Pyramid was originally ten feet (3m) higher, although today the difference is just two and a half feet (0.7m).*

It is linked to the upper passage by an ascending corridor. The lower entrance was abandoned in favour of the upper one, which begins, like that of Khufu, in the north face of the pyramid. Why a new entrance was constructed, and why it was connected to the old one, are not known, but it has been suggested that after the corridor was built the architect decided to reposition the entire pyramid a little further to the south. At the end of the upper passageway, also partly cut from the bedrock, is the chamber containing Khafre's red granite sarcophagus. Two other tunnels in the pyramid are the work of ancient tomb robbers. The remains of a subsidiary pyramid – the tomb of a queen – lie near the main pyramid.

Khafre's mortuary temple was built of local limestone and incorporates a pillared hall, two long narrow chambers, an open courtyard that may have contained a seated sculpture of the king, five statue niches and magazines (storerooms). This basic arrangement was followed by all later Old Kingdom mortuary temples. Around the temple are five boat pits. Linking the mortuary and valley temples is a causeway about sixteen hundred feet (495m) long and sixteen feet (5m) wide. The valley temple – the best-preserved building of the Fourth Dynasty – is a square building with two entrances that bear the only inscriptions in the entire temple; all that can still be deciphered of them today are the words: "Khafre Beloved of [the goddess] Bastet" and "Khafre Beloved of [the goddess] Hathor".

The smallest of the Giza pyramids, that of Menkaure, was called "Menkaure is Divine" (see illustration, p.169). The pharaoh died before his funerary complex was finished, and parts of it were completed by his son Shepseskaf (ca. 2508–2500BCE). Many additions were made to the complex during the Fifth and Sixth dynasties, indicating that, despite his untimely death, the king's cult flourished for more than three centuries.

Originally about 240 feet (73m) high, the pyramid now measures 204 feet (62.2m) on a base 357 feet (109m) wide. The first explorer to enter it in modern times was Howard Vyse in 1837. From the entrance, on the north side, a descending passage leads to the burial chamber, in which Vyse found a basalt sarcophagus that would originally have contained Menkaure's mummy. Unfortunately, the sarcophagus was lost at sea off Spain in 1838 en route to the British Museum. An anthropoid wooden coffin lid bearing Menkaure's name was also found in the pyramid.

The pyramid was never finished, but Shepseskaf completed his father's mortuary temple, causeway and valley temple in mudbrick. The causeway, about 1,995 feet (608m) long, leads to the valley temple, where a number of triad statues were discovered depicting the king with Hathor and goddesses of the Egyptian nomes (provinces). South of the main pyramid are three subsidiary pyramids, each with a mudbrick temple on its eastern side. One of these smaller pyramids may be that of Khamerernebty II, Menkaure's principal queen, who is portrayed with her husband in a fine statue found in the valley temple (see illustration, p.216).

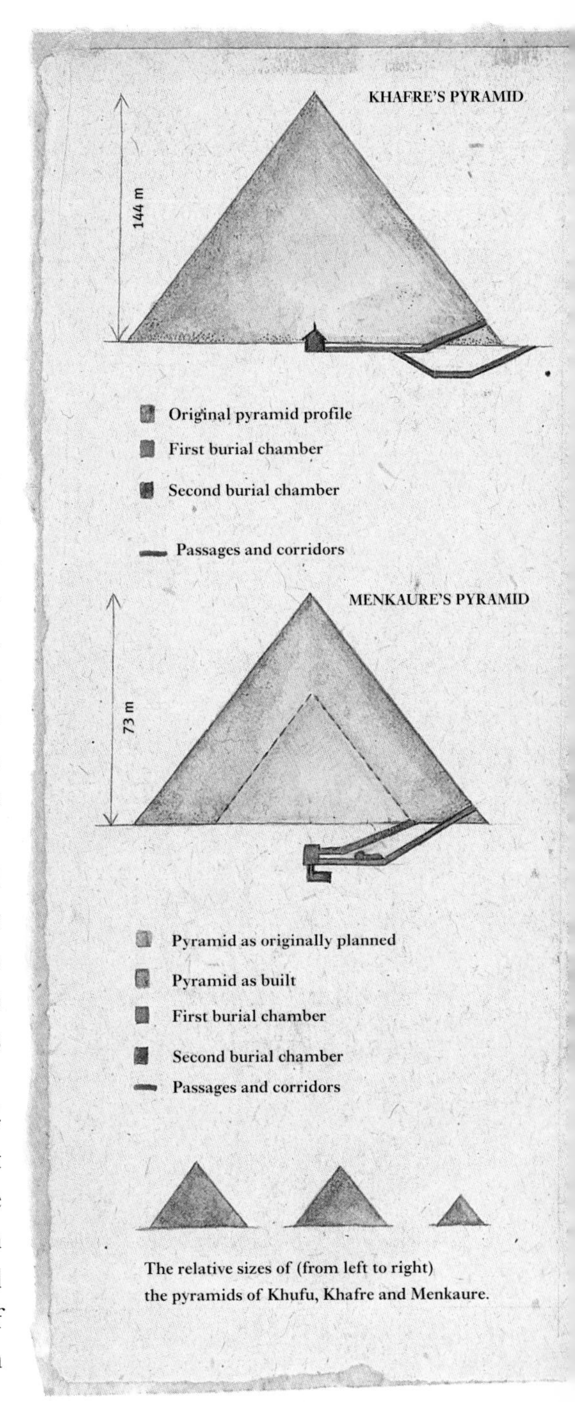


*Cross-sections of the pyramids of Khafre (top) and Menkaure (centre), showing their basic construction. Both originally had two entrances; in the case of Menkaure's pyramid, the second was constructed because the pyramid itself was increased in size. The bottom diagram illustrates the relative sizes of all three pyramids of the Giza group.*

# THE GREAT SPHINX

*The Sphinx, viewed through the remains of its temple, with Khafre's pyramid in the background. Built on an east–west axis, the temple may have been associated with the solar cult. The east and west walls of its main hall each have six recesses that were perhaps used in rites of the rising and setting sun, and twenty-four pillars in the hall may represent the hours of the day and night.*

**TUNNELS UNDER THE SPHINX**
**Many legends, some going back to ancient times, claim the existence of secret passages under the Sphinx. The Italian explorer Captain Giovanni Battista Caviglia, who investigated the Sphinx in 1817, hoped – in vain – to discover a tunnel leading from the Sphinx to Khafre's pyramid. According to a more modern myth, the Sphinx is the sole remnant of an advanced civilization otherwise lost to archaeology. A few people believe that the entire records of this civilization are located under the Sphinx's right paw, but there is no archaeological evidence to support such claims.**

**In co-operation with the American Egyptologist Mark Lehner, the Egyptian Antiquities Organization excavated in and around the Sphinx and located three tunnels under the statue. They had been found and entered before by an archaeologist who, however, had never published his findings.**

**The first tunnel is located behind the Sphinx's head. It goes inside the body of the statue for a distance of about five metres. The second, in the Sphinx's tail, is about thirty feet (9m) long. The third was opened by M. R. Baraize in 1926 and is located on the north side of the statue. All the evidence shows that these passages date to pharaonic times, but their purpose remains unknown.**

The Great Sphinx has stirred the imaginations of poets, scholars, adventurers and tourists for centuries. The first colossal statue of pharaonic Egypt, it presides majestically over the Giza necropolis. It has a lion's body and the head of a king, which wears the royal *nemes* headcloth and false beard. Although badly weathered, the features of King Khafre (see illustration, p.24) are recognizable. The Greek word "sphinx" may derive from the Egyptian *shesep-ankh*, "living image". This magnificent creature is carved from a knoll of rock that is largely of poor-quality limestone. Although some scholars believe that the knoll was a part of the quarry for Khufu's pyramid that was not exploited owing to its inferior stone, the location of the Sphinx in relation to Khafre's mortuary complex suggests that its site was carefully chosen. For example, Khafre's valley temple and the Sphinx temple are almost in exact alignment, and both buildings are constructed of large limestone blocks faced with harder red granite.

The function of the Sphinx is much debated. Lions were guardian figures in ancient Egypt, and there is a theory that the Sphinx was built as the guardian of the Giza Plateau. But according to the German Egyptologist Herbert Ricke, it was associated with the solar cult and was created by its Fourth-Dynasty sculptors as the image of Hor-Em-Akhet, "Horus of the Horizon", an aspect of the sun god and the name given to the Sphinx in New Kingdom times. According to this theory, the Sphinx was envisaged as standing against the "horizon" formed by the Great Pyramid – which was called *Akhet-Khufu*, "Horizon of Khufu" – and the pyramid of Khafre. For anyone approaching Giza, the head of the Sphinx, framed by the two pyramids behind it, may have resembled the hieroglyph for "horizon" (𓈌), which represents the sun rising between two mountains. The statue is perhaps best understood as a portrait of Khafre, as Horus, paying homage to his divine father Khufu, who was identified with Re, the sun (Khafre bore the title *Sa-Re*, "Son of Re").

Because the main body of the Sphinx was carved from weaker rock, it was faced with large blocks of Tura limestone, similar in quality to that which encased Khafre's pyramid. The head and neck, carved from a stronger vein, were left unfaced. The beard, now lost, must have been sculpted together with the head, because it would have been impossible to suspend such a heavy piece from the underside of the Sphinx's chin. Giovanni Battista Caviglia found a small part of the beard in 1817. The fragment is in two pieces that are now in the Cairo and British museums.

## THE DREAM OF THUTMOSE IV

A red granite stela between the front paws of the Sphinx relates that, when he was still a prince, King Thutmose IV (ca. 1400–1390BCE), went hunting near Giza. He fell asleep in the shade of the Sphinx, which appeared to him in a dream and complained that its body had fallen into ruin. Thutmose was not heir to the throne, but the creature promised that he would one day be king if he restored the monument. The rest of the inscription is worn away, but Thutmose did indeed become king, removed the accumulated sand from the Sphinx and reset some of the facing stones that had fallen off the body. Finally he erected the stela, which, as well as giving an account of the dream, shows Thutmose making offerings to the Sphinx. Known as the Dream Stela, it was probably once the main door of King Khafre's mortuary temple.

The Sphinx underwent a number of sand clearances in ancient times. The occasional disappearance of the monument under sand may account for why certain ancient authors, such as Herodotus, fail to mention it in their accounts of Egypt. Major refacing work was carried out on the body at least twice, once probably during the Twenty-Sixth Dynasty (664–525BCE), and again in the Roman period, some time between 30BCE and 100CE.

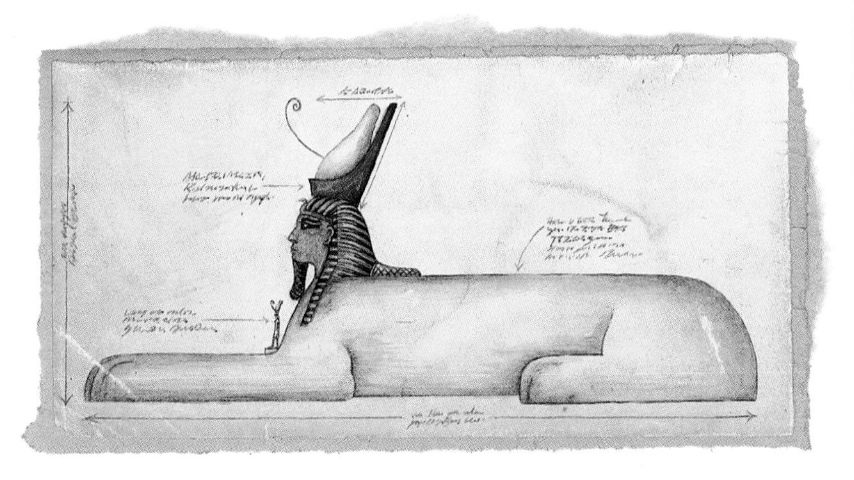

*An artist's impression of the Sphinx as it may have looked in its original painted state. A hole, now filled in, on top of the head may have been the socket for a crown. A statue of the king, of which almost nothing remains, probably once stood in front of the creature's breast.*

# PYRAMIDS OF THE "FORGOTTEN PHARAOHS"

*The pyramids of (left to right) Djoser (3rd Dynasty), Userkaf (5th Dynasty) and Teti (6th Dynasty) at Saqqara.*

The kings of the Fifth to Eighth dynasties (ca. 2500–2130BCE) do not share the renown of their Fourth-Dynasty predecessors and might justifiably be called the Old Kingdom's "forgotten pharaohs". Yet some of them built remarkable pyramids at North and South Saqqara and Abusir and introduced new features to the royal mortuary complex. For example, Unas (ca. 2371–2350BCE) was the first to inscribe the funerary "Pyramid Texts" on the inner walls of his pyramid (see pp.136–7). As notable were the sun temples about a mile (1.6km) north of Abusir, constructed, in addition to their pyramids, by six Fifth-Dynasty kings and based on the sun temple of Heliopolis. The focus of each temple was an altar before a *benben*, a squat obelisk with a pyramidal point representing the hill over which the sun rose at the beginning of creation (see pp.120–21). The *benben* stood on a low platform at one end of a court open to the sun. Remains of only two sun temples have been found, those of Userkaf (ca. 2500–2485BCE) and Nyuserre (ca. 2455–2425BCE) at Abu Ghurab. How they functioned is unclear, but like the standard pyramid complex they each have a mortuary temple, a valley temple and a connecting causeway.

*The interior of the pyramid of King Unas, which is called "Beautiful are the Places of Unas". The last king of the 5th Dynasty (ca. 2371–2350BCE), Unas built his pyramid close to the southwest corner of Djoser's Step Pyramid. The walls of the tomb chamber are covered with the first "Pyramid Texts", funerary spells and invocations that were intended to guide the dead pharaoh safely into the afterlife.*

Pyramid design also changed in the Fifth Dynasty. Within their limestone casings, pyramids were now built of rubble, debris and mudbrick rather than of solid limestone, and their state of preservation is generally poor as a result. However, it seems unlikely that this change reflects a period of economic decline, as has often been suggested, because the pharaohs of the Fifth and Sixth dynasties were more lavish than their predecessors in the inner decoration of their tombs and funerary complexes. The teams of trained artists required for such work were undoubtedly more expensive than the unskilled stone-haulers who constituted the bulk of a Fourth-Dynasty pyramid labour force.

Following the end of the Sixth Dynasty, the Old Kingdom disintegrated rapidly, a state of affairs that is reflected in the royal tombs of the Seventh and Eighth dynasties, whose rulers spanned just forty years. At South Saqqara, next to the pyramid of Neferkare Pepy II, the last significant ruler of the Sixth Dynasty, is the small and very poorly preserved pyramid of King Ibi of the Eighth Dynasty. This pyramid and its complex were built in the cheapest and least durable of all materials: mudbrick. Ruins of the mortuary temple exist, but no trace of a causeway or valley temple have been found. A number of other Eighth-Dynasty pyramids probably stood nearby. To the east of the pyramid of Teti (ca. 2350–2338BCE), the founder of the Sixth Dynasty, are the remains of another small pyramid that may date to the time of the Ninth or Tenth dynasties (ca. 2130–1980BCE), during the First Intermediate Period, when Egypt was a disunited kingdom.

## THE ABUSIR BLOCKS

Most kings of the Fifth Dynasty chose to be buried at Saqqara, but four of them – Sahure (ca. 2485–2472BCE), Neferirkare Kakai (ca. 2472–2462BCE), Neferefre (died ca. 2455BCE) and Nyuserre (ca. 2455–2425BCE) – built their pyramids further to the north, at Abusir. Of these, Sahure's complex is the best preserved, incorporating a mortuary temple, a causeway, the remains of a double-entranced valley temple, and a subsidiary pyramid just to the south.

In 1993, Egypt's Supreme Council of Antiquities decided to open the site of Abusir to visitors. In preparation for the opening, the author and his colleagues cleared Sahure's causeway and partly restored his mortuary temple. Wind-blown sand had accumulated to a depth of some ninety-eight feet (30m) over the causeway since 1907–8, when the German Egyptologist Ludwig Borchardt excavated there. The workmen assigned to remove the sand made a surprising discovery: twenty limestone blocks carved with delicate relief.

The blocks are unique: nothing remotely like them has been found at any other Old Kingdom pyramid complex. One scene, spread over several blocks, gives a fascinating glimpse of part of the pyramid-building process. It shows a gang of men hauling the pyramidion (see p.171) of Sahure's pyramid. The portion depicting the pyramidion itself is missing, but not the accompanying inscription, which reads, "bringing the pyramidion covered with fine gold [electrum] to the pyramid".

Other blocks portray the wooden sled on which the pyramidion was dragged to the pyramid. In this scene, one man pours water on the ground in front of the sled in order to reduce friction and ease its passage. A visitor to Abusir in late antiquity defaced this scene with a graffito of an archer shooting at the pyramidion.

*The 5th-Dynasty pyramid of Sahure at Abusir, with the remains of the king's mortuary temple in the foreground.*

*An inlaid bronze bust of Amenemhet III (ca. 1818–1772BCE), whose 46-year reign was the longest of the 12th Dynasty. He was probably buried in the pyramid at Dahshur (see below). The king's other pyramid, at Hawara, was famed for its now vanished mortuary temple, which has been known since antiquity as the "Labyrinth", after the legendary building of King Minos of Crete. According to the Greek historian Herodotus it had some three thousand rooms and a maze of passageways.*

# MIDDLE KINGDOM PYRAMIDS

Following the chaos of the First Intermediate Period, pyramid-building recommenced with the restoration of national unity at the beginning of the Middle Kingdom (ca. 1938–1630BCE). While still faced with limestone, the pyramids of this last great pyramid age were built for the most part with an inner core consisting almost entirely of mudbrick, and today many of them appear as little more than shapeless mounds. But their uniqueness lies within: Middle Kingdom rulers built interiors with a maze of chambers and passages to deceive tomb robbers. They also set traps, the success of which is indicated by the discovery of ancient corpses, some hanging by the legs. From Senwosret II onward, pyramid entrances are in the east or south side rather than the north, as in earlier times; this may also have been a device to foil robbers. Funerary texts now appear on coffins (the Coffin Texts), rather than inscribed on the walls, like the Pyramid Texts of the Old Kingdom.

Amenemhet I (ca. 1938–1909BCE), the founder of the Twelfth Dynasty, moved his capital to near el-Lisht, where he was buried. His pyramid complex imitates those of the Old Kingdom, with a mortuary temple, causeway and valley temple. The pyramid itself is cased with Tura limestone and incorporates blocks removed from pyramids at Giza and Saqqara. Just to the south, the pyramid of his son, Senwosret I, is notable for its ten subsidiary pyramids and fine statuary, including an avenue of Osiride statues of the king that lined a corridor on top of the causeway.

Amenemhet II was buried near the pyramid of Sneferu at Dahshur. His pyramid (the "White Pyramid") and its complex are of little special interest except for the jewelry discovered in 1894 (see box, opposite page).

*Amenemhet III's pyramid at Dahshur, known as the "Black Pyramid". When the white limestone casing was stripped for other buildings, the mudbrick core simply collapsed, leaving a misshapen stump towering above the desert. The condition of the pyramid is so dilapidated that archaeologists have been unable to calculate its original size exactly. The Dahshur pyramid contains, for the first time since King Djoser of the 3rd Dynasty, the burials of other members of the royal family in addition to the king.*

Senwosret II decided on Kahun (Lahun), near the Faiyum, for his pyramid, which is famous for its neighbouring artisans' quarter, or "pyramid town" (see p.71). His two successors returned to Dahshur for burial. Here, the plan of Senwosret III's pyramid complex appears to follow that of Djoser's Step Pyramid. His valley temple is undiscovered, but the causeway, with its beautiful reliefs, has been found, as have some more royal jewels (see box). Amenemhet III built pyramids at Dahshur and Hawara (see illustrations, opposite), but the latter is probably just a cenotaph.

The last Twelfth-Dynasty rulers, Amenemhet IV and Queen Sobekneferu, built pyramids just south of Dahshur at Mazghuna. There are two pyramids of the Thirteenth Dynasty (ca. 1759–1630BCE) at South Saqqara. For a century, in the Second Intermediate Period, no pyramids were built. Ahmose (ca. 1539–1514BCE), the first ruler of the Eighteenth Dynasty, built a pyramidal cenotaph and a dummy pyramid for his grandmother at Abydos. But the great age of pyramid-building was over.

**NUBIAN PYRAMIDS**

**Inspired by their Egyptian counterparts, Nubian kings also erected pyramids. The sandstone pyramids at Dongola and Kush are much smaller than Egyptian ones and have steep, stepped sides. Piye, the Kushite king who ruled in Egypt ca. 747–716BCE, built a pyramid near Napata, and his successors Shabaka and Taharqa, were also buried in pyramids. In the early centuries CE, the rulers of Meroe, a Nubian kingdom, likewise built stone pyramids. After Meroe fell ca. 250CE pyramid-building ceased forever in the Nile Valley.**

## ROYAL JEWELRY OF THE TWELFTH DYNASTY

After having worked for several years at Senwosret III's pyramid at Dahshur, Dieter Arnold from the New York Metropolitan Museum of Art discovered the tomb of Queen Weret under the pyramid in 1994. The daughter of Amenemhet II, wife of Senwosret II and mother of Senwosret III, Weret was greatly revered during her lifetime and afterward, and her status is reflected in her magnificent gold jewelry, discovered hidden in a passage of her tomb.

On the day of its discovery, the cache was rapidly transported to the Egyptian Museum in Cairo amid tight security. When they were found, the jewels were in many pieces that included cowrie shells, amethyst scarabs, golden lions originally from a bracelet, and some seven thousand beads of various sizes. These were all carefully restored before being put on display at the museum.

*An openwork pectoral and golden cowrie-shell belt from the jewelry cache of Princess Sat-Hathor. The pectoral, of gold* cloisons *inlaid with semi-precious stones, is in the form of a monumental pylon (gateway), with two Horus falcons flanking the* praenomen *(throne name) of Senwosret II. The shells on the belt are separated by a double row of semi-precious stone beads.*

This is the latest in a series of discoveries of fabulous jewelry of the Middle Kingdom at Dahshur. During his 1894–5 excavation season, Jacques de Morgan came across marvellous jewelry in the tombs of two princesses, Ita and Khnumit, near the pyramid of Amenemhet II, and in the tombs of Princess Sat-Hathor and Queen Mereret near the pyramid of Senwosret III.

The cache of Khnumit, a daughter of Amenemhet II, included gold bracelets with exquisite *cloisonné* clasps in the forms of hieroglyphs (see illustration, p.242). Sat-Hathor, one of Senwosret II's daughters and possibly the wife of Senwosret III, left a collection of superb cowrie necklaces, scarabs, beads, buckles, collars and pectorals (see illustration, left), which had to be painstakingly disentangled, cleaned and reassembled by restorers at the Cairo Museum. (See also p.229.)

● CHAPTER 13

# TOMBS AND TEMPLES

From the great pyramids of the Fourth Dynasty to the massive temples of the New Kingdom and Ptolemaic Period, the Egyptians honoured the dead and their gods with some of the world's most monumental architecture. For royalty and commoners alike, the tomb was not merely a repository for their bodies but a habitation which would cater for all their needs in the afterlife. Temples were institutions where the gods, upon whom the nation's well-being depended, were appeased and given sustenance. These vital functions ensured Egypt's temples a central role in the life of the country.

▲



ABOVE: *One of the two hypostyle, or columned, halls in the temple of Sety I (ca. 1290–1279BCE) at Abydos in Upper Egypt. The building, which contains some of the finest Egyptian temple decoration, celebrates the royal ancestor cult, and its many reliefs include a list of seventy of Sety's predecessors.*

## THE DWELLING PLACES OF THE DEAD

Among the essential requirements for a "successful" afterlife, as far as the Egyptians were concerned, was a tomb that could serve as both a secure final resting-place and a home to various aspects of their deceased persona. Over more than three millennia, the Egyptian tomb evolved from the simplest of pit graves to the most elaborate monumental stone structures, filled with a treasury of artefacts associated with the belief in the afterlife. The tombs of royalty were accompanied by mortuary temples serving the cult of the deceased (see pp.210–11).

With some exceptions, the Egyptians built their tombs and necropolises on the west bank of the Nile: the dead, like the setting sun, were believed to enter the underworld in the west. There are basically two lines of development in tomb architecture and decoration, one royal and the other non-royal, or "private". The royal tombs were built using the most expensive materials and often involved great innovations. Many aspects of royal tombs, such as the mortuary texts intended to guide the deceased through the netherworld (see pp.136–7), and the pyramidal tomb shape (see previous chapter), eventually filtered down to the tombs of non-royal classes. However, the finest surviving examples of mortuary architecture reflect the highest echelons of Egyptian society, and are not representative of the population as a whole.

Regardless of the materials that were used (mudbrick, limestone, sandstone), or of how they were constructed (free-standing or hewn from the living rock), most tombs contained a superstructure, which was accessible to the living, and a substructure, which housed the mummified body and was sealed for eternity. The earliest superstructures were mudbrick mounds or rectangular enclosures, possibly showing Mesopotamian influence of and often containing a series of niched façades in imitation of domestic architecture. The superstructure incorporated a tomb chapel, in which relatives would place food offerings to the deceased before a "false

ABOVE: *Predynastic and Protodynastic burials, such as this one of ca. 3500BCE from Naqada (excavated by the author), were simple oval pits, sometimes lined with matting or wood. The tightly flexed body was typically accompanied by just a few vessels and cosmetic items. Throughout ancient Egyptian history, most of the population were probably buried in a simple pit tomb with a small number of artefacts.*

LEFT: *In dynastic times, the élite were buried with an array of finely crafted grave goods that catered to the needs of the dead in the afterlife. The body itself might lie within one or more elaborate sarcophagi bearing the portrait of the deceased in the prime of life. This Late Period basalt lid covered the sarcophagus of Sisobek, vizier under Psamtik I (664–610BCE).*

door", an inscribed niche that served as a symbolic entrance to the netherworld (see p.197). As the Old Kingdom progressed, tomb superstructures became increasingly elaborate. Members of the élite social classes came to expect a number of rooms and a diverse repertoire of wall decoration, all magically endowed to serve the needs of the deceased in eternity. Laudatory inscriptions, scenes depicting daily life and the rich resources of the tomb owner's estate, representations of loving family members, and a vast display of offering scenes assured continued abundance and prosperity in the afterlife.

The substructure was the actual burial place, containing the mummy, canopic equipment (for the preserved essential organs) and other mortuary objects. From the mastaba tombs of the Old Kingdom (see p.197), via the rock-cut chambers of Middle and New Kingdom sepulchres, to the great Theban funerary "palaces" of the Late Period, these basic elements were constant. To usurp a tomb, recarve an inscription, or chisel out the figures on a wall carving was to attack the legacy, and the eternal survival, of the tomb's occupant (see pp.144–5).

*A view of the Valley of the Kings in Western Thebes, the royal burial ground of New Kingdom pharaohs from the time of Thutmose I, the third ruler of the 18th Dynasty. The stonework at the entrance to each tomb is modern.*

# ROYAL TOMBS

As dynasties rose and collapsed, capitals and necropolises shifted, and tomb robbery increased, the tombs of the Egyptian royal families underwent a complex evolution. By the Old Kingdom proper (Fourth to Sixth dynasties, ca. 2625–2170BCE), the construction of mastabas and step pyramids had crystallized into true pyramid complexes, which included subsidiary pyramids (for queens or for ritual purposes); pyramid temples; long causeways to the Nile Valley; and valley temples (see pp.171–3). Middle Kingdom pyramids at sites such as el-Lisht and Dahshur followed this pattern, if somewhat more modestly than their Old Kingdom counterparts (see pp.190–91).

By the New Kingdom, much had changed. Egypt had experienced its first domination by foreigners (see pp.26–7) and a Theban family had liberated the country and founded the Eighteenth Dynasty, shifting the capital once again to Upper Egypt. For the next few centuries, Thebes was the principal burial site for Egypt's rulers, beginning with Thutmose I (ca. 1493–1482BCE). His chief architect, Ineni, found a secret location for his tomb, a wadi (dry riverbed) behind the cliffs on the west bank of the Nile. So began the long history of the Valley of the Kings. Without superstructures (or rather, with the Theban Peak serving as a pyramidal superstructure for all the burial places at its foot), these tombs were excavated out of the living rock, and contained staircases, corridors, storerooms, shafts and burial chambers. Some of the larger tombs boast seemingly countless chambers and penetrate more than one hundred yards (90m) into the cliffside. The limestone walls were smoothed by the ancient craftsmen living in the nearby village of Deir el-Medina, who then carved and painted them. Scenes from the underworld, spells from the Book of the Dead (replacing the Middle Kingdom Coffin Texts, which in turn replaced the Old Kingdom Pyramid Texts; see pp.136-7), royal names and titles and images of protective deities cover the walls in polychrome splendour. Many of these are well preserved, although flash floods, climatic changes and tourism threaten the valley as never before.

As the necropolis grew from the Eighteenth to the Twentieth dynasties (ca. 1539–1075BCE), a contingent of guards must have been stationed around the valley to protect its royal inhabitants and the precious objects buried with them. However, despite the fact that the sepulchres of Egypt's mightiest pharaohs were no longer marked by individual pyramids, they still proved an easy target, and most were robbed in antiquity. During the Twenty-First Dynasty (ca. 1075–945BCE), priests eventually gathered the royal mummies out of their plundered tombs and hid them from further damage. A secret tomb south of the terraced temple of

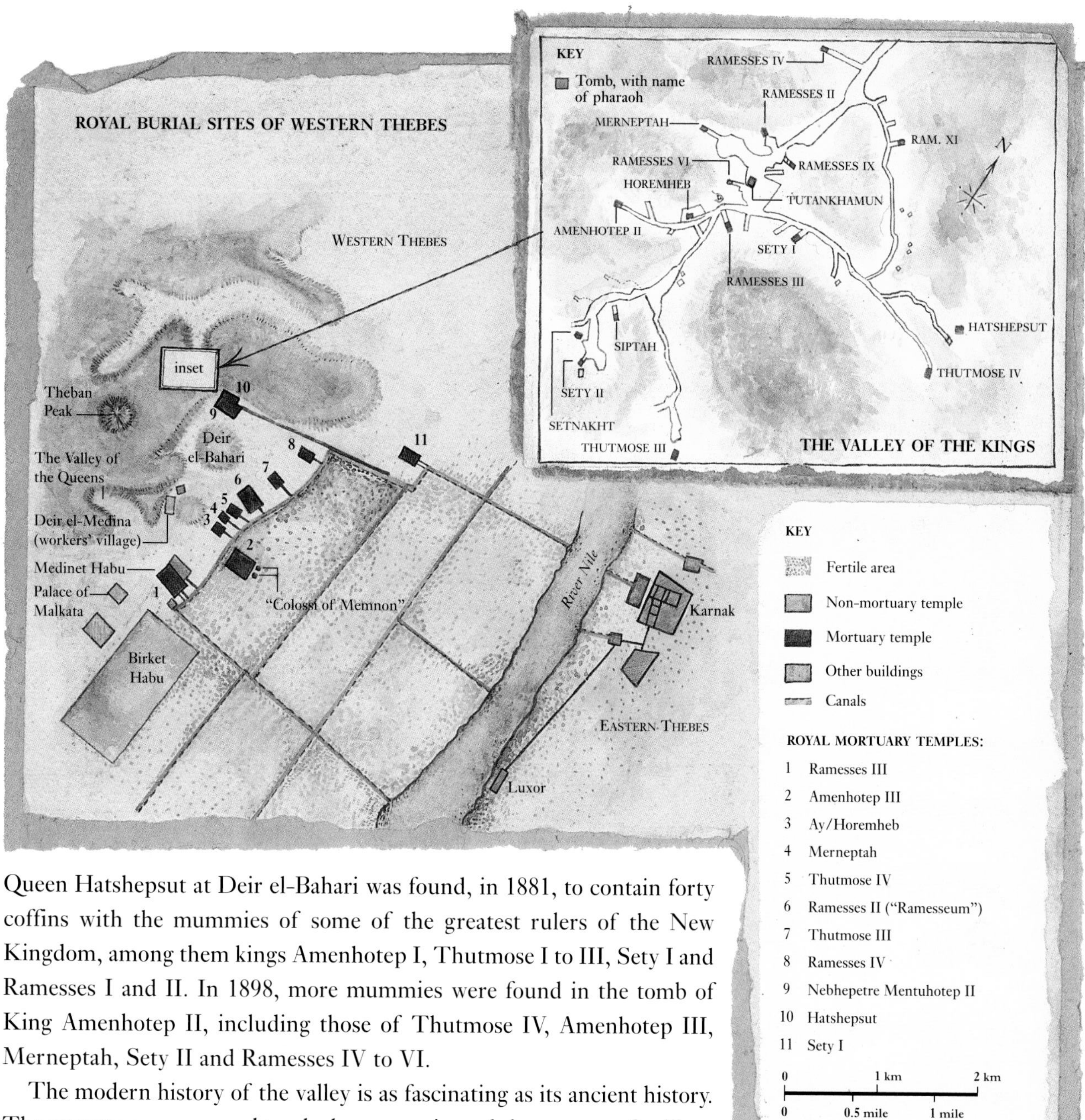


*This plan of Western Thebes and the Valley of the Kings shows clearly how the New Kingdom pharaohs built their mortuary temples (see pp.210–11) on the edge of the Nile floodplain, away from the rich burials in the valley. A series of canals and at least one artificial harbour, at Medinet Habu, linked the temples to the Nile, and thence with the great temples of Karnak and Luxor on the opposite bank of the river.*

Queen Hatshepsut at Deir el-Bahari was found, in 1881, to contain forty coffins with the mummies of some of the greatest rulers of the New Kingdom, among them kings Amenhotep I, Thutmose I to III, Sety I and Ramesses I and II. In 1898, more mummies were found in the tomb of King Amenhotep II, including those of Thutmose IV, Amenhotep III, Merneptah, Sety II and Ramesses IV to VI.

The modern history of the valley is as fascinating as its ancient history. The seventy or more royal tombs here experienced the passage of millennia in subterranean silence, until in the early 1800s they were discovered by European "explorers" such as Giovanni Belzoni, who used a battering ram to open some of the tombs. They withstood the hordes of tourists surrounding Howard Carter's meticulous clearance of the tomb of Tutankhamun in the 1920s (see p.196), and later enjoyed conservation by the Egyptian authorities. In 1987 and 1995 they even offered up another spectacular discovery: KV5 (first found in 1825, but mistaken for an unfinished tomb), apparently the tomb of many of the sons of Ramesses II.

Farther south lies the counterpart to the kings' valley, known as the Valley of the Queens, which houses the tombs of queens and princes of the late New Kingdom. The most famous of these belongs to Queen Nefertari, the wife of Ramesses II. Brightly coloured wall paintings

**TOMB ROBBERY**

**It was no secret that, as the burial process grew more elaborate, so did the value of the grave goods interred with both royal and non-royal mummies. Gilded coffins, amulets of precious stones, exotic imported artefacts all proved too tempting for thieves. When embalmers began to include protective amulets of precious stones, gold or silver, within the mummy wrappings, even the deceased's corpse came under threat. Robbers probably attacked royal tombs soon after the king's funeral, and there is evidence of corruption among the necropolis employees charged with protecting the tombs.**

depicting the queen in the company of the gods, surrounded by countless hieroglyphic texts, each one a minor masterpiece of intricate detail, provide one of the best glimpses into the original splendour that once adorned hundreds of Egyptian sepulchres that are now lost or destroyed. In recent years, the fragile frescoes have received painstaking cleaning and restoration by an international team.

While the royal tombs themselves were hidden in the Valley of the Kings, the cult of the king was still maintained through a mortuary temple (see pp.210–11). But a major difference distinguishing New Kingdom royal tombs from their earlier counterparts is the physical separation of burial and mortuary temple (see plan on p.195).

Later dynasties adopted different burial customs. The kings of the Twenty-First and Twenty-Second dynasties (ca. 1075–712BCE) built their sepulchres within the temple precinct at their Delta capital of Tanis. Other capital cities of the Third Intermediate Period and Late Period, such as Bubastis, Herakleopolis, Hermopolis, Leontopolis and Sais, probably also saw royal burials, but few have survived as well as those of the New Kingdom at Thebes.

## THE TOMB OF TUTANKHAMUN

The British Egyptologist Howard Carter (1874–1939) spent many relatively fruitless years excavating in the Valley of the Kings at Thebes on behalf of his patron, Lord Carnarvon (1866–1923). But on November 4, 1922, the first step of a carved descending staircase was unearthed. After debris was removed, a long corridor and a sealed door were uncovered. When Carter looked through a hole pierced through the wall, all he could make out was "gold. Everywhere the glint of gold". The antechamber he saw was packed with objects, ceremonial couches in the shape of bizarre animals, boxes of linen, chariots and life-sized statues. Three more chambers were also found: the burial chamber, the treasury and an annexe.

Tutankhamun (ruled ca. 1332–1322BCE) lived during a fascinating period of religious upheaval at the end of the Eighteenth Dynasty, when the cult of the Aten had only recently been abandoned (see pp.128–9). Many of the objects found in the tomb must span this era of transition.

*The gold sarcophagus that contained Tutankhamun's mummy.*

# PRIVATE TOMBS

THE MASTABA TOMB

chapel
false door
burial shaft
burial chamber
sarcophagus

PLAN

chapel
burial shaft
false door
burial chamber and sarcophagus

ELEVATION

false door
chapel
burial shaft
burial chamber
sarcophagus

ABOVE: *A typical mastaba tomb. Limestone blocks masked a rubble and debris core, through which a burial shaft was sunk into bedrock. Construction could take 15 months.*

For all their emphasis on the netherworld, non-royal tombs, built for the élite classes and their families, provide a wealth of insights into Egyptian society. Their wall decoration, re-creating everything from fishing scenes to carpentry to house parties, provides frozen glimpses of daily life along the Nile.

Although some features of royal tombs (for example, their architectural forms and the use of mortuary texts) were gradually incorporated into the private tomb, the latter for the most part followed its own course of development. In the formative decades (late fourth to early third millennium BCE), the private tombs of the élite were usually concentrated around "royal" mortuary complexes at sites such as Abydos and Hierakonpolis (see p.69) in the south and Saqqara in the north. Simple mudbrick pit graves were covered with unadorned superstructures. Decoration was sparse; small, crudely inscribed name stelae often provided the only inscriptions. Grave goods consisted of stone and ceramic vessels, cosmetic implements, slate palettes and articles of jewelry.

Mudbrick architecture never fell completely out of use, but with the rise of the Third and Fourth dynasties the élite classes increasingly built their tombs in stone. The typical Old Kingdom private tomb is known as a mastaba, the Arabic for "bench", from its resemblance to the solid benches outside a typical rural Egyptian house. A mastaba is a rectangular limestone or mudbrick superstructure with sloping sides. The elaborate niching, or "palace façade", of earlier tombs was reduced to a north

RIGHT: *The entrance to an Old Kingdom tomb near the pyramid of Khafre at Giza.*

and south niche on the exterior of the east wall. These niches became the indispensable "false doors" through which the spirit of the deceased magically passed to partake of food offerings left by the living. "Slab stelae" at Giza – often the only inscribed surface in these early Old Kingdom tombs – carved with the tomb owner's name, titles and image, were incorporated into the design of the false door, along with additional inscriptions and prayers. For protection, the Egyptians later moved their false doors and offering scenes into the previously solid core of the mastaba itself, forming an offering chapel. By contrast, statue chambers (called *serdabs*), vertical burial shafts (replacing stairways) and underground burial chambers were inaccessible to the living. Decoration soon spread beyond the false door to adorn all four walls of the chapel. Late Old Kingdom private tomb architecture culminated in the Fifth- and Sixth-Dynasty tombs at Saqqara (see box, below). These tombs were larger than their predecessors, and rooms were added, often for other members of the tomb owner's family.

## MASTABAS AT SAQQARA

The most elaborate private tombs from the Old Kingdom period are to be found at Saqqara. This was the primary necropolis of the Egyptian capital, Memphis, and it was used by almost every dynasty up to the First Intermediate Period. The superstructures here contain sometimes twenty or more chambers, with multiple false doors and burial shafts for family members, and thousands of hieroglyphic inscriptions and multicoloured wall scenes. Not surprisingly, Saqqara remains a primary tourist attraction to this day.

Perhaps the greatest variety of wall scenes may be found in the Fifth- and Sixth-Dynasty tombs of high-ranking administrators such as Ti, Ptahhotep, Mereruka, Kagemni and others. In the Egyptian mind there was no difference between actual objects (such as people and offerings) and paintings or models of them. Each of these offering scenes would therefore have been endowed with reality in the afterlife, the people, objects and activities depicted serving as provisions for the posthumous life of the deceased's spirit. Large-scale figures of the deceased, signifying his importance, dominate the walls of many of the tombs. Other representations include workshop scenes displaying craftsmen at work, boating games in the marshes, and ritual pilgrimages to holy sites such as Abydos. Specialized representations, including scenes of circumcision (which occur, for example, in the tomb of the physician Ankhmahor), and the touching depiction of a cow concerned for her calf being carried across a canal (in the tomb of Ti), enhance the vitality and uniqueness of each monument.

*A wall relief showing a herdsman leading bulls, from the tomb chapel in the mastaba of Ptahhotep, a government official, at Saqqara. Old Kingdom, late 5th Dynasty, ca. 2380BCE.*

Rock-cut tombs, containing chambers cut into the bedrock or high up on the Nile Valley cliffside, are also found in the Old Kingdom, often built by influential high officials. These rock-cut tombs proliferated in regional necropolises such as Deshasheh, Meir and Sheikh Said, and became the preferred type of burial structure following the collapse of the Old Kingdom and the First Intermediate Period (see pp.26–7). Middle Kingdom rock-cut tombs appeared throughout Middle and Upper Egypt at sites such as Beni Hasan, Thebes, Deir el-Bersheh and Asyut. They often contained elaborate chambers with statue niches and columns, fronted by courtyards, porticoes and causeways leading down to the riverbank. Reflecting the political conditions of the age, military themes often featured prominently on the walls. In the early Middle Kingdom, wall scenes depicting the activities of daily life were transferred to three-dimensional wooden models, buried with the deceased.

*A view looking toward the Nile from the Theban cliffs across the necropolis area of Western Thebes known as the Asasif, east of the temple of Hatshepsut at Deir el-Bahari. This is the site of the monumental private tombs of the 26th Dynasty, such as those of Montuemhat (see p.37 and also illustration, p.200), Ankh-hor, Basa, Pedamenophis and Pabasa. Only their mudbrick superstructures are visible, because most of the chambers in the tombs are subterranean.*

After the Second Intermediate Period, in the early New Kingdom, elaborate tombs once again appeared. As Thebes grew to prominence throughout the Eighteenth to Twentieth dynasties, its necropolises rivalled those of the Old Kingdom at Saqqara and Giza. Theban rock-cut tombs usually contained a T-shaped layout (a transverse hall and corridor), which corresponded to the chapels built into the superstructures of earlier tombs. Their shafts and burial chambers were cut deep into the rock below. Wall carving and especially painting reached new heights under the early Thutmosid rulers, particularly in liveliness of colour and experimentation with secular themes (parties, fieldwork, trades). Later in the New Kingdom, these themes were replaced with funerary subjects such as scenes of mummification and funerals, and vignettes from mortuary literature. During this period, burial chambers were also decorated. Particularly good examples are the modest tombs of the necropolis craftsmen at Deir el-Medina. Small pyramids, no longer used in the royal mortuary complex, now sometimes covered private sepulchres.

The Third Intermediate Period was a time of mass burials, and coffin decoration rather than tomb architecture became the main focus. But in the Twenty-Fifth and Twenty-Sixth dynasties came a final renaissance. Important Theban officials of this period combined elements of both royal and private, tomb and temple architecture, in great funerary "palaces" built near the temple of Hatshepsut at Deir el-Bahari. Consisting of massive mudbrick entrance pylons, causeways, large sunken courts and countless underground chambers, the tombs of priests and administrators, such as Montuemhat (see p.37), are among the largest Egyptian burial structures ever conceived. Despite sporadic highlights in later eras, such as the tomb of Petosiris, high priest of Thoth at Hermopolis (ca. 320BCE), which resembled a Late Period temple, or the classically inspired catacombs of Kom el-Shukafa near Alexandria (second century CE), never again were such resources concentrated on private tombs in Egypt.

**NECROPOLIS MANAGEMENT AT THEBES**

**From ancient Thebes come some tantalizing papyrus records dealing with necropolis administration. Here we gain glimpses into disputes between the administrators of the east and west banks of the Nile, records of the trials of tomb robbers, and inspection tours of the necropolis. However, much more research is still required before we can answer such questions as who apportioned the individual "plots" for tomb construction; how long the mortuary cult of an individual lasted before the family members and financial resources ran out; who settled disputes over access routes to the tomb chapel; and how legal was the usurpation of tombs by subsequent generations.**

# CITIES OF THE DEAD

The cemeteries dating from the First and Second dynasties at Abydos in the south and Saqqara in the north set patterns of necropolis development that were followed for millennia. Wherever royal mortuary structures were built, the tombs of family members, viziers and high officials were located nearby. Even in the New Kingdom, when pharaohs separated their actual tombs from their mortuary temples, hiding the former in the Valley of the Kings, private necropolises developed in the same general region of the Theban cliffs.

Early Old Kingdom private necropolises centred on the Memphis region, and recent excavations suggest that sites along the desert's edge may have been linked by more continuous settlements and burials than has been previously imagined. At North Saqqara may be found some of the earliest monumental private mastaba tombs. In fact, with some interruptions, Saqqara boasts an almost unbroken range of private tombs, stretching at least through the First Intermediate Period, into the Middle Kingdom and continuing into the New Kingdom (see p.198).

The Fourth Dynasty marks the major break from Saqqara. Private tomb development evolved further at sites such as Meidum, with three small private necropolises of mastaba tombs in the general vicinity of the pyramid of King Sneferu (ca. 2625–2585BCE). Additional early Old Kingdom mastaba fields are also located near the pyramids of Sneferu at Dahshur. Sneferu's successors chose the Giza plateau, farther to the

ABOVE: *The sunken court of the Western Theban tomb of Montuemhat (fl. ca. 655BCE), mayor of Thebes, governor of Upper Egypt and Fourth Prophet of Amun at the temple of Karnak (see p.37).*

RIGHT: *The excellently preserved paintings in the 19th-Dynasty tomb of Sennedjem at Deir el-Medina in Western Thebes include scenes of the deceased and his wife in the afterlife, reaping and ploughing in the paradisial "Field of Reeds" (see p.116).*

north, as the site for their mortuary complexes, and their families and officials followed suit. Giza represents the first example of private necropolis layout to preconceived, organized plans on a massive scale. Cemetery units with street after street of aligned mastaba tombs appeared on both the east (for the royal family proper) and west sides of the Great Pyramid of Khufu. With minor exceptions, the rest of the Fourth Dynasty expanded upon the Giza pattern set by Khufu. Later in the Old Kingdom, the streets of the various mastaba fields became choked with subsidiary burials, suggesting a breakdown in the original maintenance of the necropolis.

Several kings of the Fifth Dynasty built their pyramids at Abusir, just north of Saqqara. The largest private tomb found here so far belongs to Ptahshepses, the vizier and son-in-law of King Nyuserre (ca. 2455–2425BCE). Additional tombs, belonging to viziers and other high officials, have been unearthed in recent years. But the most impressive Old Kingdom private tombs are to be found back at Saqqara, dating from the Fifth and Sixth dynasties. These tombs reflect the increasing power of the country's highest administrators, and utilize all the skills of royal artisans from the Memphite court. The massive superstructures and multiple chambers of some of these far outmatch their much smaller counterparts from the Fourth Dynasty at Giza. Additional rock-cut tombs hidden beneath the causeway of Unas's pyramid also reveal multiple chambers and vibrant colours. It is difficult to establish any organized pattern of development at this huge necropolis, although many tombs from all periods still await discovery.

Toward the end of the Old Kingdom, the "nomarchs" (see p.27) ceased to aspire to a Memphite burial, and instead chose sites closer to home for themselves and the members of their ruling house. Thus provincial necropolises developed throughout the First Intermediate Period up to the reunification of the country under the Theban Eleventh Dynasty.

It was not until the New Kingdom that dense private necropolises comparable to the vast Old Kingdom cemeteries of Saqqara and Giza were built once again. The governing classes of the Eighteenth to Twentieth dynasties constructed their sepulchres in the western cliffs of Thebes, between the royal tombs and the royal mortuary temples, on the edge of the floodplain. Even so, the regulated layouts of the Old Kingdom have so far not been discernible. The most unified and compact private necropolis of this period is probably that of Deir el-Medina, site of the village of artisans employed in constructing and decorating the royal tombs.

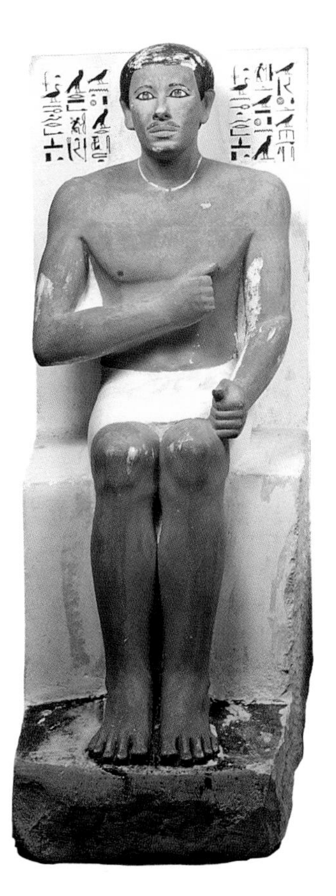

*Statues of Prince Rahotep and Princess Nofret from Meidum (see below). The prince, who sports the thin moustache that was popular during the Old Kingdom, may have been a son of King Sneferu (ca. 2625–2585BCE), the founder of the 4th Dynasty. Unlike Nofret, Rahotep has a tanned complexion from working out of doors.*

**A PRINCELY COUPLE**

**Among the most striking portraits of the Old Kingdom are the painted limestone statues of Prince Rahotep and his wife Nofret (see illustration, above) from the couple's mastaba tomb near the pyramid of Sneferu at Meidum. The paint is almost perfectly preserved on these four-foot (122cm) high sculptures, which are made more realistic by the use of inlaid crystal and amethyst for the eyes. In 1871, when Auguste Mariette discovered the tomb, the Egyptian workmen who first came across the statues were so startled by the eyes glinting in the light of their candles that they fled in fear.**

# THE TEMPLE

*"Trajan's Kiosk", the ceremonial landing stage built by the Roman emperor Trajan at the temple of Isis at Philae, an island in the Nile near Aswan. Isis was worshipped at Philae from ca. 675BCE until well into the Christian era. Far from Egypt's capital, Alexandria, the temple defied (or ignored) Roman decrees from the late 4th century CE ordering all pagan temples to close. The last functioning temple of the old religion, it was finally abandoned in the mid-6th century CE. The temple was moved from Philae to higher ground on a nearby island, Agilkia, following the construction of the Aswan High Dam in the late 1960s.*

*King Menkaure (ca. 2532–2510BCE) flanked by the goddess Hathor (left) and a goddess of one of the Egyptian nomes; a statue from the king's mortuary temple at Giza (see p.185).*

Temples were a vital force in Egyptian society on both local and national levels, playing a critical role in spiritual and secular aspects of the country. They served a multitude of different purposes. However, in contrast to, for example, Christian churches, Egyptian temples were not places of congregation for the masses. Ordinary Egyptian citizens never saw the inner sanctuaries, where the statue of the temple's god dwelt in seclusion, serviced only by the highest priests in the temple hierarchy.

There are basically two major types of Egyptian temple: those dedicated to local or state deities (see pp.206–9), and those specifically built for the mortuary cult of the pharaoh (see pp.210–11). In addition to these major structures, the Egyptians also built solar temples devoted to the cult of the sun god; barque stations (ritual stopping points for the god's sacred boat on processional routes); "birth houses", dedicated to the creation myths of specific deities; and small temples and shrines for local deities.

Through its architectural forms (pylons, courtyards, columns in plant form and sacred enclosures), its decoration (representations of deities, royalty and rituals, and the inclusion of texts) and its location at a sacred site, the Egyptian temple as a whole re-created part of the Egyptian universe on earth. Throughout the evolution of the non-mortuary temple, almost every building incorporates one basic Egyptian concept: that of the primeval mound – the sacred hill, which first arose from the chaotic waters of darkest antiquity (see p.120). This idea was manifested in the elevation of the innermost sanctuary above the level of the outer chambers of the temple.

Some of the earliest temples were little more than mudbrick enclosure walls with a shrine (a tent or box-shaped structure) in the centre and pennants hung from tall poles (the origin of the flag-like hieroglyph 𓊹 for *netjer*, "god"). Remains of some of the oldest temples, dating from the early

## TEMPLES OF THE HERETIC KING

The paradigm of the Egyptian temple, with its succession of courts and chambers leading to the mysterious darkness of the innermost sanctuary, experienced a sort of inversion during the reign of Akhenaten (see pp.128–9). At the king's new capital of Akhetaten (modern el-Amarna) a massive area was reserved for the Mansion of the Aten temple, in the centre of the city. The temple consisted of two major structures. First came a frontal temple with six pylons and seven discrete sections. Unlike the covered courts and chambers at Thebes, most of these sections were open, and were filled with hundreds of altars. Many more altars surrounded the building. Cult chapels stood at the back of the structure, and a round-topped stela rested on a platform in the temple court. The rear temple, divided into two major sections, served as the cult sanctuary, with a central altar and numerous statues of Akhenaten and Nefertiti.

A second, smaller temple lay farther to the south. It was enclosed by a wall, and once again there were numerous altars, with the main sanctuary in the third and rearmost court. Finally, two "garden temple" structures were added at the southern edge of the city. The decoration scheme of Amarna temples consisted primarily of figures of the king and queen, along with the royal family, making offerings in the presence of life-giving rays from the sun's disc.

*Part of a reconstructed wall of Akhenaten's temple to the Aten at Karnak. The king's temples employed small blocks known by the Arabic word* talalat. *When the temples were torn down by Akhenaten's successors, the blocks were used as infill for their own monuments, such as new pylons at Karnak. Thousands of* talalat *have been recovered from the ruins of these later buildings, and are currently being matched and pieced together by Egyptologists with the aid of computers.*

*The rock-cut temple of Nefertari at Abu Simbel, built by Ramesses II (ruled ca. 1279–1213BCE) in Egypt's Nubian frontierland, outside ancient Egypt proper. Like the temple of Philae (see p.202) and many other monuments in this region, Nefertari's temple and its companion (dedicated to Ramesses II himself) were dismantled and rebuilt on higher ground in the 1960s to prevent their submersion by Lake Nasser, the reservoir created by the Aswan High Dam.*

third millennium BCE, exist at sites such as Hierakonpolis and Abydos in Upper Egypt and Buto and Heliopolis in Lower Egypt.

The transition from mudbrick to stone construction, which took place in the Old Kingdom, is seen in the royal mortuary temples built near the pyramids for the cult of the dead pharaoh. However, the temples of the gods were built in more populous areas closer to the Nile floodplain and often in more perishable materials, and they have for the most part disappeared. With a few exceptions, much the same situation prevailed during the Middle Kingdom. It was not until the New Kingdom that the more durable sandstone replaced limestone in non-mortuary temple architecture, and consequently more structures from this era survive.

Sandstone allowed bigger building blocks – over ten feet (3m) long – resulting in ever-larger temple structures. Mudbrick ramps and scaffolding were used in the construction of successive courses of masonry in walls and columns, which were then dressed, carved and painted, usually from the top down. Where temples were cut from the living rock, the same quarrying and polishing techniques were employed as for rock-cut tombs. From the New Kingdom onward, temple entrances were flanked by the characteristically Egyptian massive gateways, or pylons, which would have been adorned with flagstaffs and pennants. Also from this period are the first soaring obelisks and hypostyle (columned) halls.

Some of the more famous temples may be found at Karnak (see pp.208–9), Luxor, Tanis and even many sites in ancient Nubia (present-day southern Egypt and Sudan) and farther south. Construction expertise hardly disappeared in the Greco-Roman period, but flourished at sites such as Dendera, Esna, Edfu, Kom Ombo and Philae (see p.202).

## TEMPLE ADMINISTRATION

The bureaucracy associated with major Egyptian temples could rival that of a present-day governmental or religious institution. In addition to fulfilling the ritual needs of the temple's cult statue by ensuring that festivals took place and daily food offerings were made, the priesthood also administered their temple's portfolio of properties. Service rotas were arranged for groups of priests called "phyles", each one serving for approximately two months per year.

Much of what we know about how mortuary temples were run comes from documents discovered at Abusir, part of the necropolis of Memphis. The first group of papyri, discovered in 1893, concern the funerary property of the Fifth-Dynasty pharaoh Neferirkare Kakai (reigned ca. 2472–2462BCE). The papyri deal with the transfer of revenue to his estate, offerings to his mother, Queen Khentkawes, and various other activities.

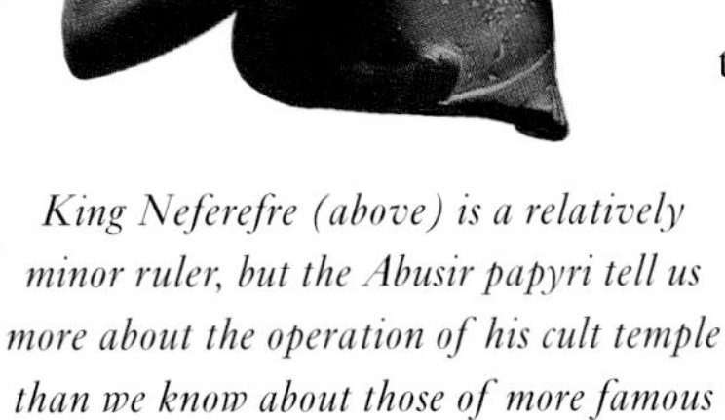

*King Neferefre (above) is a relatively minor ruler, but the Abusir papyri tell us more about the operation of his cult temple than we know about those of more famous Old Kingdom rulers, such as Khufu.*

In 1982, Czech archaeologists working in the mortuary temple of King Neferefre, or Reneferef (died ca. 2455BCE), a minor king of the Fifth Dynasty, discovered a further two thousand additional fragments of administrative papyri, which shed more light on the organization of a typical Old Kingdom mortuary temple. They describe the daily rites performed by the priests who served the cult of the deceased pharaoh – rites that might be maintained for many generations after the king's death.

Every day, a procession of priests would march around the royal pyramid, and the cult statue in the nearby mortuary temple would be anointed with scented oils, painted, dressed and "fed" (that is, it would be presented with food offerings). After the statue was deemed to have "eaten", the offerings would be removed and distributed among those who were responsible for the maintenance of the royal cult.

In the 1960s and 1970s, the ancient Egyptian monuments of Nubia, such as the temples of Abu Simbel and Philae, were the subject of a massive international collaborative effort to remove them from the path of the rising waters of Lake Nasser. Under the auspices of the UN, many monuments were painstakingly dismantled and rebuilt on higher ground.

Despite the ruinous monochromatic appearance of most temples today, in ancient times they would have been colourful, even gleaming structures. White-painted backgrounds set off the thousands of polychrome hieroglyphs and brightly coloured offering scenes depicting colossal figures of the pharaoh and the gods. Such texts and images were always of primary cultic significance, never merely decoration.

The inner sanctuary, deep inside the temple, would have been a dark and quiet place, visited by only the most privileged of priests. In contrast, the outside precincts were probably noisy and bustling with worshippers and others. In addition to their complex hierarchies of priests, temples employed all manner of part-time and full-time personnel, from farmers and carpenters to scribes, jewellers and keepers of livestock.

# HOUSES OF THE GODS

*A view of the hypostyle hall of the temple of the goddess Hathor at Dendera, one of the best-preserved temples of the Late Period. The tops of the columns are decorated with what is usually interpreted as a representation of the goddess's head.*

Although the oldest temples dedicated to the gods have long since disappeared, two forms are known from hieroglyphic representations: the shrine of the northern crown (Per-neser, or Per-nu), and that of the southern crown (Per-wer). The northern (Lower Egyptian) structure, known from the Delta site of Buto, consisted of a small, probably wooden room with a vaulted roof, surrounded by a simple enclosure wall. The southern (Upper Egyptian) shrine took on more of the form of a mythic creature: a skeletal framework with mats of animal skins draped over the roof, tusks protruding over the entrance, and totems or flagstaffs erected in front of the entire structure.

The preserved remains of Old Kingdom non-mortuary temples are slight compared with those of later ages. Memphis, the location of the

pharaonic palace and administration throughout much of Egyptian history, certainly boasted a number of impressive temples, in addition to the precinct of Ptah, creator god and patron deity of the region. But foundation walls and ruins are all that remain today of both the city and its temples. Middle Kingdom temple remains are not much better; the few notable exceptions include an undecorated temple at Qasr el-Sagha and the pavilion of Senwosret I at Karnak.

The New Kingdom provides our best evidence for understanding the architecture and inner workings of temples to the gods. In Upper Egypt, the Theban temples of Karnak and Luxor are not only among the largest, but are accompanied by the greatest supplementary material. For example, most of the élite classes of priests, administrators, even the "overseers of the cattle of Amun", were buried across the river from the temple, and these tombs tell us much about the responsibilities that these people bore, and the rituals and festivals that they supervised. Minor deities worshipped in small, regional shrines were more oriented toward local concerns, such as overseeing a plentiful inundation and a fruitful harvest.

The average Egyptian probably entered only the outermost courtyards to witness the processions and ceremonies on specially designated festival days. However, minor structures often stood outside the temple enclosure, dedicated to the petitions and common prayers of the populace. In addition, many private households contained a small shrine or cult place for offerings to local protective deities (see pp.84–5). Egyptians of means could also commission and dedicate statues to be erected in the temple's interior, in order to benefit from the god's beneficence.

**STATUE BURIAL CACHES**

**From time to time, the temple would become so choked with votive statues that the priests gathered and buried many of the older pieces. Some of the most exciting archaeological discoveries have been the location of such burial "caches". One of these was found in 1903–4 on the north–south axis of the temple of Karnak, and contained 800 statues and stelae, along with 17,000 smaller objects. A smaller cache turned up in the temple of Luxor at Thebes in 1989, and contained many life-sized masterpieces of royal sculpture from the Eighteenth Dynasty and later.**

## THE FORTUNES OF THE TEMPLES

While certain deities within the pantheon remained fundamental to the Egyptian world-view throughout the millennia, others achieved a greater prominence at specific periods. As their cults grew so did the resources allocated to them – among them, the construction of temples. For example, at Abydos, Osiris, god of the resurrection, subsumed the aspects of his predecessor deity Khenty-amentiu ("foremost of the Westerners"; that is, the dead), and rose in popularity from the Old Kingdom onward. In a similar way Re, the sun god, achieved special status in the mid- to late Old Kingdom, as demonstrated by the sun temples of Abu Ghurab and other sites (see pp.188–9), and by the many kings whose names included that of the god (Sahu*re*, Nyuser*re*, and so on). With the rise of the Theban Eighteenth Dynasty came the cult of the state god Amun and the consequent prominence of the Karnak temple (see pp.208–9). Akhenaten's failed attempt to supplant Amun at the end of the dynasty (see pp.128–9) involved a new architectural plan for the Aten's temples at el-Amarna (see p.203).

Many temples, particularly during the New Kingdom, accrued vast land holdings, numerous personnel, cities and even jurisdiction over foreign prisoners, and became politically influential at national level.

# The Temple of Karnak

Perhaps the most impressive Egyptian monument after the pyramids of Giza is the temple of Karnak on the east bank at Thebes. Larger than St Peter's in Rome, Karnak is in fact a series of temples built by a succession of pharaohs and dedicated primarily to the state god Amun (see pp.126–7). It is best understood from the inside out, beginning with the earliest Middle Kingdom sanctuaries and expanding outward with almost each succeeding reign, through the New Kingdom and beyond. A multitude of pylons (monumental gateways), avenues of sphinxes and sacred lakes mark two main axes: east–west, following the daily course of the sun and aligning with the royal monuments of Western Thebes (see plan, p.195), and north–south, the natural axis of the country itself, linking Karnak with the complex of the goddess Mut and that of Luxor, farther south. With countless historical texts, rituals, hymns and prayers recorded on its walls, Karnak remains a primary source for all aspects of ancient Egyptian culture after the Old Kingdom.

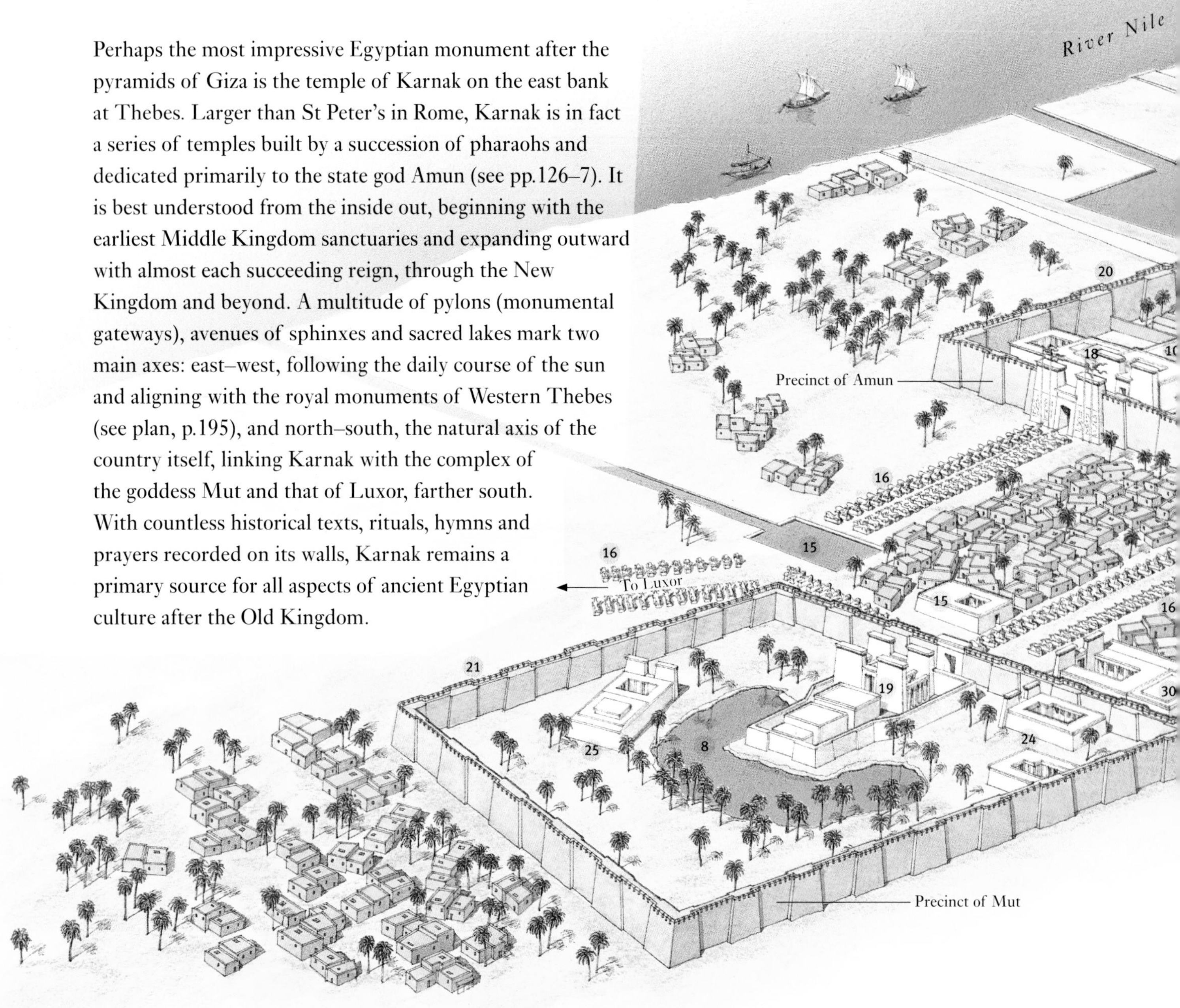

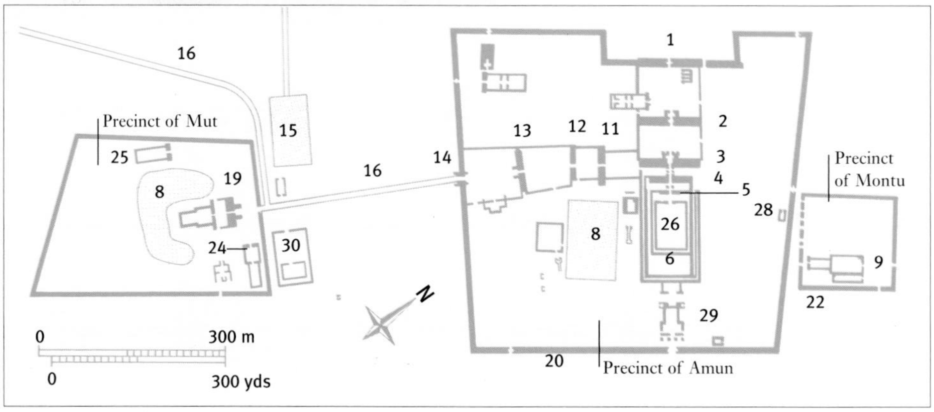

*A plan of the three main precincts that make up the Karnak temple complex.*

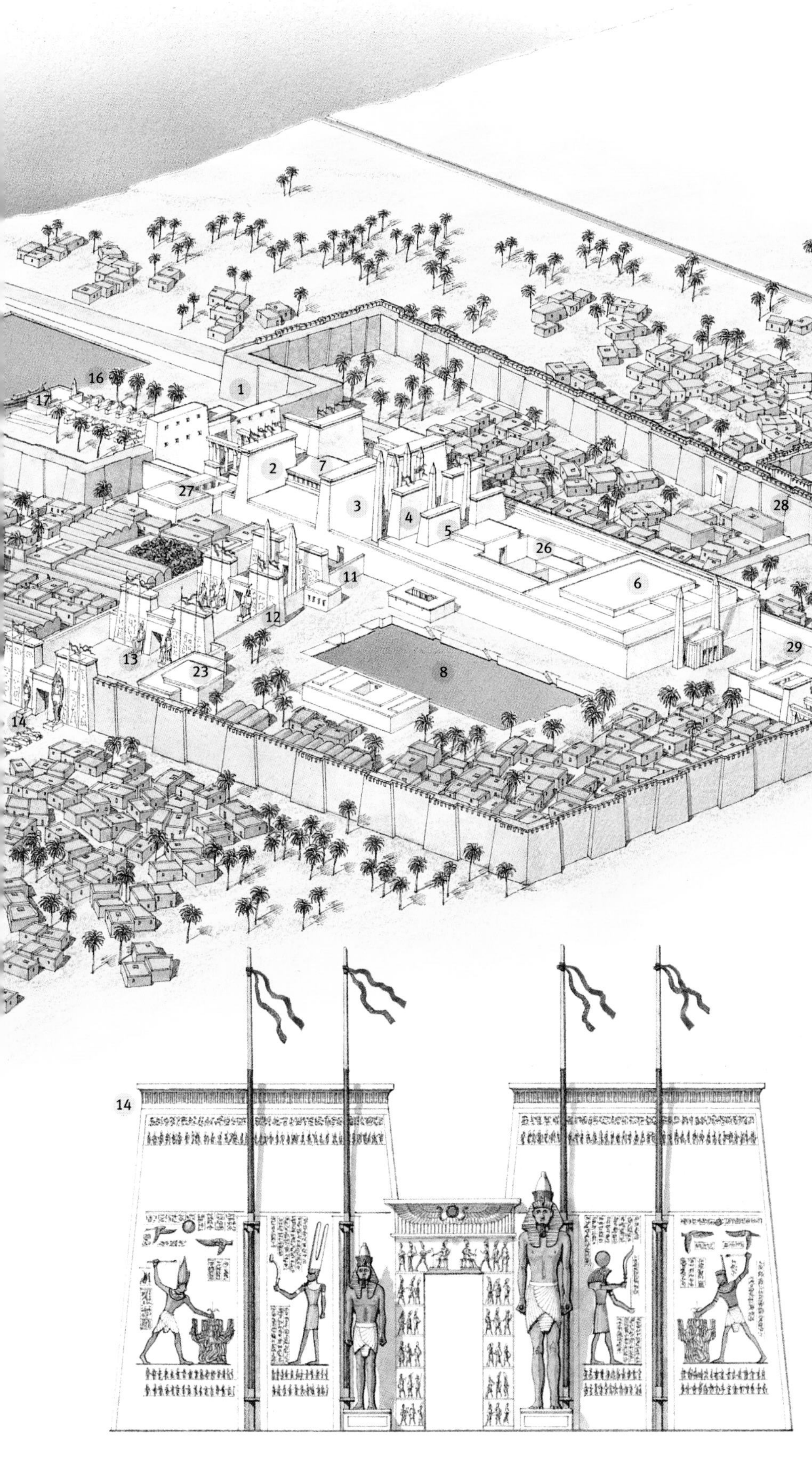


*A reconstruction of the Tenth Pylon at Karnak, showing how striking the external decoration of a New Kingdom temple would have appeared. On each wing of the pylon, the main scene shows the pharaoh crushing his enemies before the god Amun, who marches "out of" the temple to reward the king with order, justice, victory over the forces of chaos, and the rejuvenation of the monarchy and the land.*

**KEY TO RECONSTRUCTION AND PLAN**

1 First Pylon (Nectanebo I)
2 Second Pylon (19th Dynasty)
3 Third Pylon (Amenhotep III)
4 Fourth Pylon (Thutmose I)
5 Fifth and Sixth pylons (Thutmose I and III)
6 Memorial temple of Thutmose III
7 Great Hypostyle Hall
8 Sacred lakes
9 Temple of Montu
10 Temple of Khonsu
11 Seventh Pylon (Thutmose III)
12 Eighth Pylon (Hatshepsut)
13 Ninth Pylon (Horemheb)
14 Tenth Pylon (See also reconstruction, left)
15 Barque stations
16 Avenues of sphinxes
17 Departure quay for the barque of Amun
18 Temple of Opet
19 Temple of Mut
20 Enclosure wall of the precinct of Amun
21 Enclosure wall of the precinct of Mut
22 Enclosure wall of the precinct of Montu
23 Temple of the *Sed* (Jubilee) festival of Amenhotep II
24 Temple of Khonsu Pa-Khered (Khonsu the Child)
25 Temple of Ramesses III
26 Middle Kingdom court and inner sanctuary
27 Temple of Ramesses III
28 Temple of Ptah
29 "Temple of the Hearing Ear" of Ramesses II
30 Sanctuary of Amun Kamutef ("Bull of his Mother")

*The entrance obelisk at Luxor displays the deeply incised carvings of the Ramesside era which successfully prevented usurpation. But the Ramessides were not themselves above usurping monuments: the statue behind the obelisk bears the name of Ramesses II but was in fact erected by an earlier pharaoh.*

# ROYAL MORTUARY TEMPLES

Unlike the temples to deities of the Egyptian pantheon, royal mortuary temples served the cult of a single pharaoh. The king usually decreed a number of construction projects during his reign, but the construction of the royal tomb and its accompanying mortuary temple complex must have held the greatest personal significance for him, for this was the institution intended to maintain his cult after his death.

The earliest "royal" mortuary temples known are at sites such as Abydos and Hierakonpolis (see p.69) in Upper Egypt, and Saqqara in Lower Egypt, where there is evidence of massive mudbrick enclosure walls. The evidence is much fuller for the Old Kingdom proper, when part of the royal pyramid complex included several temples. One of the most significant and enigmatic constructions is the Step Pyramid complex of Djoser (Third Dynasty) at Saqqara (see pp.178–9). In this first major stone building in Egypt (and possibly all of the ancient Near Eastern world), the "mortuary temple" is a whole complex surrounded by an enclosure wall niched in imitation of a palace façade. This surrounds both the pyramid and a host of dummy buildings representing shrines and palaces of Upper and Lower Egypt, together with ritual structures designed to renew the vital force of the king's persona. In succeeding reigns, the typical Old Kingdom royal mortuary temple included a pyramid temple in close proximity to the pyramid itself, a long causeway leading down to the Nile Valley, and a valley temple, as well as satellite pyramids, boat pits and ceremonial structures. Limestone, granite and even basalt were the preferred construction materials, as these buildings were intended to last through all eternity. Lands and personnel were endowed specifically for the temple's activities, and many were still in operation generations after the death of the king for whom they were originally built.

A new type of royal mortuary temple appeared in the Eleventh Dynasty, under Nebhepetre Mentuhotep II (ca. 2008–1957BCE). In the centre of a natural bay, called Deir el-Bahari, in the cliffs of Western Thebes, this pharaoh constructed a low, terraced temple with approach ramps, engaged pillars in the form of the god Osiris and a mound, pyramid or rectangular structure on top. Other Middle Kingdom pharaohs returned to the north in the Twelfth Dynasty (ca. 1938–1759BCE), following the Old Kingdom pyramid model at sites such as Lisht and Dahshur, albeit on a more modest scale.

The height of royal mortuary temple construction dates from the New Kingdom, when a string of sandstone buildings dotted the desert's edge on the west bank at Thebes. By the Eighteenth Dynasty the pharaohs had physically separated the royal burials from the mortuary temples them-

**USURPATION**

**Hieroglyphic texts and figural representations held a magical significance for the Egyptians, and the alteration or destruction of texts and images was equally meaningful. Throughout Egyptian history all types of wilful destruction, recarving or other alteration took place. In some cases, names or images were chiselled out by the enemies of the deceased to ritually "kill" him or her. There are also many cases in which monuments were "usurped" by changing the name on the statue, in order to divert all its divine benefits to the new "owner". From the Ramesside Period, kings ordered that hieroglyphs be cut up to five inches (12.5cm) deep into the wall, making it much harder for anyone to erase or alter an inscription.**

ABOVE: *The "Colossi of Memnon", two massive statues of King Amenhotep III that are all that remains of the king's mortuary temple in Western Thebes. Now much eroded, the statues once flanked the main gateway in the temple pylon.*

LEFT: *The terraced mortuary temple of Hatshepsut at Deir el-Bahari, which has been undergoing careful restoration in recent decades (see also illustration, p.8).*

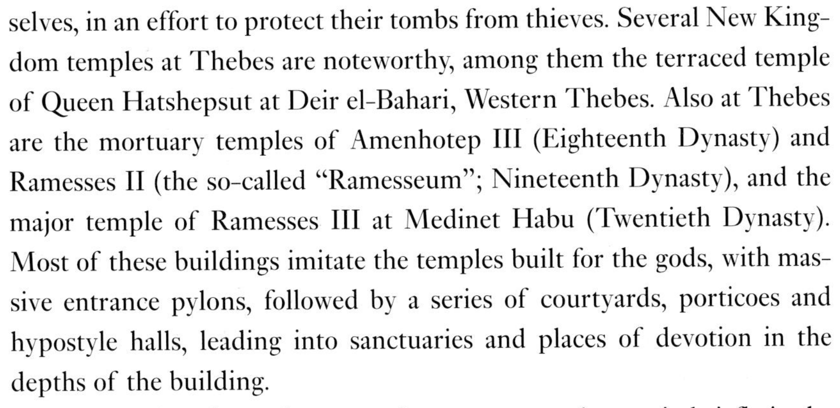

selves, in an effort to protect their tombs from thieves. Several New Kingdom temples at Thebes are noteworthy, among them the terraced temple of Queen Hatshepsut at Deir el-Bahari, Western Thebes. Also at Thebes are the mortuary temples of Amenhotep III (Eighteenth Dynasty) and Ramesses II (the so-called "Ramesseum"; Nineteenth Dynasty), and the major temple of Ramesses III at Medinet Habu (Twentieth Dynasty). Most of these buildings imitate the temples built for the gods, with massive entrance pylons, followed by a series of courtyards, porticoes and hypostyle halls, leading into sanctuaries and places of devotion in the depths of the building.

Royal tomb and temple seem to have come together again briefly in the Third Intermediate and Late periods. For example, the royal families of the Twenty-First and Twenty-Second dynasties were interred in the north at Tanis, in family sepulchres within the temple precinct.

### MEDINET HABU TEMPLES

**One of the best-preserved temple complexes may be seen at Medinet Habu in Western Thebes (see plan, p.195). Here an Eighteenth-Dynasty temple was begun by Hatshepsut. The walls bear the evidence of successive restorations and expansions by many subsequent rulers up to the Ptolemaic and Roman periods. During the Twentieth Dynasty, Ramesses III chose a nearby site for his mortuary temple and added a ceremonial palace adjoining the temple on the south side. Excavations have revealed a throne-room, offices and sleeping chambers. Other buildings within the walls of this "city" included storerooms, the houses of high administrators, and tombs and chapels added during the Twenty-Sixth Dynasty.**

● CHAPTER 14

# EGYPTIAN ART

Just a few centuries separate the simple polished stoneware and burnished pottery of the fourth millennium BCE – itself of high artistic value – from the magnificent sculpture, relief and painting that already characterized the earliest dynasties. Mastering techniques in rock, wood, metals (especially gold), semi-precious stones and other media, Egyptian artists worked on a large and small scale, in accordance with the conventions of a formalized aesthetic, to produce one of the world's most astonishing artistic legacies.

▲



ABOVE: *The conventions of Egyptian art were often disregarded in the representation of labourers and foreigners. The scruffy beard and curly hair of this carpenter from an 18th-Dynasty tomb at Deir el-Medina, Western Thebes, would never have been shown on a picture of the tomb's owner.*

## THE POTENT IMAGE

Judged by modern standards, ancient Egyptian artisans produced some of the most spectacular art ever made, yet their culture had no word for art and no concept of art for art's sake. For the Egyptians the activity that we today would consider art served a greater purpose: to embody life. Although the vast majority of statues were created as idealized likenesses of their subjects rather than actual portraits, they were brought to life through the "Opening of the Mouth" ceremony (see p.141). Tomb statuary served as the eternal repository for the *ka* (spirit) of the deceased. Similarly, a representation of a deity became the deity itself, and images of provisions and activities carved and painted on the walls of tombs or temples were imbued with all the qualities of the real thing.

The idea that images had the same properties as their real-life counterparts governed styles of representation. At times this becomes humorously apparent, especially in two-dimensional works where, for example, an animal is depicted from the side, but more is shown than would normally be visible in profile. For example, a First Intermediate Period painting from Gebelein, near Thebes, attempts to portray a donkey wearing two pouches, one on each side of its back; the pouch on the side normally invisible in profile was flipped up so that it, too, could be seen.

Certain formulae dictated the manner in which a tomb owner and his family were depicted. The poses of the main figures tended to be stiff and formal, each artist having a relatively limited repertoire from which to choose and little opportunity for individual creativity. But no such constraints governed the depiction of figures of lesser importance, who were customarily shown smaller than their master. Servants carrying out their daily tasks cover tomb walls, and while each activity had to incorporate a number of standard elements, fine details were left to the individual artist.

Despite the splendour of their works, colossal or miniature, in soft cypress wood or hard granite, the artisans of ancient Egypt have revealed remarkably little about themselves. Most artisans remain anonymous and can only rarely be associated with a specific work. All that we know comes

*In this scene from the 18th-Dynasty tomb of the priest Khamuas at Western Thebes, the stiff and formal figure of the deceased (left), his wife kneeling at his feet, oversees the activities of his servants, who are drawn with less rigidity and smaller than their master, in accordance with their status.*

from the artisans' tombs, representations of them at work, and references in texts to their training. On large projects such as the painted or relief-decorated tombs, it is clear that artisans worked cooperatively. The Deir el-Medina artisans (see p.73) who prepared the vast royal tombs of the New Kingdom at Western Thebes worked in teams, each one comprising line draughtsmen, relief sculptors and scribes (see p.223). It is sometimes possible to differentiate between the work of a master artisan producing images of the tomb owner and his family and that of less-skilled apprentices showing figures of lesser importance. Different artisans may have worked on sculpture and relief. Among those who worked on sculpture, there was probably a further specialization by material.

Egypt's rich natural resources provided a readily available supply of many types of raw materials (see pp.64–5), some of which were exploited even prior to the beginning of Egyptian dynastic history. Of the softer stones, limestone and sandstone were readily available in large quantities and were employed both for sculpture and architecture. Harder stones such as granite, quartzite, diorite and basalt were in shorter supply, which limited their use largely to sculpture. In general, limestone and sandstone statuary was completely painted, while on sculptures of harder stone often only details such as the eyes and lips were highlighted in paint.

Some sculpture was made from wood and metal, although relatively little has survived. The first metal used in sculpture was copper, either solid or hollow (hammered over a wooden core). Among the finest examples of the hollow type are life-size, and lifelike, sculptures of the Sixth-Dynasty kings Pepy I and his son Merenre. By the Middle Kingdom, Egyptian artists had mastered the technique of hollow casting. In the New Kingdom, bronze replaced copper as the favoured metal for sculpture.

*Realism and an overall attention to detail are the outstanding features of this rare wooden statue of the steward Ka-aper from Saqqara. Carved from pieces of sycamore and assembled by tongue-and-groove joinery, it was then partly plastered and painted. Fifth Dynasty, ca. 2490BCE.*

*This Predynastic clay head of ca. 5000BCE is the oldest three-dimensional Egyptian work of art. It is believed to have served a ritual function.*

# PREDYNASTIC AND EARLY DYNASTIC ART

Almost as soon as techniques of agriculture and animal husbandry made permanent settlement possible along the Nile, Egyptians began to transform utilitarian items into objects of beauty. Sites in the Delta, the Faiyum and the northern reaches of the Nile Valley have yielded information from ca. 5000BCE. However, most of our knowledge of Predynastic art comes from Upper Egypt, and largely from grave goods. Major phases of Predynastic culture are named after the vast Upper Egyptian cemetery site of Naqada. The earliest phase, Naqada I (also called Amratian after the cemetery of el-Amrah), was a time of great artistic creativity. Objects such as combs and slate palettes (used to grind eye-paint) might be carved in animal forms, and colourful semi-precious stones found in the desert were polished and worn as jewelry. Pottery was produced in an extraordinary variety of shapes, wares and surface treatments, some of which even interfered with their practical purposes. Seldom were two vessels identical.

The Naqada II (or Gerzean) period was a time of bigger settlements spread over a broader geographical area, increased contact between the settlements, and larger and more numerous graves. These circumstances seem to have fostered the mass production of artefacts, and a corresponding diminution in individual creativity. Palettes and combs were simplified into more abstract, less innovative forms than appeared in Naqada I, and the same types occur frequently. A similar decline in quality is seen in pottery: while a grave might have more numerous and larger pots, fewer would be decorated; and those that displayed any form of decoration tended only to have repetitive designs.

*The first royal sculptures identified by name are two seated images from Hierakonpolis of the 2nd-Dynasty King Khasekhemwy (ca. 2675BCE). The face was skilfully executed with considerable attention to detail. The neck, a narrow and potentially vulnerable area, was strengthened by extending the back of the king's Upper Egyptian crown down to his shoulders.*

The final phase of Naqada culture, Naqada III or Dynasty "0", is marked by a growth in the size and complexity of settlements, and by gradual unification under strong local leaders (see pp.106–7). Many of the same types of objects were manufactured, but their decoration took on a new sophistication. Palettes became larger, creating space for historical or mythological narratives carved in relief on their soft surfaces; writing also started to appear on palettes and on other types of object.

By the beginning of dynastic Egypt (ca. 3000BCE), many of the traditions and styles associated with Egyptian art for the next three millennia were already in place. One such element was the method of rendering a three-dimensional figure in two dimensions so that it incorporated what the Egyptians considered to be its most important aspects, often without regard for realism. The Narmer Palette, a large slate tablet celebrating the victory of King Narmer over a people of the Delta (see illustration, p.23), provides many instances of the manipulation of the human figure to accommodate Egyptian ideals. The main image on the palette is dominated by a large-scale representation of King Narmer himself, whose name ("Baleful Catfish") appears in a *serekh* frame. His head is rendered in true profile, but the eye is represented frontally, in part because this effect was easier to achieve successfully, but also because the eye was considered the most important element of the face. The broad shoulders are also shown frontally, to emphasize the king's strength and power. However, the legs and feet are shown in profile, with the left foot marching forward. Such combinations of profile and frontal views became the norm for royalty, commoners, servants and foreigners alike. Size indicated importance, and the king is consequently the largest figure on the Narmer Palette.

In their methods of rendering the human figure, Egyptian artisans succeeded in conveying a large amount of information in a limited space, and at the same time in creating a harmonious composition that rarely seems overcrowded. Other aspects of ancient Egyptian art already in evidence at this time are the use of hard stone, monumentality, and the reduction of an image to its essential elements. Royal sculpture of the Early Dynastic Period, and the private sculpture that imitated it, are both characterized by a compact, block-like appearance that persisted until the artists achieved a greater comfort in working with hard materials and greater skill in liberating the human figure from its stone matrix.

## FOREIGN INFLUENCES

Against a background of increasing prosperity based on abundant natural resources, the early third millennium BCE saw the Egyptians developing a more extensive trade network. The Levant, Sinai, Mesopotamia and Persia to the east, Nubia in the south and Libya to the west, supplied new ideas, raw materials and finished products to the developing cultural centres of the Nile Valley. Mesopotamian ships, distinguished by their high prows and sterns, are represented – together with a figure dressed in a turban and a long belted robe, a typical Mesopotamian garment – in a tomb at Hierakonpolis and on a hippopotamus-tusk ivory knife handle from Gebel el-Arak in Middle Egypt. Cylinder seals found in Egypt must have been brought in by traders or travellers, because they are characteristic of Mesopotamia and Persia. Lapis lazuli, a blue mineral mined in Afghanistan and probably brought to Egypt by way of ancient Persia, Mesopotamia and Syria–Palestine, also attracted the attention of Egyptian artists. As well as materials, artists also borrowed foreign iconography, such as winged griffins and serpopards (serpent-neck leopards; see illustration).

*Two serpopards from the Predynastic Narmer Palette (on the reverse side from that shown on p.23). The necks of the animals, tethered with ropes held by two figures, form a round depression that was probably used for grinding eye-paint, the original function of such palettes.*

*King Menkaure (ca. 2532–2510BCE) and Queen Khamerernebty II, a statue from the king's mortuary temple at Giza. Perfectly proportioned according to the Egyptian artistic canon, the king strides forward, with his queen, almost as tall, by his side.*

# CLASSICAL EGYPT: THE OLD KINGDOM

Noteworthy for its refined forms and mastery of stoneworking techniques on a large scale, the Old Kingdom (ca. 2625–2130BCE) is justifiably considered the classical age of Egyptian sculpture. With increasing prosperity came a greater demand for sculpture, both royal and private. Artists gained more experience in working with stone and became more confident in their own abilities to carve it. Approximately two-thirds lifesize, the schist statue of the Fourth-Dynasty king Menkaure and his queen (see illustration, left) found in the king's mortuary temple in his pyramid complex at Giza conveys the Old Kingdom ideal. The statue is not a realistic portrait but an idealized likeness of the human image – the epitome of royalty. Both figures exhibit bodies which represent the youthful ideal for their sex, the king with broader shoulders and slightly more developed musculature than his wife. Although the queen embraces the king, with an arm around his waist and a hand on his arm, the expressions worn by both figures are rather impassive. Ageless and emotionless, the royal pair gaze ahead at an eternal beyond.

At times, Fourth-Dynasty artisans were commissioned to depict specific individuals, usually members of the pharaonic bureaucracy, in what may be considered true portraits. One fine example is the statue of the vizier Hemiunu. Life-size and lifelike, his stern face with its set mouth, full jowls, double chin and decidedly hooked nose can leave little doubt that an individual rather than a "type" is represented. Further, his more-than-ample torso with its rolls of fat spilling over the waistband of his kilt, evocative of a prosperous career, provides a dramatic contrast to the taut form typical of representations of royalty.

*A woman brushes her hair back to keep it from the flames as she stokes a fire. A terracotta figurine from a 5th-Dynasty tomb at Giza.*

Artistic challenge is often a crucible for originality, and there is no better example of this in Egyptian art than the statue of the dwarf Seneb and his family (see p.84). By depicting Seneb seated with his short legs crossed beneath him and his two children standing where the subject's legs would usually be, the artist has tempered, although not hidden altogether, Seneb's infirmity. The paint is perfectly preserved, allowing the viewer to notice at once that Seneb's wife and daughter have pale skin, untouched by the sun, reflecting their social status as high-ranking females who work mainly indoors. In contrast, the tanned, ruddy skin of Seneb and his son is typical of Egyptian depictions of men of all ranks, who were more accustomed to working outdoors. Despite their different ages and roles, all four members of Seneb's family have similarly bland faces, illustrating a style that was to continue into the Fifth Dynasty.

The growth and prosperity of the bureaucratic classes in the Fifth and Sixth dynasties meant a greater demand for tombs and for statuary with which to furnish them, particularly at Giza and Saqqara. In general at this time, the mass production of sculpture left little room for artistic individuality. Exceptions are figurines of servants engaged in a variety of domestic tasks, which it was believed they would also perform for the tomb owner in the afterlife.

An important official's tomb might contain more than two dozen rooms, each decorated from floor to ceiling with painted scenes of the owner, his family, and people at work on his estate. The smaller vignettes, in particular, display considerable variety and even humour, presenting a remarkable picture of daily life in the age of the great pyramids.

## THE CANON OF PROPORTION

At least as early as the Old Kingdom, the Egyptians perfected a way of replicating their ideal concept of the human image in sculpture and relief. In wall paintings in the Fifth-Dynasty tomb of Perneb at Saqqara, some of the technical devices used by artisans to achieve symmetry and the desired proportions are still in place. For example, a line runs vertically down a number of servant figures, exactly bisecting their torsos. Horizontal lines intersect the central axis line at the hairline, base of neck, armpit, elbow, base of buttocks and top of knee. By the Middle Kingdom, these guidelines were fleshed out both horizontally and vertically into a full grid of eighteen squares that extended from the hairline to the bottom of the feet. The grid was further subdivided during the Twenty-Sixth Dynasty into twenty-one squares, which allowed even greater control. The Egyptians could lay out these lines on any surface to reproduce identically-proportioned figures of whatever size they wanted.

*A wooden gesso-surfaced drawing board showing the gridmarked figure of Thutmose III (ca. 1479–1425BCE), perhaps the work of a trainee artist. Next to the king are his names and trial hieroglyphs.*

The divisions of the body and the size of the grid units were based on a fixed ratio of body parts. The basic unit was the cubit, a unit based on the distance between the elbow and the tip of the thumb. The cubit was subdivided into six handbreadths, with one handbreadth equalling the width of four fingers measured across the knuckles. In the Twenty-Sixth Dynasty, the "royal cubit" became the standard measure of the human figure. As its name suggests, it was grander than the standard cubit and consisted of seven rather than six handbreadths.

# REGIONAL STYLES AND ART OF THE IMPERIAL AGE

With the loss of royal power at the end of the Old Kingdom, regional capitals and local rulers replaced the central authority of Memphis. In the absence of court sponsorship, fewer tombs were constructed in the royal necropolises of Giza and Saqqara, and artistic activity there slowed considerably. Local officials were buried in their home cities, and studios of local artisans sprang up to meet the consequent need for tomb sculpture and decoration. Memphite styles were adapted to local conditions and available resources. Understandably, the greater the distance from Memphis and the more time that elapsed, the weaker the adherence to the models. During the First Intermediate Period (ca. 2130–1980BCE), many places developed their own distinctive styles.

A stela from the tomb of Wadj-Setji at Naga ed-Deir, approximately halfway between Memphis and Thebes, demonstrates one local interpretation of Memphite style. Beneath a hieroglyphic formula, Wadj-Setji and his wife Merirtyef stand beside a pile of food and drink offerings. The subject-matter is typical of the Old Kingdom, although the style is very

## TEMPLE SCULPTURE OF THE MIDDLE KINGDOM

One of the major innovations of the Middle Kingdom was the large-scale use of sculpture in temples, and the new types of sculpture that arose to meet this need. Life-size figures of King Senwosret III depict him with his hands placed flat on his triangular kilt in an attitude that signified reverence for his predecessor, Nebhepetre Mentuhotep II, at whose funerary temple at Deir el-Bahari in Western Thebes they were erected for all to see. In these sculptures, like those of Senwosret's successors in the Twelfth and Thirteenth Dynasties, the dispassionate countenance characteristic of Old Kingdom royal portraits (see illustration, p.216) is replaced by a much sterner facial expression. Although not new to the Middle Kingdom, sphinxes were also set up outside temples, usually in pairs.

"Block-statues", so called from the solid, cuboid appearance that arose from the squatting posture of the subject, also appear for the first time during the Middle Kingdom. The slightly abstracted form of such sculptures created flat surfaces that proved ideal for inscription, and the block statue became a popular way of advertising its owner's personal qualities in temple precincts.

*This powerfully realistic portrait of Senwosret III (ca. 1836–1818BCE) was clearly intended to inspire awe in the onlooker. The authority of the pharaoh is conveyed by the stern gaze and the downturned, almost grimacing, mouth.*

different. The stick-like figures have spindly arms which hang limply or bend awkwardly to hold walking sticks. An extremely prominent nose and a large eye take up most of the face, while the neck is all but absent. Heads are small, waists high and legs long. A central vertical line and several horizontal lines incised to the sides of the main figures reflect an awareness of the rules of proportion (see box, p.217), but no true understanding of their application. The distinctive features of this piece and of others from the same site – the treatment of figures, broad border of alternating coloured rectangles, and the distinctive forms of the hieroglyphs – amount to a definite local style. Regional idiosyncrasies also developed throughout the Nile Valley and oases at this period.

After Egypt was reunified in the Eleventh Dynasty by Nebhepetre Mentuhotep II of Thebes (ca. 2008–1957BCE), and a central authority again controlled the land, it was understandably the Theban style that prevailed at the beginning of the new era. Bold, massive sculpture conveyed the power of the conqueror. Freedom to travel made the Old Kingdom monuments of the Memphite area again accessible, and they served as models for both sculpture and relief – especially after Amenemhet I, the first king of the Twelfth Dynasty, moved the capital back to the north. Works from the beginning of the Middle Kingdom may be distinguished only in subtle details from those of the Old Kingdom. As in the Old Kingdom, official private sculpture generally copied royal models in the depiction of facial features, and many non-royal statues display the stern visages seen in statues of kings from Senwosret III onward (see illustration, opposite). By the Thirteenth Dynasty, when a strong administrative class produced a series of weak kings, some of the finest sculptures were those of powerful private individuals. A major innovation of the Middle Kingdom was the development of hollow casting for metal sculpture.

*A wooden sculpture from Dahshur representing the* ka *(spirit) of King Auyibre Hor. The naked statue has inlaid eyes of bronze, crystal and quartz, and was originally stuccoed and painted. On the king's head are two raised hands, the hieroglyph for* ka. *Middle Kingdom, 13th Dynasty, ca. 1340BCE.*

The contents of the tomb of Djehutynakht, a nomarch (local ruler) of the town of Deir el-Bersha, offer a glimpse of provincial life during the early Middle Kingdom. Farming, weaving, baking, brewing, slaughtering and carpentry – the main activities on the nomarch's estate – are represented in almost three dozen wooden models, rather than sculpted on tomb walls as during the Old Kingdom. Crowded around the coffins of Djehutynakht and his wife were nearly sixty boat models, including funerary boats, sailing boats, rowing boats and papyrus skiffs, testifying to the importance of the river in both religion and daily life. Painted in exquisite detail on the inner walls of the nomarch's two coffins are his personal effects, including garments, jewelry and weapons. An assortment of wooden models of male and female servants, and processions of male and female offering-bearers, further ensured Djehutynakht's comfort and wellbeing in the afterlife.

*A variety of beautifully crafted faience animals have been found in tombs of the Middle Kingdom. This hippopotamus is decorated with the aquatic plants in which it bathed.*

From around the same time, a glimpse of the life of an Egyptian queen is provided by the exquisitely carved limestone reliefs on the sarcophagus

*A scene from the limestone sarcophagus of Queen Kawit, who was buried in the funerary precinct of her husband, Nebhepetre Mentuhotep II, at Deir el-Bahari in Western Thebes. A servant dresses the hair of the queen, who holds a mirror in one hand and a bowl of milk in the other. A male attendant pours milk into a bowl with the words: "For your* ka *[spirit], O mistress". Other scenes show Kawit wearing similarly fine jewelry and attended by servants who fan her and offer her jars of ointment. Middle Kingdom, 11th Dynasty, reign of Nebhepetre Mentuhotep II (ca. 2008–1957BCE).*

of Kawit, one of the wives of King Nebhepetre Mentuhotep II, who is depicted at her toilet attended by male and female servants.

The expulsion of the Asiatic Hyksos rulers at the end of the Seventeenth Dynasty (see pp.31–2) inaugurated a new period in Egypt's history and art, marked by an attempt to recapture the spirit of the past by copying its imagery. However, the mood of imitation changed when Hatshepsut proclaimed herself king (see p.89). Whether they show her as a woman or a man, her portraits have a decidedly feminine quality. In the mid-Eighteenth Dynasty both royal and private sculptures and reliefs were characterized by heart-shaped faces, arched eyes and eyebrows, and a sweet smile.

Dramatic changes took place during the reign of Amenhotep III (ca. 1390–1353BCE), when international contacts inspired new ideas and economic prosperity brought additional resources to a growing bureaucratic class. An opulence pervades both sculpture and the decorative arts. Elaborately curled wigs, diaphanous garments and an attention to detail, particularly in the rendering of jewelry, characterize representations of both sexes. Men and women alike tend to have cherubic faces, elongated eyes and benignly smiling mouths, highlighted by vermillion lines.

Amenhotep III's successor, Akhenaten, took liberties in the depiction of the human body (see box, opposite), but his eccentricities were all but forgotten by the beginning of the Nineteenth Dynasty. Rulers such as Sety I and Ramesses II rendered the human form more conservatively, while maintaining the aura of elegance. Ramesses II's long and prosperous reign (ca. 1279–1213BCE) inspired quantity in art and, at times, quality. Throughout Egypt and Nubia, temples and sculptures commemorated his military exploits and festivals. No later ruler matched this productivity.

*Masks for covering and protecting the head of a mummy formed an important part of Egyptian funerary equipment from the Old Kingdom onward. This sumptuous example belonged to Thuya, the mother-in-law of Amenhotep III (ca. 1390–1353BCE), and is made of gilded* cartonnage *(linen stiffened with plaster) inlaid with semi-precious stones and glass.*

## THE AMARNA REVOLUTION

When King Amenhotep IV abandoned Egypt's traditional gods in favour of the sun disc, the Aten, and changed his name to Akhenaten to reflect his devotion to this deity, he also profoundly altered the way in which artists depicted both divine and human images. No longer shown anthropomorphically, the sun god was depicted as an abstract disc emitting rays which presented *ankh*s (☥, the hieroglyph for "life") to the king and royal family who stood under its protection.

*One of the daughters of Akhenaten and Queen Nefertiti nibbles delicately on a roasted duck on this unfinished relief fragment from el-Amarna. The relief has been sketched out but only partly carved. The girl's relaxed pose and elongated body are typical features of the art of the Amarna Period.*

Perhaps influenced by the new religious philosophy, artisans emphasized new and different themes in art. For example, household stelae show the king and his queen, Nefertiti, kissing and caressing their royal daughters in a natural, human way. Previously, any display of intimacy was alien to Egyptian art, and royal children were not generally depicted in the company of their parents. Formal poses now gave way to lounging figures. Another innovation of the "Amarna style" (named after el-Amarna, the site of Akhenaten's capital, Akhetaten) was the representation of temple and palace interiors, affording rare glimpses of the activities that took place within.

Some of the most extraordinary pieces found at Akhetaten come from the studio of the sculptor Thutmose, the only such workshop known from ancient Egypt. Unquestionably, the most famous piece found here was the bust of Queen Nefertiti (see p.88), where the eccentricities of the Amarna style were abandoned in favour of a classical elegance. The bust was a highly unusual form for Egyptian art, and the function of this piece may never be known; it was perhaps used for teaching painters and sculptors. Thutmose's studio also produced a unique series of plaster casts of the faces of men and women, vividly bringing to life the ordinary people of ancient Egypt.

Painted scenes had adorned tombs since Predynastic times, but in the New Kingdom tomb painting really came into its own. In Western Thebes alone, scenes of ritual and daily life covered the walls of up to a thousand royal and private tombs (see illustration, p.88). Texts from the village of Deir el-Medina, home to the workers who decorated the tombs, indicate that the position of artist was highly prized and often passed from father to son. The artists worked in teams, often on opposite walls. One man laid a grid on a smooth limestone surface and outlined the main figures in black ink. Before the paint was applied, another artist checked the outlines, marking corrections in red ink. Paint was made by mixing water and gum arabic with ground pigment. Black, blue, green, yellow, red and white were the most common colours.

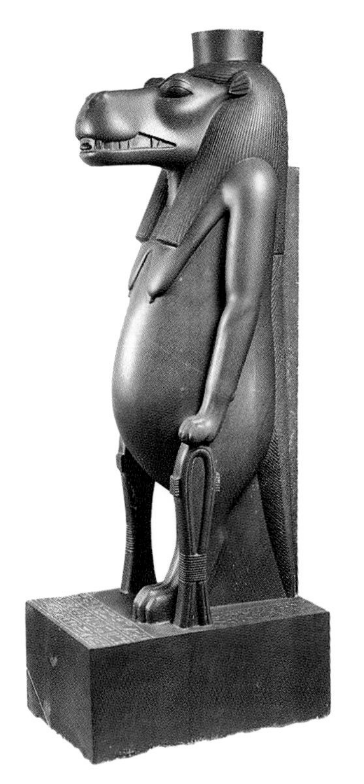

ABOVE: *A schist statue of Taweret, the goddess of childbirth, as a hippopotamus. The high polish of the stone is characteristic of the Late Period. From Karnak; 26th Dynasty, reign of Psamtik I (664–610BCE).*

BELOW: *Great numbers of small bronze votive statues of gods were deposited in shrines during the Late Period. This bronze cat inlaid with gold and silver represents the goddess Bastet; ca. 575BCE.*

# ART OF THE LATE PERIOD

The Third Intermediate Period and the Late Period are marked historically by an alternation between native Egyptian and foreign rule, and the art of this era reflects the opposing forces of tradition and change. The bland faces and insipid smiles of Third Intermediate Period sculptures represent an attempt to recapture the halcyon days of the early Eighteenth Dynasty through an imitation of its art. In some instances, the rulers of the Third Intermediate Period simply usurped earlier monuments. For example, there is so much usurped statuary of the Ramesside era (Nineteenth and Twentieth dynasties, ca. 1292–1075BCE) at Tanis, the capital of the Twenty-First Dynasty (ca. 1075–945BCE), that people in later periods assumed it was a Ramesside city. The Kushite invaders of the Twenty-Fifth Dynasty identified themselves with earlier periods, and copied them in their art, but at the same time they also promoted a brutal realism that showcased the skill of Egypt's artisans.

When the Kushites were pushed back to their southern homeland, the native Egyptian rulers who took over (as the Twenty-Sixth Dynasty) at times preferred more placid, youthful faces with a faint smile. Life-size figures of non-royal individuals embracing an image of a deity express a personal piety, perhaps aimed at excelling the Kushites. The dynasty is also noted for its emphasis on the fine details of human anatomy and the high polish that its artisans achieved on hard stone. Large-scale bronzes, made by the hollow-cast technique and often inlaid with gold and silver, are another characteristic of the Late Period, as is its superb jewelry, some of which incorporated older pieces. The prosperity of the time is also reflected in art on a smaller scale, particularly bronze statuettes of deities, vessels of Egyptian faience and tiny amulets intricately carved from a variety of materials.

The contribution made by the Persians during their first occupation of Egypt (referred to as the Twenty-Seventh Dynasty, ca. 525–405BCE) is represented by the appearance of new gestures and garments. In their admiration for Egypt and its artisans, the Persians commissioned, most likely from Egyptian artists, a life-size sculpture in Egyptian stone of their king, Darius I (522–486BCE), for erection in their homeland. Darius wears a typically Persian garment, the front of which bears an inscription in Persian, Babylonian and Egyptian hieroglyphs, the key languages of diplomacy in the Near East in the first millenium BCE.

Egypt experienced its final renaissance under native rule in the Thirtieth Dynasty (ca. 381–343BCE), when the last ethnically Egyptian pharaohs revived the arts. Temple building commenced on a large scale. For example, extensive programmes of construction and decoration were carried out by King Nakhtnebef (Nectanebo I, 381–362BCE) at temples throughout the country. Among his many projects, he constructed some of the earliest parts of the temple of Isis at Philae, erected a massive enclosure wall around the temple complex of Karnak and began the construction of its First Pylon, which was to remain unfinished. Nakhtnebef's successor, Nakhthoreb (Nectanebo II, 362–343BCE) erected a temple of Isis at Behbeit el-Hagar in the Delta, and also undertook new building activity at the temple of the cat goddess Bastet at Bubastis and at other sites associated with the animal cults that enjoyed great popularity during the period (see p.163).

Painters and sculptors must have been in great demand as a consequence of such building activity. A new plasticity of modelling is evident in both sculpture and relief of the time, and it was this style that the Greek Ptolemaic Dynasty – following another period of Persian rule and the conquest of Egypt in 332BCE by Alexander the Great of Macedon – adopted as their own. Sculpture and relief from the last days of Egypt under its native kings are often indistinguishable from those of the early years of the Ptolemies.

## THE ART OF THE KUSHITES

Of all the foreigners who ruled Egypt in the Late Period, only the Kushites left a lasting impact on Egyptian art and culture. They shared with the Egyptians a devotion to the god Amun, and saw themselves as keepers and restorers of Egypt's ancient temples. Part of their temple renovation programme involved erecting statues of their own kings, which blended Egyptian and Kushite styles and iconography.

In keeping with their reverence for the Egyptian gods, the Kushites at times also copied Egyptian art from earlier periods, including poses, wigs, garments and body modelling. So skilful were they that often only subtle differences distinguish works dating to the Twenty-Fifth Dynasty from their Old, Middle and New Kingdom prototypes.

Some of the most expressive sculptures of the Kushite dynasty depict private individuals. Unfettered by strict conventions, artisans were apparently free to depict their subjects more realistically.

Kushite sculptures of women are especially distinctive. The lean figures of previous dynasties give way to fuller breasts and broader hips. This new standard of beauty was continued in the Meroitic art of the Sudan long after the Kushites were forced out of Egypt, and it may even have set the pattern for the full-figured representations characteristic of Ptolemaic Egypt.

*This sphinx portrait of the Kushite king Taharqa (690–664BCE) is more massive than traditional Egyptian examples. The two* uraei *(cobras) are another Kushite trait; native rulers usually wore only one.*

*In the Roman Period, techniques of decorative glass production reached high standards. To make brightly coloured plaques such as this one of the god Horus, rods of glass were heated and fused together then cut into sections.*

*Antinous, the favourite of the Roman emperor Hadrian, is depicted in pharaonic* nemes *headdress and kilt in this 2nd-century* CE *sculpture from Hadrian's villa at Tivoli in Italy. The figure's muscular torso and naturalistic distribution of weight make it clear that the statue was created by a Roman, not Egyptian, artist.*

# ART OF THE PTOLEMAIC AND ROMAN PERIODS

When Alexander the Great and the Macedonian Greeks conquered Egypt, they ended forever native pharaonic rule but not pharaonic culture. At least since the Eighteenth Dynasty, the Greeks and the Egyptians had been frequent trading partners and, particularly in the Late Period, each borrowed and benefited from the other's artistic expertise. No monument better illustrates the juxtaposition of Greek and Egyptian influences than the tomb of Petosiris at Tuna el-Gebel in Middle Egypt, carved during either the final days of pharaonic rule or the beginning of the Greco-Roman Period. Throughout the tomb the subject-matter is Egyptian, but the artists used both Greek and Egyptian styles. Rarely, however, were the two styles combined. Exceptions are the relief-carved figures in the outer chamber, shown engaged in farming, animal husbandry and manufacturing. They are often represented in Greek-inspired naturalistic poses and wear flowing garments anchored at the shoulder or gathered at the waist – the traditional Greek cloak (*himation*) and tunic (*chiton*) respectively. By contrast, in the inner chamber of the tomb, which features traditional scenes from the Egyptian funerary cult, the manner in which the participants are represented, as well as their garments and accoutrements, are completely Egyptian.

The two styles continued to exist side by side for the next several hundred years. Works created for Egyptians continued to echo pharaonic style and those made for the Greek population followed contemporary Hellenistic models, particularly at Alexandria, the Mediterranean port city founded by Alexander the Great on the site of the native town of Raqote. In royal sculpture, the same Ptolemaic ruler might be depicted in purely Hellenistic style, with an expressive face and natural curls, or impassively as a pharaoh in the traditional royal *nemes* headcloth that dated back nearly three millennia. Recent discoveries beneath the sea off modern Alexandria are shedding new light on the monuments of the ancient capital, much of which was submerged following an earthquake. The greater part of the Ptolemaic and Roman city lies buried beneath the thriving modern city and is as yet unexplored.

Although they rarely blended the two traditions in one work, native Egyptian sculptors were influenced by contact with the Hellenistic world. The tendency toward softer, fleshier body modelling and the depiction of natural hair and draped garments are probably attributable to Greek influence, although all these features have Egyptian precursors.

Temple-building continued on a large scale under the sponsorship of the Ptolemies, no doubt to secure the support of Egypt's powerful priesthood and of the population in general. In their layout and decoration they followed Egyptian models. Reliefs depicted Ptolemaic rulers dressed as pharaohs in traditional style demonstrating their devotion to the traditional Egyptian gods. There is no more beautiful example of Ptolemaic temple construction than the sanctuary of Isis that originally occupied almost the entire island of Philae in the Nile near Aswan. Although the earliest known buildings date to the Thirtieth Dynasty (see p.225), most of the extant structures are of Ptolemaic and Roman date. Following a basic traditional Egyptian plan that is also seen at Luxor – begun more than a thousand years earlier – two massive pylons lead to a courtyard, hypostyle hall and inner sanctuary. On the outermost pylon the king is depicted in the act of smiting his enemies, a scene that first appears at the very beginning of Egyptian dynastic history (see p.155). The entire complex was removed to the neighbouring island of Agilkia in the 1970s in order to preserve its buildings, reliefs and sculptures from the rising waters of the Aswan High Dam.

*This beautiful 2nd-century CE encaustic (coloured beeswax) portrait of a woman was painted during her lifetime and incorporated into her mummy wrappings at death. In the Roman Period, such images – the finest surviving examples of painted portraiture from anywhere in the Roman world – took the place of the funerary masks of earlier times. The form and technique prefigure the style of later Byzantine icons.*

When Augustus conquered Egypt for Rome in 30BCE, he and his descendants continued the tradition of erecting Egyptian-style temples throughout Egypt and Nubia, thereby establishing themselves in the eyes of the people as the rightful heirs of the pharaohs. In general, however, Rome exploited the Egyptians, whose standard of living declined markedly under imperial rule. The Romans saw Egypt as an appealingly exotic land, and exported its obelisks and other monuments, as well as its fine hard stones, to Rome. Nile scenes were depicted in mosaics (see illustration, p.56) and Egyptian sculpture was copied (usually with an inauthenticity that is almost comic). In his re-creation of his vision of Egypt and the Nile for his villa at Tivoli, the second-century Roman emperor Hadrian commissioned fanciful statues of his idea of Egyptian priests and priestesses that bore little relationship to either Egyptian or Roman works. Statues produced to commemorate Hadrian's favourite, Antinous, who drowned in the Nile during the emperor's visit to Egypt in 130–31 CE, depict Egyptian pharaonic regalia on a muscular physique that often holds a characteristically Roman pose – an ill-fitting combination at best.

Back in Egypt, it was largely in the funerary arts that the pharaonic tradition was maintained. Masks or entire coffins of *cartonnage* (stiffened linen overlaid with plaster) or wood were brightly painted with age-old iconography but in a style reminiscent of the curvaceous reliefs of Ptolemaic Egypt. These idealized representations existed simultaneously with strikingly lifelike portraits painted on wood in encaustic (coloured beeswax). The nature of this medium gives the portraits a remarkably thick texture that somewhat resembles modern oil painting. The image, which was probably hung in the home during the subject's lifetime, served as a mask for a person's mummy after death. Most surviving examples have been recovered from the Faiyum region, and they are hence known as "Faiyum portraits".

# THE ART OF ADORNMENT

*Beauty and function are combined in this mirror, which consists of a silver disc and a handle of obsidian inlaid with gold, semi-precious stones and faience. The face of the fertility goddess Bat-Hathor and the handle in the form of an open papyrus symbolize rebirth, as does the mirror disc itself, which recalls the sun. From the tomb of the princess Sat-Hathor-Iunet at Kahun; 12th Dynasty, reign of Amenemhet III (ca. 1818–1772BCE).*

Egyptian artisans devoted as much care and attention to the minor arts as they did to large-scale statuary and relief. No object was too small or too insignificant to be made beautiful. Observers are often surprised at the scale of Egyptian objects when they see them, because from a photograph everything looks monumental, so minute is the detail.

The love of beauty and the urge to decorate any available surface is apparent as early as Predynastic Egypt; and at times decoration completely transformed an object's function. On a pottery cup, for example, a row of three-dimensional cows trot around the outcurved rim; perhaps whimsical or perhaps a symbolic expression of the owner's desire for plentiful herds, the rim decoration would have made drinking from the cup almost impossible.

In many cases, function necessarily governed the choice of material. Items made for everyday use in perishable materials such as basketry or wood were reproduced in more permanent substances for funerary purposes, because they had to last for eternity. Reeds bound together and tied at the ends might serve as a tray during life, but for an Early Dynastic tomb a duplicate would be carved in stone. Function might dictate that the shape alone be copied, but the artisan went further, carving into stone each individual reed and stitch, and each twist of rope binding the ends.

Objects that to the untrained observer have a purely aesthetic appeal might have been potent sources of symbolic magic for the Egyptian viewer. For example, a charming sculpture of a woman tenderly holding a child is in fact a vessel that was designed to contain the milk of a mother who has borne a son: drinking milk from such as container was believed to ensure the health of a sick child.

The appeal of colour to Egyptian artisans was undeniable, and realism was often sacrificed in its favour. Bright blue faience was used to make a whole menagerie of animals, including ducks, monkeys, hedgehogs, rabbits, fish, lions and hippopotamuses (see illustration, p.219). When a vein of amethyst was discovered during the Middle Kingdom, it was exploited so extensively for jewelry, stone vessels, scarabs, amulets and small sculpture that the resource was soon exhausted, and post-Middle Kingdom amethyst objects are rare.

Scarcity made some materials more desirable than others, and artists were not above making a cheaper material look more expensive through the skilful application of paint or veneer. Accordingly, pottery could be turned into stone, common limestone into granite, or soft woods into hard ebony. Gilding gave anything the appearance of gold. However, accompanying texts often declare that the more valuable material was used.

## JEWELRY

The Egyptians, like people of all other developed cultures of the ancient world, valued personal adornment. Some of the earliest Predynastic graves contained beads made of shell, common stones or dried mud strung together into simple necklaces or bracelets. Men and women alike wore jewelry, which was also used to adorn the statues of both kings and gods. Virtually every type of jewelry that is familiar today existed in ancient Egypt. From the evidence of New Kingdom coffins and statuary, it appears that a well-dressed man or woman of the time – a period of great affluence – might wear a diadem, earrings, an elaborate broad collar consisting of six or more strands of beads, a pectoral, bracelets, armbands and a number of rings and anklets.

*Elaborate solar and lunar symbolism characterize this luxuriant pectoral from the tomb of Tutankhamun (ca. 1332–1322BCE), one of the most extensive treasures ever discovered – there were 143 precious objects on the king's mummy alone. The focus of this pectoral is a winged scarab of translucent chalcedony.*

Gold was the most prized material for jewelry, either used by itself, or elegantly combined with brightly coloured semiprecious stones such as lapis lazuli, turquoise or carnelian. Such pieces were highly valued and often passed down through the generations. Those who could not afford precious metals or gemstones made do with substitutes or faience. Coloured glass could also be used, but it was not inexpensive. Funerary jewelry might even be made of plaster gilded to look like solid gold.

Not only were there numerous types of jewelry, but also the variety within each type was extensive. Earrings, for example, were first worn during the Second Intermediate Period, and the earliest were simple hoops inserted into pierced ears. By the end of the New Kingdom, one could choose from elaborate hoops and dangles, tube and boss earrings and ear studs. Ear plugs more than two inches (6cm) in diameter might be inserted into the earlobe. Mummies with their earlobes stretched to at least that size indicate that such enormous items of jewelry were actually worn.

Jewelry beautified the wearer but could also serve other functions. A type of collar made of tiny disc beads strung together in rows, the so-called "gold of honour", signified exalted status. Flies, the epitome of tenacity, were made in gold, ivory or semi-precious stones and awarded for acts of heroism (see illustration, p.30). Amulets were also incorporated into jewelry to provide protection or divine blessing.

Among the most impressive and elegant of all small-scale works of art of the pharaonic period are the pectorals and other jewelry found in the tombs of princesses of the late Twelfth Dynasty, such as Sat-Hathor, one of the daughters of King Senwosret II (ca. 1847–1837BCE; see p.191). Gold *cloisons* (raised compartments) inlaid with semi-precious turquoise, lapis lazuli, garnet, feldspar and carnelian form a variety of beautiful and symbolic amulets. Tiny beads of the same materials were strung together in bracelets that were ingeniously fastened by clasps or locking pins. No modern jeweler could surpass his or her ancient Egyptian counterpart in the delicacy, sensitivity to design and use of colour displayed in these exquisite examples of personal finery.

*The intricate* cloisonné *of this pectoral of the 12th-Dynasty princess Mereret fills every available space in its twin depictions of Amenemhet III smiting a symbolic enemy. The king appears under the protective outstretched wings of the vulture goddess.*

ABOVE: *The sacred* wedjat *or Eye of Horus is the central element of this gold, faience, carnelian and lapis bracelet, one of seven found on the mummy of King Shoshenq II of the 22nd Dynasty, whose name is inscribed on the inner surface of the bracelet. The* wedjat *was intended to protect the king in the afterlife.*

OPPOSITE: *Two cartouches define the shape of this golden ointment box inlaid with semi-precious stones, from Tutankhamun's tomb. Traditionally, cartouches contained the royal name, but here they enclose seated images of the king, executed in the Amarna style (see p.221). The box is surmounted by two plumes incorporating sun discs.*

● CHAPTER 15

# SIGNS, SYMBOLS AND LANGUAGE

For many centuries, Ancient Egyptian remained locked up in the mysterious hieroglyphic symbols whose secrets had been lost for a millennium and a half. Once deciphered, the rich treasury of the language could be appreciated in the countless writings, wall paintings and inscriptions found all over Egypt. Essential as a means of communication, the written language also served as a decorative tool in the considerable repertoire of Egyptian artists.

▲



ABOVE: *Hieroglyphs served an aesthetic function as well as a linguistic one. As can be seen in this relief from the Middle Kingdom tomb chapel of Ihy (ca. 1920BCE), they might be carved in detailed raised relief with as much care as the accompanying figures.*

## DISCOVERING THE KEY

Ancient Egyptian, a language recorded for more than three thousand years, is familiar today in the script that we term hieroglyphs. The script first appeared ca. 3100BCE, and consists of images derived mainly from the world in which the Egyptians lived. Illiteracy would have prevented most people from understanding the full meaning of the texts that they saw carved and painted on the temple walls. Later generations of foreign conquerors, visitors and tourists viewed the same monuments with even greater incomprehension of the enigmatic signs and symbols. Hieroglyphs ceased to be a viable means of written communication at some time during the early centuries CE – after Egypt had succumbed to foreign conquest, Hellenization and, finally, the spread of Christianity. The last firmly datable hieroglyphs were crudely carved in 394CE on the walls of the temple of Isis at Philae.

The ancient Egyptians themselves referred to their pictographic script as *medou netjer*: "words of the gods". Later, the Greeks described the script as "sacred letters" and also "sacred carvings", the term from which our word "hieroglyph" derives. Greek historians and philosophers in the first centuries CE thought, mistakenly, that hieroglyphs represented only ideas and concepts, a view that may underlie the later European belief that they were symbolic in nature. *Hieroglyphica*, a text probably belonging to the fourth or fifth century CE – purportedly written by an Egyptian called Horapollo, and then translated into Greek by a certain Philippus – explains the signs as symbols of Egyptian knowledge and gives ingenious, if often tortuous, explanations of their allegorical significance. For example: "To symbolize the cosmic God, or fate ... they draw a star, because the forethought of God preordains victory, by which the movement of the stars and the whole universe is accomplished" (Horapollo 1:13).

Horapollo's text influenced European scholars of the fifteenth century, the time of the Renaissance, and the symbolic theory of hieroglyphs persisted well into the eighteenth century. However, it was the discovery of

*The hieroglyphs in Karnak's Great Hypostyle Hall (ca. 1250BCE) were deeply incised into the stone in order to prevent their erasure by later kings (see p.210).*

**THE ROSETTA STONE**

**Near the Delta town of Rashid, whose ancient name was Rosetta, Napoleon's soldiers discovered in 1799 a monument that became the key to deciphering hieroglyphs. The Rosetta Stone (below), a slab of black basalt three feet ten inches (1.18m) high, preserved a commemorative inscription written in three scripts: hieroglyphs in the upper register, Demotic in the middle, Greek in the lowest. Dated to the ninth year of Ptolemy V Epiphanes (196BCE), the text is a copy of a decree by the priests of Memphis listing the king's honours. By comparing the Egyptian to the Greek, linguists were able to decipher the ancient Egyptian language. In 1802, under the terms of the Anglo-French treaty of Alexandria, the stone was handed over to Britain and transported to the British Museum, where it is still on display.**

the "Rosetta Stone" in 1799 that provided the real key to decipherment. Inscribed on the stone was the same text in two languages, Egyptian and Greek, and three different scripts: hieroglyphs, Demotic (a cursive Egyptian "popular" script) and Greek. For the first time, a passage of Egyptian could be compared with the same text in a known language. Scholars in many countries attempted the task of deciphering ancient Egyptian, soon concluding that the hieroglyphs were mainly phonetic, not symbolic, signs.

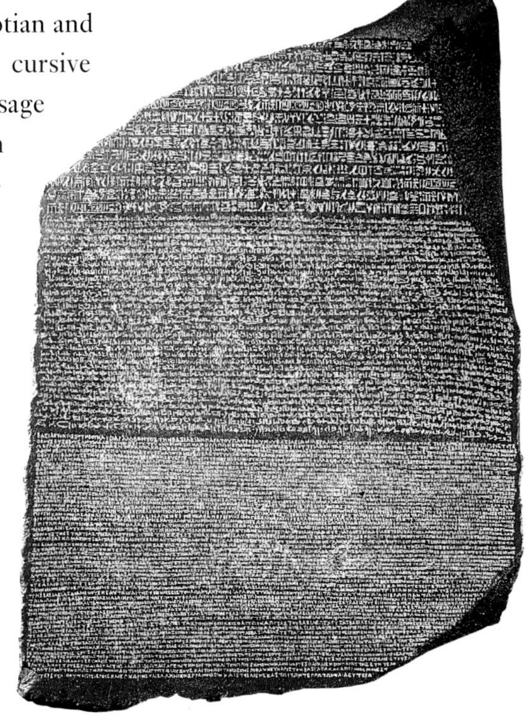

Thomas Young (1773–1829), an Englishman, was apparently responsible for much initial decipherment, but the final breakthrough is credited to the French scholar Jean-François Champollion (1790–1832), who began by identifying signs used to write royal names such as Ptolemy and Cleopatra. Eventually, drawing on his knowledge of Coptic (the last stage of the Egyptian language, used by Christians and written in Greek characters), he assigned phonetic values to other signs. After Champollion's theories were published in 1822, scholars soon made great headway in unlocking a language that had kept its secrets for more than fourteen hundred years.

# SCRIPTS AND SCRIBES

The ancient Egyptians used four scripts: hieroglyphic, hieratic, Demotic and Coptic. Hieroglyphic, the longest-lived (ca. 3100BCE to the end of the fourth century CE), may originally have been employed for most purposes, but by the early Old Kingdom (ca. 2625BCE) occurred mainly in religious, monumental or commemorative texts – principally on the walls of temples, tombs and palaces, and on stelae, statuary, utensils (such as slate palettes), coffins, sarcophagi, jewelry and amulets. Some six to seven hundred signs were in use until the Greco-Roman Period, when the repertoire expanded to around six thousand. Unlike the other scripts, hieroglyphs were for the most part recognizable images drawn from the world of the ancient Egyptians. They could be read (and could face) from left to right or right to left, and could be written in both rows and columns. Sometimes a text might run in both directions. The orientation is usually indicated by the hieroglyphs in animal and human form, which always face toward the beginning of the line, for example or .

Hieroglyphs represent the more formal style of writing, principally for use in sacred inscriptions. Hieratic, comparable to handwriting, probably appeared not long after the introduction of hieroglyphs, of which it is a simplified or shorthand form. It reads from right to left, and could be written in rows or columns, although scribes generally wrote it in rows. By

**TYPES OF SCRIPT**
**The four types of Egyptian script are depicted in the artwork below. The left-hand section shows two columns of HIEROGLYPHIC script, with the same text in the more cursive HIERATIC alongside. (The hieratic texts have been copied from what is probably a private letter; the hieroglyphic text simply transliterates these.) The artwork at top right shows DEMOTIC, and is based on a wine receipt of 145CE. Below that are two fragments of COPTIC (ca. 700CE): all the letters are Greek except for the two highlighted in blue, which are from Demotic.**

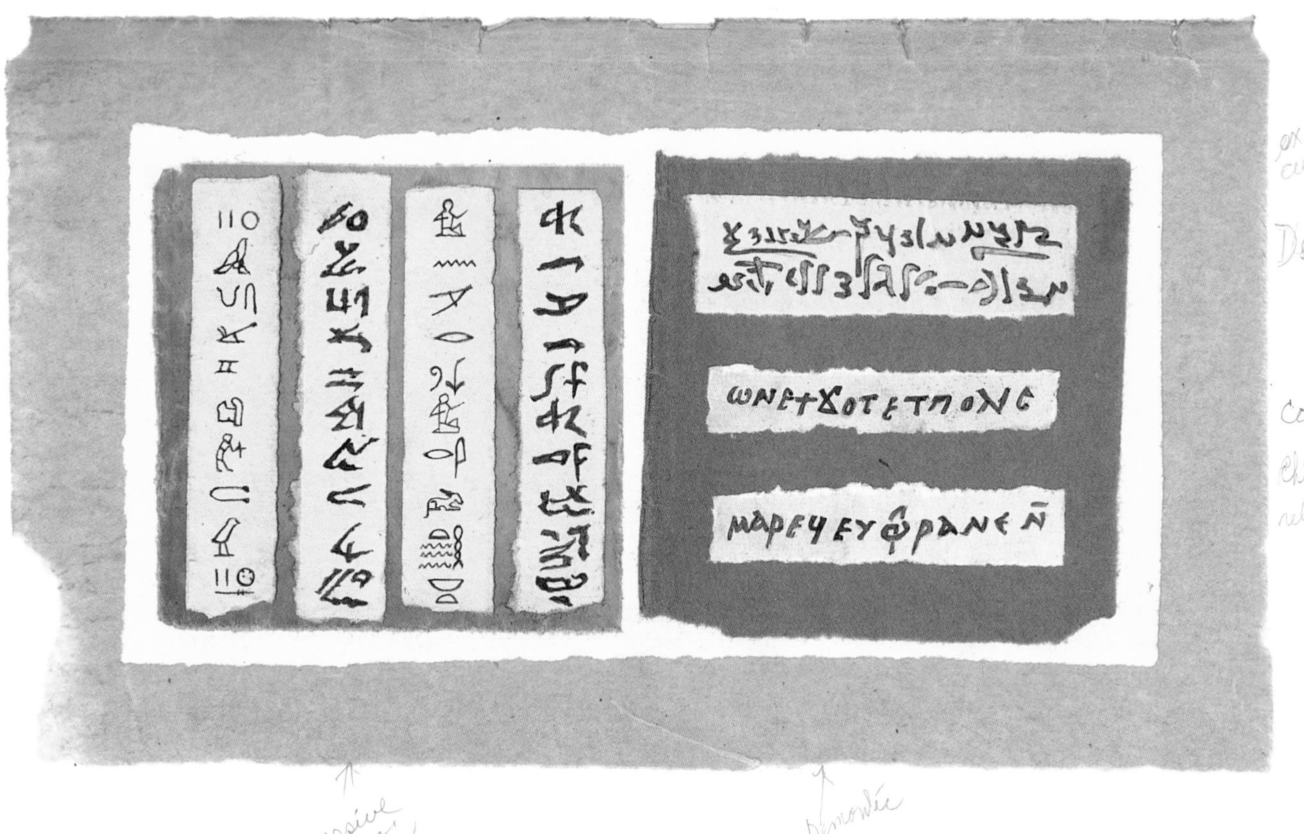

the Old Kingdom, hieratic script was used for temple accounts on papyri and was preferred for private and administrative documents. Scribes also employed it to record religious material, such as the Book of the Dead and other funerary and mythological texts. Its religious use came to predominate, hence the Greek description of this script as "priestly" (*hieratica*).

Demotic, the script that the Greeks referred to by the word *demotica* ("popular") and the Egyptians by the phrase "writing of documents", first appeared ca. 600BCE and eventually displaced hieratic. Its derivation from both hieratic and hieroglyphic is disguised to a certain extent by its extremely cursive form. At first it was confined mainly to legal and administrative documents, then later also for literary and religious texts.

Coptic (from the Greek *Aiguptia*, "Egyptian") was the last script of the Egyptian language and the only one that was fully phonetic. Its alphabet consists of twenty-four Greek letters with an additional six signs. Although employed primarily by early Christians, Coptic seems to have come into existence early in the first century CE in Egyptian magical texts to ensure the proper pronunciation of critical words, phrases and passages. The script was used for both religious and secular purposes.

*A painted limestone statue of a seated scribe from a 5th-Dynasty tomb at Saqqara, ca. 2475BCE. The profession was so respected that élite Egyptian men who were not scribes sometimes had themselves depicted in this characteristic scribal posture.*

## THE TOOLS OF THE SCRIBE

The ancient Egyptian hieroglyph for scribe, (*sesh*), depicted three important tools: a rectangular palette with two circles on it, one representing red ink, the other black; a water pot (centre); and a pen and pen holder (right). Although the sign appears unchanged on reliefs and statuary, archaeological discoveries indicate that this form of palette had probably been replaced already by the Old Kingdom by a composite version, called a *gesti*. Usually of wood (or, less frequently, ivory), the *gesti* was rectangular, with two inkwells and a slot for pens (*ar*), which were made from reeds.

Scribes also had stone mortars and pestles to grind the pigment for their inks, water pots for mixing liquid with the pigment, and knives and burnishers for preparing sheets of papyrus. As well as papyrus, they also wrote on *ostraca* (shards of pottery or flakes of limestone) and writing boards. Because the scribe often had to work outdoors, he would carry his tools in a fairly large equipment box of wood with painted decoration.

The professions of scribe and artist were closely linked. A line draughtsman was known as a *sesh kedut*, or "scribe of the drawings", and was responsible for copying hieroglyphs onto walls, which would then be painted or carved (or both) by a master artist. Painters used special palettes with multiple wells for different colours.

*A collection of ancient Egyptian scribal and painting implements, including two rectangular palettes; thin reed pens; an inkwell; two brushes (the top one made from rope, the other from twigs bound together with the ends crushed). At top is a small grinder and grinding palette and three lumps of raw pigment.*

# WORDS AND HIEROGLYPHS

**HOW WORDS WERE MADE**

**A word in hieroglyphic script is composed of one, or a series, of the principal types of sign (see main text): phonograms, logograms and determinatives. Sometimes a word can simply be a logogram with or without a determinative. For example, the logogram 𓏞 (*sesh*) signifies "writing", but with the addition of the determinative "man" (𓏞𓀀) it means "scribe". Some words consist of a single phonogram: for example, the uniliteral 𓅓 (*m*) means "in", and 𓂋𓏤, "mouth", is composed of the uniliteral 𓂋 (*r*) with a determinative stroke.**

**Other words are less simple, for example, 𓌸𓂋𓀁 (*mr*) "love", is composed of several different types of sign: the biliteral 𓌸 (*mr*), the uniliteral *r* as a "phonetic complement" (repeated sound), and a determinative. The word for "good" or "beautiful", *nefer*, can be written 𓄤𓆑𓂋, which consists of a triliteral (*nfr*) followed by two uniliterals (*f* and *r*) acting as phonetic complements. As the absence of a determinative in this word shows, its presence is not always necessary. The strictly unnecessary phonetic complement, as in 𓌸𓂋𓀁 and 𓄤𓆑𓂋, fulfils the function of a determinative by indicating that the sign 𓄤 is a phonogram, not an ideogram. In some cases the presence of such a phonetic complement can be attributed to the aesthetic sensibility of scribes, who used it to eliminate offending spaces in a text.**

Once nineteenth-century scholars had deciphered the hieroglyphs, they gained access to one of the world's oldest written languages. They concluded that the signs fall into three categories: "phonograms" (representing sounds, rather like the letters of the alphabet); "logograms" or "ideograms" (signs that represent whole words); and "determinatives" (signs that clarify the sense of a word in a particular context).

The phonograms, which are derived from the logograms, represent the consonantal sounds of Ancient Egyptian. Vowels were not recorded, perhaps because they were more likely to change in different grammatical circumstances, as in the English *sit* and *sat*. It is worth noting that Arabic and biblical Hebrew, languages which are related to ancient Egyptian, also originally wrote only consonants. There are three types of phonogram: "uniliterals" (twenty-four in number, plus two variations), "biliterals" and "triliterals" (of which there are many hundreds). Each uniliteral stands for a single consonant: for example, 𓃀 represents the sound *b*. A biliteral stands for two consonants, as 𓌸 (*mr*), while triliterals represent three, as 𓄤 (*nfr*).

Logograms (ideograms), essentially mean what they depict. For example, 𓁹 ("eye"), 𓇳 ("sun") and 𓊛 ("boat"). However, such basic pictograms developed extended meanings, so that the sign 𓇳 could also signify "day", and 𓊛 "to sail". These signs eventually also became

*The finely drawn and coloured hieroglyphs on the First Intermediate Period stela of Zezennakht are employed to their full decorative effect on what is otherwise a plain composition. Determinative hieroglyphs (see main text) can be seen, for example, in the fifth line, which ends (reading from right to left) with a seated man and woman, following the word for "brothers and sisters".*

identified with the specific sounds of the words they represented, developing into, and complementing, phonograms.

Logograms were used as determinatives, which are signs without phonetic value. Several hundred determinatives were in use in Ancient Egyptian. They helped to indicate the meaning of words in a particular context, especially in the case of "homographs" – different words that look alike because they have the same consonantal structure. For example, 𓌸 (*mr*) followed by the determinative 𓈘, indicating a context of water, means "canal"; whereas 𓌸 followed by the determinative 𓀁, indicating human emotion, means "love". (See also sidebar, opposite.)

Scribes left no spaces between words, and a determinative might also be placed at the end of a word to mark its end; an example would be a simple stroke, 𓏤. A stroke is especially common after logograms, serving to distinguish them from their phonogram derivatives. Three strokes (𓏥) were employed to signify plurals.

## EGYPTIAN NAMES

For Egyptians, names were an important aspect of the personality, and great care was taken when choosing them. A person's name can consist of a single word or a group of words. In either case, it expresses an idea, thought, wish or emotion, characterizing the individual in life and ensuring that he or she was remembered for eternity. Sometimes a name beginning with the word 𓄤𓆑𓂋 (*nfr*, "good", or "beautiful", usually written *nefer*, as in Nefertiti and Nefertari) was chosen in the hope that the child would be good or handsome. If the parents wished a name to reflect their own devotion to a deity, they might select, for example, Amenhotep (𓇋𓏠𓈖𓊵𓏏𓊪), "Amun is at Peace".

Because names usually consisted of words or phrases taken from the everyday vocabulary, their use in inscriptions was potentially confusing. For this reason, it was decided to distinguish them graphically. Egyptians might add the determinative of a seated man (𓀀) or woman (𓁐) to the end of a personal name, to indicate its sex. Some variations occur; for example, the figure might be seated on a chair, or squatting: 𓀁.

Royal names (see pp.112–13) were formed in a similar way to those of private individuals, but they do not employ the seated male or female figure determinatives. However, the sign of a standard, 𓋇, is occasionally used . The names of the king were written within an oval rope, or "cartouche" (𓍷 ; see illustration, below).

*The name of King Khufu consists of the signs for* kh, w, f *and* w *in a cartouche, above which is the royal title "Lord of the Two Lands". In the inscription to the right, the name of the 4th-Dynasty Prince Rahotep, at the bottom of the left-hand column, lacks the determinative sign of a man because this function was served by the statue on which the inscription appears (see p.201).*

# THE EVOLUTION OF THE LANGUAGE

Ancient Egyptian can be divided into five phases of evolution that broadly correspond to the standard historical divisions: Old Egyptian, Middle Egyptian, Late Egyptian, Demotic and Coptic. These phases indicate the most important version of the language used at the time, but there were considerable periods of overlap, when the later stages of one phase co-existed with the first stages of the next. Moreover, each stage could reflect more than one genre or style.

The longest-lived phase was Middle or "Classical" Egyptian, used from the First Intermediate Period onward. In the Middle Kingdom, this language was formalized and became the standard medium for religious, legal and monumental texts and some literary works. It continued in use well into the Greco-Roman Period (fifth century CE). Middle Egyptian evolved from Old Egyptian, the main language of the Old Kingdom. Inscribed on the walls of tombs and temples, Old Egyptian was used also in administrative, legal and personal documents.

## THE STRUCTURE OF THE LANGUAGE

Egyptian had many of the linguistic components of most modern languages, including grammar and syntax, with parts of speech such as nouns, pronouns, adjectives, adverbs, verbs, prepositions, participles, and so on. Different personal pronouns occur for specific uses within the sentence, indicating a concept of cases (as in English "I" and "me"). There was a complicated verbal system to indicate tense, "aspect" and "mood". Nouns, pronouns, adjectives and other words, had indications of gender and number (singular, plural and a third type called "dual").

In its simplest form, the sentence or independent clause sometimes has a verb, sometimes not. If it does, its elements follow a set pattern, with occasional modification: (1) verb; (2) subject; (3) direct object; (4) indirect object; (5) adverb (or adverbial phrase). For example, the sentence, "the man says the name to the woman in the house" is expressed in Egyptian, "says [the] man [the] name to [the] woman in [the] house" (𓄂𓀀𓂋𓈖𓈖𓂋𓁐𓅓𓉐). Generally, in non-verbal sentences the verb "to be" is absent, but understood. For example, "the woman is in the temple" is expressed simply as the subject, "woman" plus a prepositional phrase, "in [the] temple" (𓇋𓅱 𓋴𓁐 𓅓𓉟𓊹𓉐).

From these basic building blocks, the Egyptians constructed a fully developed language. As with all languages, Ancient Egyptian evolved over the centuries; conventions changed, and certain usages came into or fell out of popularity. For example, the earlier phases of the written language did not indicate the indefinite or definite article ("a", "an", "the" in English), but these were consistently written from Late Egyptian on.

Punctuation in the modern sense did not exist, but certain elements, called particles, and other words in a sentence, as well as the context, would help to indicate the beginning and end of sentences and clauses. In Old and Middle Egyptian, the particle 𓇋𓅱 at the beginning of the sentence indicates a plain statement, as in the above sentence "the woman is in the temple". However, if the particle 𓇋𓈖 is the first word, what follows is a question.

Late Egyptian, the main language in use during the New Kingdom, developed from the vernacular and was employed primarily to record secular material. Whereas the two earlier phases of Ancient Egyptian are "synthetic", Late Egyptian has a more complex, "analytical" character, with meanings expressed through separate components (the difference is roughly the same as that between Latin (synthetic) and English (analytical); in Latin a single word such as *poetae* can be rendered in English only by two or more ("of the poet", "the poet's"). In many respects Demotic was the successor of Late Egyptian. Employed principally for secular texts, Demotic continued in use well into the Christian era.

Coptic, the final phase of the language, used a script composed mainly of Greek letters with six additional signs from Demotic (see p.232). It employed many of the constructions and much of the vocabulary of the previous forms of language, and its alphabetic script included vowels for the first time – a fact that has assisted linguists in their reconstruction of the unwritten vowels of earlier Ancient Egyptian. The vowels of Coptic enabled scholars to discern many distinct dialects; these may have occurred earlier, but were perhaps not reflected in the written language.

*The hieroglyphs in this relief from the 6th-Dynasty tomb of Ti from Saqqara record the everday speech of the figures much like the "speech balloons" of a modern cartoon. In the top register, for example, an overseer on the left instructs one of the herdsmen to proceed briskly with the delivery of a calf. The precise translation is uncertain, but versions include "O herdsman, there is the problem!" and "O herdsman, pull powerfully the mother cow!"*

# THE SIGNS SPEAK

Modern scholars have followed various classifications in their approach to ancient Egyptian writings – by genre, by style, by grammar, and sometimes by script. But the boundaries of these artificial divisions often merge, and texts attributed to one category will often fall into another. Although we would expect funerary texts, which are both sacred and conservative, to be composed in the classical language and in hieroglyphs, sections of spells that appear to be direct quotations from other works might use grammar that is closer to the vernacular. Moreover, some funerary texts, such as the Coffin Texts, the Book of the Dead and later mythological papyri, are occasionally written in hieratic.

In terms of subject matter, one of the largest categories would be religious texts, ranging from the Pyramid and Coffin Texts and later texts of the same genre (such as the Amduat, the Book of Gates, the Book of Breathing) to hymns and rituals, which the Egyptians recorded on papyri, on stelae, on the walls of tombs and temples, and in various other media. Many of the cult rituals performed in temples had associated dramas, depicted on temple walls or recorded on stelae or papyri. In their homes, Egyptians would use spells and rituals to focus on fertility, childbirth, or protection against disease, demons, enemies, or any form of evil, and these spells could be written on a variety of materials. The Egyptians also recorded "scientific" information, mainly for teaching purposes, in treatises on medicine, mathematics and astronomy (see pp.92–6). In addition, there were didactic texts – compilations of wisdom often offered by a father to his son. These works were popular hundreds of years after their composition. Some appear to contain elements of political propaganda.

*Ramesses II (ca. 1279–1213BCE) in his war chariot; a relief in the temple of Abu Simbel. Elsewhere in the temple, the pharaoh recorded his battle at Kadesh in Syria against the Hittites. The conflict was inscribed in both poetry and prose and employed a variety of literary devices. Accounts of the battle were popular during the king's reign and after, although the battle was not the unequivocal success that Ramesses II claimed it to be (see p.108).*

Literature was popular through most of Egyptian history, and the remnants that we have today probably represent only a fraction of the corpus. *The Story of Sinuhe* is one of history's first epic narratives. Dealing with the wanderings of an expatriate, it has political implications, and eventually a happy ending, and was so popular that copies were in circulation centuries after it was first written. *The Contendings of Horus and Seth* relates an essentially religious theme first referred to in the Pyramid and Coffin Texts (see pp.134–5), presenting the myth to a New Kingdom audience in a raucous, popular style with passages of broad humour. *The Doomed Prince*, another Late Egyptian story, deals with humankind and fate, and it may well be the world's first recorded fairy tale.

Egyptian authors recorded the accomplishments of their rulers in monumental inscriptions that combined factual information with a tendency to glorify the pharaoh. Sometimes, as in the case of the Kadesh inscriptions of Ramesses II (see illustration, opposite page), the authors used both prose and poetry in their account.

## LOVE POEMS

Nowhere is the literary creativity of the ancient Egyptians shown to better effect than in their love poetry. This is preserved on fragments of pottery and on three different papyri, all dating from the Nineteenth and Twentieth dynasties. Like surviving personal letters between private citizens, the poems were recorded in the hieratic script in the vernacular language, and reflect the concerns of the living. They are composed in a rhythmic, lyrical style, but the liveliness, freshness and contemporary qualities of the verses make more of an impression than the sophisticated underlying structure, which is less apparent.

The poems in each collection are entitled either "sayings" or "songs", reflecting their participation in oral as well as written tradition. They are written in the first person – from the perspective of the lover, either male or female. In one collection, alternate stanzas are spoken by each of the two lovers in turn. The terms used to refer to the lovers translate as "brother" and "sister", the words normally employed to denote closeness and affection between members of the opposite sex in Egyptian society. There is a clear rhythmic structure to the verses, but because of the absence of vowels in the Ancient Egyptian written language it is impossible to determine whether rhymes or assonance were present. One love poem, among several written on fragments of a vase that is now in Cairo, reads:

"My sister has come, my heart exults,
My arms spread out to embrace her;
My heart bounds in its place,
Like the red fish in its pond.
O night, be mine forever,
Now that my queen has come!"
(translated by Miriam Lichtheim in *Ancient Egyptian Literature*, volume II, The New Kingdom, page 193).

*Female singers and dancers were often employed to entertain guests at the social functions of the élite. The decoration of this blue faience bowl, now in Leiden, Holland, depicts a woman sitting in a bower playing an instrument resembling a lute, while a pet monkey plays next to her. New Kingdom, 18th or 19th Dynasty.*

*These exquisite pendants from the jewelry cache of the 12th-Dynasty princess Khnumit (see p.191) are formed from the hieroglyphs for (left to right) "joy", "birth" and "all life and protection".*

# SYMBOL AND IMAGE

The pictographic nature of Egyptian writing (which is probably connected with the appearance ca. 3100BCE – almost simultaneously with hieroglyphs – of the characteristic two-dimensional imagery of Egyptian art) makes hieroglyphs recognizable on several levels. The illiterate mass of the ancient Egyptian population could no doubt appreciate them visually, even if they did not understand what they meant.

One aspect of the versatility of hieroglyphs is their frequent use as a decorative element in architecture, furniture and the minor arts. For example, the hieroglyph 𓊽 is found as a decorative component in the headboards and footboards of beds, as well as featuring in windows, small chests and jewelry. Although this sign is effective aesthetically, it also retained its verbal message, "endurance". Dual functions can be seen with other hieroglyphs, such as 𓎃, meaning "protection", which was also the shape of life preservers; and 𓋹 meaning "life", both of which figure prominently in jewelry and amulets. The last sign can also mean "mirror", and it is perhaps this denotation that inspired the creator of one of Tutankhamun's mirrors to use 𓋹 as the form of its case.

Often, more than one hieroglyph is used, making a message to express positive wishes. For example, 𓇼 "joy" and 𓎃𓋹𓎃 "all life and protection" occur as amulets, and figure prominently in jewelry from the Middle Kingdom (see illustration, left). Royal names sometimes form the major elements of bracelets. A highly elaborate calcite unguent vessel from the tomb of Tutankhamun (ca. 1332–1322BCE) has hieroglyphs as its centrepiece and supports, and the whole composition expresses the king's rule over the "Two Lands" of Egypt.

*A perfume vessel from Tutankhamun's tomb. It comprises two Nile deities and the hieroglyph for "Union" – a reference to the united "Two Lands" – which combines the lotus and the papyrus plants, symbols of Upper and Lower Egypt respectively.*

Sometimes, the status of imagery as text is less apparent. A statue of a deceased individual, or a relief representation of them on a stela or tomb wall, may sometimes act as a large-scale determinative – the sign of a human figure often inscribed at the end of a personal name (see p.237). On a larger scale too, one can even see the use of hieroglyphs in architecture and town planning. The hieroglyph 𓈌, "horizon", which depicts the sun rising between two hills, may well be the inspiration for the design of the pylon of an Egyptian temple. When the sun rose over the pylon, the resulting image clearly resembled the hieroglyph. The very same hieroglyph may have influenced Akhenaten (ca. 1353–1336BCE; see pp.128–9) in his choice of a site for his new capital, Akhetaten – "Horizon of the Aten [Sun Disc]" – in Middle Egypt. When the sun rose above the cliffs here, it may well have appeared to the pharaoh like a giant version of the "horizon" hieroglyph.

In some tomb scenes where the text and the accompanying scene are closely related, it is difficult to decide whether the scene acts as a determinative for the text, or whether the text is a literary complement for the scene. In a few Old Kingdom tombs, labels or short texts recording dialogue include words with determinatives (see pp.234–5) omitted, because the appropriate figures to whom they refer are actually depicted, on a larger scale, in the associated scene.

Interestingly, this unity of imagery and writing is also shown in the anthropomorphizing treatment of several hieroglyphs. Signs such as ☥, 𓊽 and 𓎿 may sprout arms or legs and become part of a decorative frieze on monumental architecture, or even active participants in scenes on the wall of a temple or tomb, or on a piece of jewelry or a fan. A lamp from the tomb of Tutankhamun has as its central element the hieroglyph ☥ with outstretched arms holding a torch.

These are just some of the many examples that illustrate the closeness of hieroglyphic symbol and visual image in Egyptian art. Artists were able to merge flawlessly a verbal message and a decorative or symbolic image in one harmonious composition.

## CRYPTOGRAPHY

Cryptographic writing – the practice of disguising the obvious interpretation of a hieroglyphic inscription – was popular in Egypt, especially during the New Kingdom, although examples occur as early as the Old Kingdom and even in the later periods. Sometimes the hieroglyphs used were freshly devised for the occasion, while in other instances the scribe gave new forms to existing hieroglyphs. By means of such devices the real meaning of a sacred text could be limited to a select few. If a name was written cryptographically, it was less likely to be appropriated or destroyed.

In a statue of the Eighteenth-Dynasty pharaoh Hatshepsut, the three elements of her personal name Ma'at-ka-Re ("Truth is the Soul of Re") are presented as a three-dimensional sculpture, whose major focus is a large snake. A statue of Ramesses II of the Nineteenth Dynasty (see illustration, right), groups the sun god in the form of a falcon, the pharaoh as a child and a heraldic plant – elements that "spell out" the king's throne name Ra-mes-su (Ramesses). On such statuary, the rulers' names were not inscribed on the surface, but were hidden within the entire composition. The same intention could lie behind the sculpture of the sun god Re above the entrance to Ramesses II's temple at Abu Simbel. On one side of the god is a staff in the shape of the triliteral hieroglyph *wsr*; on the other, the goddess Ma'at. These elements form Ramesses' personal name, User-ma'at-Re ("Powerful is the Justice of Re"). Even if attempts had been made to efface the king's name inside the temple, this disguised identification would remain intact.

*The elements of this monumental sculpture – the sun god (*Ra*, or* Re*), the king as a child (*mes*) and the sedge plant (*su*), a heraldic symbol, in the child's left hand – form the name of King Ramesses II (ca. 1279–1213 BCE).*

# GLOSSARY

***akh*** A dead person's blessed spirit, capable of manifestation as a ghost.

***ankh*** "Life", represented by the hieroglyph ☥, a sacred emblem also frequently employed as a decorative motif.

***ba*** The winged spirit of a blessed dead person. Depicted with the body of a bird and head of the deceased, the ***ba*** was able to fly from the underworld to visit, unseen, the world of the living.

**Book of the Dead, The** Term used by Egyptologists to describe a type of funerary text comprising around two hundred spells that were intended to guide the deceased safely into the afterlife. It was usually written on papyrus and succeeded the earlier **Pyramid Texts** and **Coffin Texts** (see below) in the Second Intermediate Period.

**canopic (or Canopic)** Of or pertaining to the port of Canopus in the Delta. In particular, the term refers to pottery or stone jars, often with a stopper in the likeness of the head of the deceased, in which a person's viscera were stored in a tomb. "Canopic jars" were named from their similarity to vessels from Canopus bearing the head of Osiris, which were venerated as manifestations of the god.

**Coffin Texts, The** Term used by Egyptologists to describe a type of funerary text popular in the Middle Kingdom. Inscribed or painted on coffins, the texts were drawn from more than eleven hundred spells that in turn derived from the earlier **Pyramid Texts** (see below).

**Horus-king** Any one of the early Egyptian rulers whose name is written in a **serekh** (see below) surmounted by a figure of a falcon representing the god Horus. The nomenclature so written is known as the "Horus-name".

**Hyksos** The name given to the Asiatic rulers who became kings in Egypt in the late 17th century BCE and ruled for just over a century as the Fifteenth Dynasty (ca. 1630–1523BCE). The word represents a Greek rendering of the Egyptian *Heka Khaswt* "Rulers of Foreign Lands".

***ka*** The "vital force" or "creative life energy" of a person or a god. The *ka* of a mortal was created at birth and remained with the body for life as a "spiritual double", sometimes depicted in art as a smaller version of the individual. When the body died, it was the *ka* that remained in existence and (an Egyptian hoped) successfully entered the afterlife, where it was sustained by grave goods and the votive offerings of the living.

**Kush** A kingdom in Nubia (see below) whose rulers occupied the throne of Egypt for several decades in the eighth century BCE as the Twenty-Fifth, or Kushite, Dynasty (ca. 760–756BCE).

***ma'at*** "Truth", or "right", a cosmic principle which, it was believed, should govern all human and divine actions; it was personified and revered in the form of the goddess Ma'at, daughter of the sun god Re and guardian of truth, justice and harmony.

**mastaba** (Arabic: "bench") A type of rectangular tomb common for wealthy private burials from the Old Kingdom onward, and so called from the similarity of their shape to the squat stone or mudbrick benches commonly found outside Egyptian rural houses.

**Memphite** Of or pertaining to the city of Memphis.

**natron** A naturally occurring compound of sodium carbonate and sodium bicarbonate, used in mummification (for desiccation) and also for a variety of everyday purposes, such as washing and cleaning the teeth. The main source of natron was the Wadi Natrun, northwest of the Delta.

**nomarch** The governor of a **nome** (see below).

**nome** One of the forty-two traditional provinces of ancient Egypt, known in Egyptian as a *sepat*.

**Nubia** The region immediately south of ancient Egypt, beyond Elephantine (Aswan). Often called Yam by the Egyptians, it fell geographically into two regions, Lower Nubia (the northern part, roughly the south of the modern Egyptian republic) and Upper Nubia (the southern part, roughly present-day Sudan). Within Nubia there arose at various times a number of powerful states, such as **Kush** (see above), Kerma and Meroe.

**ostracon, ostrakon** (pl. **ostraca, ostraka**) The Greek word for a shard of pottery or a limestone flake, a common material employed for writing or drawing upon, usually for brief notes and artist's sketches.

**pharaonic** Of or pertaining to the pharaohs (kings) of Egypt. The term commonly refers only to the centuries spanning the Old Kingdom to the Late Period, when Egypt was usually governed by kings of native origin, rather than the subsequent Ptolemaic and Roman periods, when the country was under permanent rule by foreigners.

**Pharaoh** The king of Egypt, a Greek term derived from the Egyptian *per-aa* or *per-ao* ("Great House"), which originally referred to the royal palace but from the New Kingdom onward was also used to mean the ruler.

**Ptolemaic** Of or pertaining to the Greek dynasty founded by Ptolemy I in 310BCE and ending in 30BCE with the murder (on Roman orders) of Ptolemy XV, the son of Cleopatra VII and Julius Caesar.

**pylon** The monumental entrance of an Egyptian temple or palace.

**Pyramid Texts** A term used by Egyptologists to describe the first known Egyptian funerary texts, recorded on the interior walls of royal pyramids from King Unas (ca. 2371–2350BCE) onward and comprising around eight hundred spells.

**Ramesside (Ramessid)** **(1)** *adj.* Of or pertaining to the "Ramesside" period, the Nineteenth and Twentieth dynasties (ca. 1292–1075BCE), spanning the reigns of Ramesses I to Ramesses XI. **(2)** *n.* A pharaoh of the Ramesside period.

**serdab** (Arabic: "cellar") A chamber in a **mastaba** (see above) in which was placed a statue representing the dead person's ***ka*** (see above).

***serekh*** A hieroglyphic sign probably representing the façade of a royal palace ( ), in which the name of an early **Horus-king** (see above) was customarily written.

***shawabti*** ("Persea-wood figure"), a magical figurine placed in a tomb to act as a servant for the deceased in the afterlife. Later, probably through linguistic confusion, called an ***ushebti*** ("answerer").

***ushebti*** *See* ***shawabti***.

**wadi (wady)** A dry or seasonal river bed.

# BIBLIOGRAPHY


### GENERAL BIBLIOGRAPHY

Baines, John and Jaromír Málek. *Atlas of Ancient Egypt*. New York: Facts On File, 1993.

Faulkner, Raymond O. *The Ancient Egyptian Coffin Texts*. 3 vols. Warminster: Aris & Phillips, 1973–78.

Faulkner, Raymond O. *The Ancient Egyptian Pyramid Texts, translated into English*. Oxford: Oxford University Press, 1969.

Faulkner, Raymond O., Ogden Goelet, Carol Andrews and James Wasserman. *The Egyptian Book of the Dead*. San Francisco: Chronicle Books, 1994.

Gardiner, Sir Alan. *Egypt of the Pharaohs*. Oxford: Oxford University Press, 1961.

Hoffman, M. A. *Egypt Before the Pharaohs: The Prehistoric Foundations of Egyptian Civilization*. New York: Alfred Knopf, 1979.

Hornung, Erik. *Idea into Image: Essays on Ancient Egyptian Thought*. Translated by Elizabeth Bredeck. New York: Timken Publishers, 1992.

Lichtheim, Miriam. *Ancient Egyptian Literature, a Book of Readings*. 3 vols. Berkeley: University of California Press, 1980.

Shaw, Ian and Paul Nicholson. *The British Museum Dictionary of Ancient Egypt*. London: British Museum Press, 1995.

Strouhal, Eugen. *Life in Ancient Egypt*. Cambridge: Cambridge University Press and Norman: University of Oklahoma Press, 1992.

Wente, Edward F. *Letters from Ancient Egypt*. Society of Biblical Literature Writings from the Ancient World 1. Atlanta: Scholars Press, 1990.

### CHAPTER 1 THE GIFT OF THE NILE and CHAPTER 4 THE WEALTH OF THE LAND

**Fekri Hassan**

Butzer, K. W. *Early Hydraulic Civilization in Egypt: A Study in Cultural Ecology*. Chicago and London: University of Chicago Press, 1984.

Clark, J. D., and S. A. Brandt, eds. *From Hunters to Farmers: The Causes and Consequences of Food Production in Africa*. Berkeley: University of California Press, 1984.

Hassan, F. A. "Population, Ecology and Civilization in Ancient Egypt" in Carole L. Crumley, ed. *Historical Ecology*. School of American Research: Santa Fe, New Mexico, 1993.

James, T. G. H. *Ancient Egypt: The Land and its Legacy*. Austin: University of Texas Press, 1988.

Shaw, T., P. Sinclair, B. Andah and A. Okpoko, eds. *Food, Metals and Towns in Africa's Past*. London: Routledge and Unwin Hyman, 1995.

Wendorf, F., and F. A. Hassan. "Environment and Subsistence in Predynastic Egypt" in J. D. Clark and S. A. Brandt, eds. *Causes and Consequences of Food Production in Africa*. Berkeley: University of California Press, 1984.

### CHAPTER 2 THREE KINGDOMS AND THIRTY-FOUR DYNASTIES

**William J. Murnane**

Aldred, Cyril. *Akhenaten, King of Egypt*. London: Thames & Hudson, 1988.

Bagnall, Roger S. *Egypt in Late Antiquity*. Princeton: Princeton University Press, 1993.

Bowman, Alan K. *Egypt after the Pharaohs, 332BC–AD642, from Alexander to the Arab Conquest*. Berkeley: University of California Press, 1986.

Breasted, James H. *Ancient Records of Egypt*. 5 vols. Chicago: The Oriental Institute, 1906.

Clayton, Peter A. *Chronicle of the Pharaohs*. London: Thames & Hudson, 1994.

Gardiner, Sir Alan. *Egypt of the Pharaohs*. Oxford: Oxford University Press, 1961.

Grimal, Nicolas. *A History of Ancient Egypt*. Translated by Ian Shaw. Oxford: B. H. Blackwell, 1992.

James, T. G. H. *Pharaoh's People: Scenes from Life in Imperial Egypt*. Oxford: Oxford University Press, 1985.

Kemp, Barry J. *Ancient Egypt: Anatomy of a Civilization*. London and New York: Routledge, 1989.

Kitchen, K. A. *The Third Intermediate Period*. 2nd ed. Warminster: Aris & Phillips, 1986.

Murnane, William J. *Texts from the Amarna Period in Egypt*. Society of Biblical Literature Writings from the Ancient World 5. Atlanta: Scholars Press, 1995.

Quirke, Stephen. *The Administration of Egypt in the Late Middle Kingdom*. New Malden: SIA Publishing, 1990.

Redford, Donald B. *Egypt, Canaan and Israel in Ancient Times*. Princeton: Princeton University Press, 1992.

Spencer, A. J. *Early Egypt: The Rise of Civilization in the Nile Valley*. London: British Museum Press, 1993.

Strudwick, Nigel. *The Administration of Egypt in the Old Kingdom*. London: Kegan Paul International, 1985.

Trigger, B. G., B. J. Kemp, D. B. O'Connor and A. B. Lloyd. *Ancient Egypt: A Social History*. Cambridge: Cambridge University Press, 1983.

### CHAPTER 3 EGYPT AND THE WORLD BEYOND

**Donald Redford**

Adams, W. Y. *Nubia, Corridor to Africa*. London: Allen Lane, 1977.

Boardman, J. *The Greeks Overseas*. New York: Thames & Hudson, 1980.

Davies, W. V., ed. *Egypt and Africa: Nubia from Prehistory to Islam*. London: British Museum Press, 1991.

——. *Egypt, the Aegean and the Levant*. London: British Museum Press, 1995.

Dothan, T. and M. *People of the Sea: The Search for the Philistines*. New York: Macmillan, 1992.

Drewes, R. *The End of the Bronze Age*. Princeton: Princeton University Press, 1993.

Emery, W. B. *Egypt in Nubia*. London: Hutchinson, 1965.

Giveon, R. *The Impact of Egypt on Canaan*. Göttingen: Vandenhoeck and Ruprecht, 1978.

Groll, S., ed. *Pharaonic Egypt: The Bible and Christianity*. Jerusalem: Magnes Press, 1985.

Harris, J. R. *The Legacy of Egypt*. 2nd ed. Oxford: Clarendon Press, 1971.

Leahy, A., ed. *Libya and Egypt c.1300–750*. London: School of Oriental and African Studies, 1990.

O'Connor, D. B. *Ancient Nubia: Egypt's Rival in Africa*. Philadelphia: University Museum Press, 1993.

Redford, D. B. *Egypt, Canaan, and Israel in Ancient Times*. Princeton: Princeton University Press, 1992.

Sanders, N. K. *The Sea Peoples*. New York: Thames & Hudson, 1985.

Smith, W. S. *Interconnections in the Ancient Near East*. New Haven: Yale University Press, 1965.

Trigger, B. *Nubia Under the Pharaohs*. New York: Thames & Hudson, 1976.

Ward, W. *Egypt and the East Mediterranean World*. Beirut: American University of Beirut, 1971.

### CHAPTER 4 THE WEALTH OF THE LAND

**Fekri Hassan**

See under Chapter 1

### CHAPTER 5 THE SETTLED WORLD

**Ian Shaw**

Bietak, M. *Avaris: the Capital of the Hyksos*. London: British Museum Press, 1996.

Dunham, Dows and J. M. A. Janssen. *Second Cataract Forts*. 2 vols. Boston: Museum of Fine Art, 1963.

Emery W. B. et al. *The Fortress of Buhen*. 2 vols. London: Egypt Exploration Society, 1977–9.

Frankfort, H. et al. *The City of Akhenaten II*. London: Egypt Exploration Society, 1933.

Jeffreys, D. G. *The Survey of Memphis I*. London: Egypt Exploration Society, 1933.

Peet, T. E. and C. L. Woolley, *The City of Akhenaten I*. London: Egypt Exploration Society, 1923.

Pendlebury, J. D. S. et al. *The City of Akhenaten III*. 2 vols. London: Egypt Exploration Society, 1951.

Petrie, W. M. F. *Kahun, Gurob, Hawara*.

London: Egypt Exploration Society, 1890.
Petrie, W. M. F. *Illahun, Kahun, Gurob*. London: Egypt Exploration Society, 1891.
Smith, W. Stevenson. *The Art and Architecture of Ancient Egypt*. Harmondsworth: Pelican, 1981.
Trigger, B. G. et al. *Ancient Egypt: A Social History*. Cambridge: Cambridge University Press, 1983.
Uphill, E. *Egyptian Towns and Cities*. Princes Risborough: Shire Publications, 1988.

**CHAPTER 6 WOMEN IN EGYPT**
**Gay Robins**

Cerny, J. "The will of Naunakhte and the related documents" in *Journal of Egyptian Archaeology*: 31, 29–53, 1945.
Eyre, C. J. "Crime and adultery in ancient Egypt" in *Journal of Egyptian Archaeology*: 70, 92–105, 1984.
Fischer, H. G. *Egyptian Women of the Old Kingdom and of the Heracleopolitan Period*. New York: Metropolitan Museum of Art, 1989.
Friedman, F. "Aspects of domestic life and religion" in L. H. Lesko, ed. *Pharaoh's Workers, the Villagers of Deir el Medina*. Ithaca: Cornell University Press, 1994.
Janssen, R. M. and J. J. Janssen, *Growing Up in Ancient Egypt*. London: Rubicon Press, 1990.
Pestman, P. W. *Marriage and Matrimonial Property in Ancient Egypt*. Leiden: E. J. Brill, 1961.
Robins, G. "The god's wife of Amun in the 18th Dynasty in Egypt" in A. Cameron and A. Kuhrt, eds. *Images of Women in Antiquity*. Rev. ed. London: Routledge, 1993.
Robins, G. *Reflections of Women in the New Kingdom: Ancient Egyptian Art from the British Museum*. San Antonio, Texas: Van Siclen Books, 1995.
Robins, G. "While the woman looks on: gender inequality in the New Kingdom" in *KMT* 1 no.3: 18–21, 64–65, 1990.
Robins, G. *Women in Ancient Egypt*. London: British Museum Press, 1993.
Troy, L. "Good and bad women" in *Göttinger Miszellen* 80: 77–82, 1984.
Troy, L. *Patterns of Queenship in Ancient Egyptian Myth and History*. Uppsala: University of Uppsala, 1986.
Tyldesley, J. *Daughters of Isis: Women of Ancient Egypt*. Harmondsworth: Penguin, 1994.
Ward, W. *Essays on Feminine Titles of the Middle Kingdom and Related Subjects*. Beirut: American University, 1986.

**CHAPTER 7 THE BOUNDARIES OF KNOWLEDGE**
**Christopher Eyre**

Andrews, Carol. *Amulets of Ancient Egypt*. London: British Museum Press, 1994.
Borghouts, J. F. *Ancient Egyptian Magical Texts*. Leiden: E. J. Brill, 1978.
Killen, Geoffrey. *Egyptian Woodworking and Furniture*. Princes Risborough: Shire Publications, 1994.
Nunn, John F. *Ancient Egyptian Medicine*. London: British Museum Press, 1996.
Parkinson R. B. *Voices from Ancient Egypt: An Anthology of Middle Kingdom Writings*. London: British Museum Press, 1991.
Parkinson, R. B. and Stephen Quirke. *Papyrus*. London: British Museum Press, 1995.
Pinch, Geraldine. *Magic in Ancient Egypt*. London: British Museum Press, 1994.
Robins, G. and Charles Shute. *The Rhind Mathematical Papyrus: An Ancient Egyptian Text*. London: British Museum Press, 1987.
Sasson, Jack M. et al., eds. *Civilizations of the Ancient Near East*. 4 vols. London: Macmillan and New York: Simon & Schuster, 1995.
Scheel, Bernd. *Egyptian Metalworking and Tools*. Princes Risborough: Shire Publications, 1989.

**CHAPTER 8 THE LORD OF THE TWO LANDS**
**David P. Silverman**

Emery, Walter B. *Archaic Egypt*. Harmondsworth: Penguin, 1961.
Hayes, William C. *Most Ancient Egypt*. Chicago and London: University of Chicago Press, 1965.
Hoffman, Michael A. *Egypt Before the Pharaohs: The Prehistoric Foundations of Egyptian Civilization*. New York: Alfred A. Knopf, 1979.
Morenz, Siegfried. *Egyptian Religion*. Translated by Ann. E. Keep. London: Methuen, 1973.
Murnane, William. "Ancient Egyptian Co-regencies". *SAOC*. Vol. 40. Chicago: The Oriental Institute, 1977.
O'Connor, David B. and David P. Silverman, eds. "Ancient Egyptian Kingship" in *Problèmes d'Egyptologie*. Vol. 9. Leiden: E. J. Brill, 1995.
Redford, Donald B. *History and Chronology of the Egyptian Eighteenth Dynasty: Seven Studies*. Toronto: University of Toronto Press, 1967.
Rizkana, Ibrahim and Jürgen Seeher. Maadi. 4 vols. "Excavations at the Predynastic Site of Maadi and Its Cemeteries Conducted by Mustapha Amer and Ibrahim Rizkana on Behalf of the Department of Geography, Faculty of Arts of Cairo University, 1930–1953". *AV*. Vols. 64, 65, 80, 81. Mainz: Philip von Zabern, 1987–1990.
Robins, G. "A critical examination of the theory that the right to the throne of ancient Egypt passed through the female line in the 18th dynasty" in *Göttinger Miszellen*: 62, 66–77, 1983.
Silverman, David P. "Deities and Divinity in Ancient Egypt" in Byron E. Shafer, ed. *Religion in Ancient Egypt: Gods, Myths and Personal Practice*. Ithaca: Cornell University Press 1991.
Spencer, A. Jeffrey. *Early Egypt: The Rise of Civilisation in the Nile Valley*. London: British Museum Press, 1993.
Trigger, Bruce G., et al. *Ancient Egypt: A Social History*. Cambridge: Cambridge University Press, 1983.
Troy, Lana. "Patterns of Queenship in Ancient Egyptian Myth and History" *Boreas*. Vol. 14. Uppsala: University of Uppsala, 1986.
Wildung, Dietrich. *Egyptian Saints: Deification in Pharaonic Egypt*. Hagop Kevorkian Series on Near Eastern Art and Civilization. New York: New York University Press, 1977.

**CHAPTER 9 THE CELESTIAL REALM**
**James P. Allen**

Allen, James P. *Genesis in Egypt: The Philosophy of Ancient Egyptian Creation Accounts*. Yale Egyptological Studies 2. New Haven: Yale Egyptological Seminar, 1988.
Forman, Werner, and Stephen Quirke. *Hieroglyphs and the Afterlife in Ancient Egypt*. Norman: University of Oklahoma Press, 1996.
Hornung, Erik. *Conceptions of God in Ancient Egypt: The One and the Many*. Translated by John Baines. Ithaca: Cornell University Press, 1982.
Quirke, Stephen. *Ancient Egyptian Religion*. London: British Museum Press, 1992.
Simpson, William K., ed. *Religion and Philosophy in Ancient Egypt*. Yale Egyptological Studies 3. New Haven: Yale Egyptological Seminar, 1989.

**CHAPTER 10 THE CULT OF THE DEAD**
**Robert K. Ritner**

D'Auria, Sue, Peter Lacovara and Catherine H. Roehrig, eds. *Mummies and Magic, the Funerary Arts of Ancient Egypt*. Boston: Museum of Fine Arts, 1988.
Griffiths, J. Gwyn. *Plutarch's "De Iside et Osiride"*. Swansea: University of Wales Press, 1970.
Lesko, Leonard H. "Death and the Afterlife in Ancient Egypt" in Jack M. Sasson, ed. *Civilizations of the Ancient Near East III*. New York: Charles Scribners Sons, 1995.
Ritner, Robert K. *The Mechanics of Ancient Egyptian Magical Practice*. Chicago: The Oriental Institute, 1993.
Silverman, David P. "The Curse of the Curse of the Pharaohs" in *Expedition*. Vol. 29/2, 1987.
Simpson, William K., ed. *The Literature of Ancient Egypt*. New Haven and London: Yale University Press, 1973.
Spencer, A. J. *Death in Ancient Egypt*. New York: Penguin Books, 1982.

**CHAPTER 11 THE LIFE OF RITUAL**
**Emily Teeter**

Bierbrier, Morris. *The Tomb Builders of the Pharaohs*. London: British Museum Press, 1982.
Bleeker, C. *Egyptian Festivals*. Leiden: E. J. Brill, 1967.
Bleeker. C. *Hathor and Thoth*. Leiden: E. J. Brill, 1973.
Cauville, S. *Edfou*. Cairo: IFAO, 1984.
David, Rosalie. *Religious Ritual at Abydos*. Warminster: Aris & Phillips, 1981.
Decker, Manfred. *Sports and Games in Ancient Egypt*. New Haven: Yale University Press, 1992.
Epigraphic Survey. *The Festival Procession of Opet in the Colonnade Hall. Reliefs and Inscriptions at Luxor*. Vol. 1. Chicago, 1994.
Martin, G. T. *The Sacred Animal Necropolis at North Saqqara*. London: Egypt Exploration Society, 1981.

Piccioni, Peter. "Sportive Fencing as a Ritual for Destroying the Enemies of Horus" in *Gold of Praise: Studies on Ancient Egypt in Honor of E.F. Wente*. Chicago: University of Chicago Press, 1997.

Ritner, Robert. *The Mechanics of Ancient Egyptian Magical Practice*. Chicago: The Oriental Institute, 1993.

Sauneron, Serge. *The Priests of Ancient Egypt*. New York: Grove Press, 1980.

Touny, A. and S. Wild. *Sport in Ancient Egypt*. Leipzig, 1969.

Wente, E. F. and J. Harris. eds. *An X-Ray Atlas of the Pharaohs*. Chicago: University of Chicago Press, 1980.

**CHAPTER 12 THE PYRAMIDS**
**Zahi Hawass**

Badawi, A. *A History of Egyptian Architecture*. Vols 1-3. Berkeley: University of California Press, 1954–68.

Edwards, I. E. S. *The Pyramids of Egypt*. Harmondsworth: Penguin, 1995.

Fakry, A. *The Pyramids*. Chicago: University of Chicago Press, 1960.

Hawass, Zahi A. *The Pyramids of Ancient Egypt*. Pittsburgh: Carnegie Museum of Natural History, 1990.

Lauer, J.-P. *The Royal Cemetery of Memphis: Excavation and Discoveries since 1850*. London: Thames & Hudson, 1976.

Maragioglio., V. and C. A. Rinaldi, *The Architecture of the Memphite Pyramids*. Vols. 2–8. Turin and Rapallo, 1963–77.

Reisner, G. A. *A History of the Giza Necropolis I*. Cambridge, Mass.: Harvard University Press, 1942.

Reisner, G. A. Mycerinus: *The Temples of the Third Pyramid at Giza*. Cambridge, Mass.: Harvard University Press, 1942.

Verner, M. *Forgotten Pharaohs, Lost Pyramids at Abusir*. Prague: Academia Skodaexport, 1994.

Watson, Philip J. *Egyptian Pyramids and Mastaba Tombs of the Old and Middle Kingdoms*. Princes Risborough: Shire Publications, 1987.

**CHAPTER 13 TOMBS AND TEMPLES**
**Peter Der Manuelian**

Arnold, Dieter. *Building in Egypt: Pharaonic Stone Masonry*. Oxford: Oxford University Press, 1991.

Baines, John. "Palaces and Temples of Ancient Egypt" in Jack M. Sasson, ed. *Civilizations of the Ancient Near East I*. New York: Charles Scribners Sons, 1995.

Clarke, Somers and R. Engelbach. *Ancient Egyptian Construction and Architecture*. New York: Dover Publications, 1990 (reprint of 1930 Oxford University Press ed: *Ancient Egyptian Masonry*).

Dodson, Aidan. *Egyptian Rock-cut Tombs*. Princes Risborough: Shire Publications, 1991.

Emery, Walter B. *Archaic Egypt*. Baltimore: Penguin Books, 1961.

Hornung, Erik. *The Valley of the Kings*. Translated by David Warburton. New York: Timken Publications, 1990.

Reisner, George A. *The Development of the Egyptian Tomb Down to the Accession of Cheops*. Cambridge, Mass.: Harvard University Press, 1936, reprint Brockton, Mass.: John William Pye Rare Books, 1996.

Robins, G. *Egyptian Painting and Relief*. Princes Risborough: Shire Publications, 1986.

Romano, James F. *Death, Burial and Afterlife in Ancient Egypt*. Pittsburgh: The Carnegie Museum of Natural History, 1990.

Stadelmann, Rainer "Builders of the Pyramids" in Jack M. Sasson, ed. *Civilizations of the Ancient Near East II*. New York: Charles Scribners Sons, 1995.

Thomas, Angela P. *Egyptian Gods and Myths*. Princes Risborough: Shire Publications, 1986.

**CHAPTER 14 EGYPTIAN ART**
**Rita M. Freed**

Aldred, Cyril. *Middle Kingdom Art in Ancient Egypt*. London: Academy Editions, 1956.

Aldred, Cyril. *New Kingdom Art in Ancient Egypt During the Eighteenth Dynasty, 1570 to 1320BC*. London: A. Tiranti, 1961.

Aldred, Cyril. *Egypt to the End of the Old Kingdom*. London: Thames & Hudson, 1965.

Aldred, Cyril. *Jewels of the Pharaohs*. London: Thames & Hudson, 1971.

Aldred, Cyril. *Egyptian Art*. London: Thames & Hudson, 1980.

Badaway, A. *A History of Egyptian Architecture*. Berkeley: University of California Press, 1968.

Brooklyn Museum. *Cleopatra's Egypt: Age of the Ptolemies*. Brooklyn: Brooklyn Museum, 1988.

——. *Egyptian Sculpture of the Late Period*. Brooklyn: Brooklyn Museum, 1960.

Fazzini, R. *Images for Eternity: Egyptian Art from Berkeley and Brooklyn*. San Francisco: Fine Arts Museums of San Francisco, 1975.

James, T. G. H. *Egyptian Painting and Drawing in the British Museum*. London: British Museum Press, 1985.

James, T. G. H. and W. V. Davies *Egyptian Sculpture*. Cambridge, Mass.: Harvard University Press, 1983.

Mekhitarian, A. *Egyptian Painting*. New York: Rizzoli, 1979.

Michalowski, K. *Art of Ancient Egypt*. New York: N. H. Abrams.

Museum of Fine Arts Boston, 1969. *Egypt's Golden Age: The Art of Living in the New Kingdom, 1558–1085BC*. Boston, 1981.

Peck, W. H. *Egyptian Drawings*. New York: Dutton, 1978.

Robins, G. *Egyptian Painting and Relief*. Aylesbury: Shire Publications, 1986.

Russman, E. *Egyptian Sculpture*. Austin: University of Texas Press, 1989.

Schafer, H. *Principles of Egyptian Art*. Oxford: Clarendon Press, 1978.

Sauneron, Serge. *The Art and Architecture of Ancient Egypt*. 2nd ed. New York: Penguin Books, 1981.

Spanel, D. *Through Ancient Eyes: Egyptian Portraiture*. Birmingham, Alabama: Birmingham Museum of Art, 1988.

**CHAPTER 15 SIGNS, SYMBOLS AND LANGUAGE**
**David P. Silverman**

Andrews, C. *The Rosetta Stone*. London: British Museum Press, 1981.

Davies, W. V. *Reading the Past: Egyptian Hieroglyphs*. London: British Museum Press, 1987.

Davis, N. *Picture Writing in Ancient Egypt*. Oxford: Oxford University Press, 1958.

Fischer, H. G. *Ancient Egyptian Calligraphy*. New York: Metropolitan Museum of Art, 1979.

Gardiner, A. *Egyptian Grammar*. Oxford: Oxford University Press, 1973.

Harris, J. R., ed. *The Legacy of Egypt*. Oxford: Oxford University Press, 1971.

Quirke, S. *Hieroglyphs and the Afterlife in Ancient Egypt*. London: British Museum Press, 1996.

Ray, J. D. "The Emergence of Writing in Egypt" in *World Archaeology* 17 no. 3: 307–16, 1986.

Silverman, David P. "Writing" in *Egypt's Golden Age: The Art of Living in the New Kingdom*. Boston: Museum of Fine Arts. Boston, 1982.

Silverman, David P. *Language and Writing in Ancient Egypt*. Pittsburgh: Carnegie Museum of Natural History, 1990.

Zausich, K. T. *Hieroglyphs Without Mystery*. Austin: Texas University Press, 1992.

# INDEX

Page references to main text, box text and margin text are in roman, to captions in *italic*.

## A

## H



## I



## J

## K



## L



## M

## N



## O

## P

## Q



## R



## S

## U



## V



## W



## Y



## Z

# PICTURE CREDITS

*The publishers wish to thank the photographers and organizations for their kind permission to reproduce the following photographs in this book:*

**KEY**

**t** top; **b** bottom; **c** centre; **l** left; **r** right

| | |
|---|---|
| **BAL:** | Bridgeman Art Library |
| **BM:** | The British Museum |
| **CM:** | The Egyptian Museum, Cairo |
| **ICL:** | Images Colour Library |
| **JL:** | Jürgen Liepe/The Egyptian Museum, Cairo |
| **RHPL:** | Robert Harding Picture Library |
| **TSI:** | Tony Stone Images |
| **WFA:** | Werner Forman Archives |

**1** AKG; **2** RHPL; **3** ICL; **4** JL (JE46725); **5** WFA/Christie's; **6** James Davis Travel Photography; **8** Zefa; **9** BM (EA 37982); **10** Hutchison Library; **11** Science Photo Library; **12** Graham Harrison; **14-15** Britstock-IFA; **16** Graham Harrison/BM; **17** Akademie der Bildenden Kuenste, Vienna/AKG; **18** Staatliche Museen zu Berlin/AKG; **19c** WFA/Christie's; **19b** John G. Ross; **20** WFA/BM; **21** Graham Harrison; **23** JL (JE32169=CG14716); **24** JL (JE10062=CG14); **25** JL (JE36143); **26** JL (JE36195); **28** JL (JE20001=CG395); **29** JL (JE46694=CG52702); **30** JL (JE4694=CG52671); **31** JL (JE15210=CG394); **32** Graham Harrison/BM; **33t** Graham Harrison/BM; **33b** JL (JE44861); **34** CM/AKG; **36** JL (CG560); **37** JL (JE36933=CG42236); **38** WFA/BM; **39** JL (JE54313); **40** JL (JE 36457); **41t** BM (EA 921); **41b** WFA/CM; **42** Akademie der Bildenden Kuenste, Vienna/AKG; **43** JL (JE4673=CG52645); **44** WFA; **47** Graham Harrison; **48** JL (JE14276, JE89661); **51** JL (JE30986=CG258); **52** BM (EA 191); **53** BM (EA 3799); **54** JL (JE62949); **55t** Scala, Italy; **55b** e.t. archive/BM; **56** Museo Archeologico, Palestrina/Scala, Italy; **57** Museum of London (CL96/822); **58** BAL/Louvre; **59t** e.t. archive/CM; **59b** RHPL; **60** Dr Paul T. Nicholson; **61** Graham Harrison; **62** BAL/British and Foreign Bible Society; **63** BAL/BM; **64** Museo Egiziano, Turin/AKG; **66t** Graham Harrison/BM (EA41573); **66b** Graham Harrison/BM (EA 37978); **67** Graham Harrison/BM (EA 9999); **68** BM (EA 32610); **69b** JL (JE27434=CG14238); **70** BM; **71** BAL/Louvre; **72** BAL/BM; **73** JL (JE46724); **75** TSI; **76** Ashmolean Museum, Oxford; **78** BM (EA1972); **80t** ICL; **80b** BM (EA 37984); **81** Graham Harrison; **82t** Graham Harrison; **82b** JL (JE66624); **84** JL (JE51280); **85t** Staatliche Museen zu Berlin/Bildarchiv Preussischer Kulturbesitz; **85b** BM (EA 59418); **86** BM (EA104.70/6); **87** BM (EA 7876); **88t** Staatliche Museen zu Berlin/WFA; **88b** AKG; **89** JL (JE56259A and 56262); **90** John G. Ross/Louvre; **91** ICL; **92** John G. Ross; **93bl** BM (EA 11143); **93b** e.t. archive/CM; **94** e.t. archive/BM; **95** Graham Harrison; **96** BM (EA 43215); **97** JL (JE60686); **98t** JL (JE46723); **98b** e.t. archive/BM; **99** JL (JE32158=CG14717 and CG52701); **100** JL (RT 15.1.25.44); **101t** John G. Ross/Berlin Museum; **101b** WFA; **102** JL (JE28504=CG1533); **103** JL (JE4872); **104** ICL; **105** TSI; **106** ICL; **107t** TSI; **107b** BM (EA 720); **108t** Graham Harrison; **108b** BM (EA 9999/24); **110t** BM; **110b** ICL; **111** Zefa; **112t** WFA/CM;**112b** TSI; **113** AKG; **114** AKG; **115** BAL/Louvre; **116** AKG/BM; **117** BM (EA 6705); **118** ICL; **119** BAL/CM; **121** Akademie der Bildenden Kuenste, Vienna/AKG; **122** BM (EA 10554/81); **124** BM (EA 498); **125** Spectrum; **126** RHPL; **127** Hirmer Fotoarchiv, Munich; **128** BAL/Louvre; **129** Staatliche Museen zu Berlin/AKG; **130** Graham Harrison; **131** BM (EA 10470); **132** AKG/BM; **133t** JL (RT 23.11.16.12); **133b** JL (CG48406); **134** BAL; **135** BAL; **136** JL (SR 11488); **137** BM (EA 9901/3); **138** Graham Harrison; **139** Frank Spooner Agency; **140** BM (EA 22332); **141** WFA/E. Strouhal; **143** BM (EA 10470/17); **144** James Davis Travel Photography; **145t** Peter Clayton; **145b** Peter Clayton; **146** Griffith Institute, Ashmolean Museum, Oxford; **147** JL (JE27303); **148** ICL; **149t** S. Purdy Matthews/TSI; **149b** BAL/Louvre; **150** Oriental Institute, University of Chicago (pl. 40); **151** JL (JE32018=CG70018); **152** BM (EA 21810); **153** BM (EA 1198); **154** JL (JE61467); **155t** Graham Harrison/BM; **155b** WFA; **156t** WFA/CM; **156b** AKG; **157** JL (JE62028); **158t** WFA/Cheops Barque Museum, Giza; **158b** Zefa; **159** Staatliche Museen zu Berlin/Bildarchiv Preussischer Kulturbesitz; **160** BM (EA 8056); **161** RHPL; **162** JL (JE43566); **163t** Ancient Art & Architecture; **163b** JL (CG29712); **164t** WFA/BM; **164b** The Oriental Institute, University of Chicago (OIMN 18510); **165** BM (EA 994); **166** e.t. archive/CM; **167** TSI; **168** P. Der Manuelian; **169b** Spectrum; **170t** RHPL; **170b** BAL/Louvre; **174** e.t. archive/Egiziano Museo, Turin, Italy; **178** Graham Harrison/BM; **179l** RHPL; **181l** P. Der Manuelian; **181r** RHPL; **183** RHPL; **184** RHPL; **186** RHPL; **188t** WFA; **188b** WFA; **189** WFA; **190t** AKG/Collection of George Ortiz; **190b** Peter Clayton; **191** JL (JE30857=CG52001 (pectoral) and JE30858=CG53123 (belt)); **192** Zefa; **193l** Graham Harrison; **193r** P. Der Manuelian; **194** Zefa; **196** JL (JE60671); **197** P. Der Manuelian; **198** Graham Harrison; **199** P. Der Manuelian; **200l** P. Der Manuelian; **200r** Graham Harrison; **201** JL (CG3 and CG4); **202l** JL (JE40679); **202r** ICL; **203** Fred J. Maroon; **204** TSI; **205** JL (JE98171); **206** Graham Harrison; **210** Gavin Hellier/RHPL; **211l** TSI; **211r** Eye Ubiquitous; **212** Staatliche Museen zu Berlin/WFA; **213t** BM; **213b** WFA/CM; **214t** JL (JE97472); **214b** John G. Ross; **215** BAL/CM; **216t** Museum of Fine Arts, Boston (E7426); **216b** Museum of Fine Arts, Harvard Expedition 1920 (21.2600); **217** BM (EA 5601); **218** BM; **219t** JL (JE30948=CG259); **219b** JL (JE21365); **220t** BAL/CM; **220b** JL (JE95254=CG51009); **221** JL (JE48035); **222t** JL (CG39194); **222b** BM (EA 64391); **223** BM (EA 1770); **224t** WFA; **224b** G. Dagli Orti; **225** BM (EA 65316); **226** JL (JE 44920=CG 52663); **227t** JL (JE61884); **227b** JL (JE 30875=CG52002 and 52003); **228** Henri Stierlin; **229** Henri Stierlin; **230** David P. Silverman; **231t** Graham Harrison; **231b** BM (EA24); **233t** JL (JE30272=CG36); **233b** BM (EA 5547); **234** Toledo Museum of Art, Ohio (3/93.2); **237** Graham Harrison; **238** Graham Harrison; **239** Rijksmuseum van Oudheden, Leiden; **240t** JL (JE31113-6=CG52920-21/26-27/29-30, 35-36/5556/58/59-74 and 53018); **240b** JL (JE62114); **241** JL (JE64735).

*Every effort has been made to trace copyright holders. However, if there are any omissions we will be happy to rectify them in future editions.*

**Captions for illustrations on pp.1–5:**
**1. (Half-title)** *Wall painting from the tomb of Sennefer; reign of Amenhotep II (ca. 1426–1400BCE).* **2. (Opposite title)** *Colossal statue of Ramesses II (ca. 1279–1213BCE) at Abu Simbel.* **3. (Title)** *The mummy of the deceased and the god Horus, from the Book of the Dead of Hunefer (ca. 1285BCE).* **4. (Contents)** *Wooden model of a woman bearing offerings of wine jars and a duck, from the tomb of Meketre, ca. 2000BCE.* **5. (Contents)** *Roman Period coloured glass plaque depicting the god Horus.*